AQA Science

Exclusively endorsed and approved by AQA

9/06

BERKSHIRE EDUCATION
ST. CRISPIN'S
SCHOOL
COMMITTEE

Ann Fullick

Series Editor: Lawrie Ryan

GCSE Biology

Nelson Thornes

a Wolters Kluwer business

Published in 2006 by:
Nelson Thornes Ltd
Delta Place
27 Bath Road
CHELTENHAM
GL53 7TH
United Kingdom

06 07 08 09 10 / 10 9 8 7 6 5

A catalogue record for this book is available from the British Library

ISBN 0 7487 9641 X

Cover photographs: snail by Gerry Ellis/Digital Vision LC (NT); embryo
by Biophoto/Science Photo Library; *E. coli* bacteria by Dr Gary
Gaugler/Science Photo Library
Cover bubble illustration by Andy Parker
Illustrations by Bede Illustration, Beverly Curl, Kevin Jones Associates
and Roger Penwill
Page make-up by Wearset Ltd

Printed and bound in Slovenia

GCSE Biology Contents

Welcome to AQA Biology

How to use this book

This textbook will help you throughout your GCSE course and to prepare for AQA's exams. It is packed full of features to help you to achieve the best result you can.

Some of the text is in a box marked HIGHER. You have to include these parts of the book if you are taking the Higher Tier exam. If you are taking the Foundation Tier exam, you can miss these parts out.

The same applies to any Learning Objectives, Key Points or Questions marked [Higher].

HIGHER

a) What are the yellow boxes?

To check you understand the science you are learning, questions are integrated into the main text. The information needed to answer these is on the same page, so you don't waste your time flicking through the entire book.

LEARNING OBJECTIVES

By the end of the lesson you should be able to answer the questions posed in the learning objectives; if you can't, review the content until it's clear.

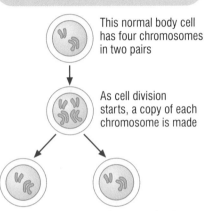

This normal body cell has four chromosomes in two pairs

As cell division starts, a copy of each chromosome is made

The cell divides in two to form two daughter cells. Each daughter cell has a nucleus containing four chromosomes identical to the ones in the original parent cell.

Figure 1 Key diagrams are as important as the text. Make sure you use them in your learning and revision.

Key words

Important scientific terms are shown like this:

observation or **anomalous**

You can look up the words shown like this – **bias** – in the glossary.

GET IT RIGHT!

Avoid common mistakes and gain marks by sticking to this advice.

KEY POINTS

If you remember nothing else, remember these! Learning the key points in each lesson is a good start. They can be used in your revision and help you summarise your knowledge.

PRACTICAL

Become familiar with key practicals. A simple diagram or photo and questions make this feature a short introduction, reminder or basis for practicals in the classroom.

Figure 2 These test tubes show clearly the importance of protein-digesting enzymes *and* hydrochloric acid in your stomach. Meat was added to each tube at the same time.

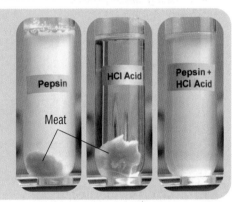

Pepsin

HCl Acid

Pepsin + HCl Acid

Meat

DID YOU KNOW?

Curious examples of scientific points that are out of the ordinary, but true…

FOUL FACTS

Some science is just too gruesome to ignore. Delve into the horrible yet relevant world of Foul Facts.

At the start of each unit you will find a double-page introduction. This reminds you of ideas from previous work that you will need. The recap questions and activity will help find out if you need some revision before starting.

SCIENCE @ WORK

When will you ever use science in 'real life'? Check this feature to find out.

SUMMARY QUESTIONS

Did you understand everything? Get these questions right, and you can be sure you did. Get them wrong, and you might want to take another look.

The ideas in 'How Science Works' are covered in the first chapter. You will need to refer back to this chapter as you work through the course.

This first chapter looks at 'How Science Works'. It is an important part of your GCSE because the ideas introduced here will crop up throughout your course. You will be expected to collect scientific evidence and to understand how we use evidence. These concepts will be assessed as the major part of your internal school assessment. You will take one or more 45-minute tests on data you have collected previously plus data supplied for you in the test. These are called Investigative Skills Assignments. The ideas in 'How Science Works' will also be assessed in your examinations.

What you already know

Here is a quick reminder of previous work with investigations that you will find useful in this chapter:

- You will have done some practical work and know how important it is to keep yourself and others safe.
- Before you start investigating you usually make a prediction, which you can test.
- Your prediction and plan will tell you what you are going to change and what you are going to measure.
- You will have thought about controls.
- You will have thought about repeating your readings.
- During your practical work you will have written down your results, often in a table.
- You will have plotted graphs of your results.
- You will have made conclusions to explain your results.
- You will have thought about how you could improve your results, if you did the work again.

RECAP QUESTIONS

Kiah wrote this account of a practical she did:

I wanted to find out how photosynthesis might be affected by temperature. I put five pieces of pondweed into five different test tubes and covered them with water. I put the test tubes into water baths at different temperatures and measured the gas coming off. I thought that the hotter the water the more gas would be produced.

One tube was left at room temperature which was 18°C and it made 15 mm³ in five minutes. The second was in 6°C water and gave 2 mm³. At 30°C there were 30 mm³, at 40°C there were 28 mm³ of gas. The test tubes were all left for thirty minutes.

1 What was Kiah's prediction?

2 What was the variable she chose to change? (We call this the independent variable.)

3 What was the variable she measured to judge the effect of varying the independent variable? (We call this the dependent variable. Its value *depends* on the value chosen for the independent variable.)

4 Write down one variable that Kiah controlled.

5 Write down a variable that Kiah did not say she had controlled.

6 Make a table of her results.

7 Draw a graph of the results.

8 Write a conclusion for Kiah.

9 How do you think Kiah could have improved her results?

How science works for us

Science works for us all day, every day. You do not need to know how a mobile phone works to enjoy sending text messages. But, think about how you started to use your mobile phone or your television remote control. Did you work through pages of instructions? Probably not!

You knew that pressing the buttons would change something on the screen (*knowledge*). You played around with the buttons, to see what would happen (*observation*). You had a guess at what you thought might be happening (*prediction*) and then tested your idea (*experiment*).

If your prediction was correct you remembered that as a *fact*. If you could repeat the operation and get the same result again then you were very pleased with yourself. You had shown that your results were **reliable**.

Working as a scientist you will have knowledge of the world around you and particularly about the subject you are working with. You will observe the world around you. An enquiring mind will then lead you to start asking questions about what you have observed.

Science moves forward by slow steady steps. When a genius such as Einstein comes along then it takes a giant leap. Those small steps build on knowledge and experience that we already have.

Each small step is important in its own way. It builds on the body of knowledge that we have. In 1675 a German chemist tried to extract gold from urine. He must have thought that there was a connection between the two colours. He was wrong, but after a long while, with an incredible stench coming from his laboratory, the urine began to glow.

He had discovered phosphorus. A Swedish scientist worked out how to manufacture phosphorus without the smell of urine. Phosphorus catches light easily. That is why most matches these days are manufactured in Sweden.

DID YOU KNOW?

The Greeks were arguably the first true scientists. They challenged traditional myths about life. They set forward ideas that they knew would be challenged. They were keen to argue the point and come to a reasoned conclusion.

Other cultures relied on long established myths and argument was seen as heresy.

Thinking scientifically

Figure 1 Tropical beach

ACTIVITY

Once you have got the idea of holidays out of your mind – look at the photograph in Figure 1 with your scientific brain.

Work in groups to *observe* the beach and the plants growing on it. Then you can start to think about why the plants can grow (*knowledge*) so close to the beach.

One idea could be that the seeds can float for a long while in the sea, without taking in any water.

You can use the following headings to discuss your investigation. One person should be writing your ideas down, so that you can discuss them with the rest of your class.

- What prediction can you make about the mass of the coconut seed and the time it spends in the sea water?
- What would be your independent variable?
- What would be your dependent variable?
- What would you have to control?
- Write a plan for your investigation.
- How could you make sure your results were reliable?

H2

Fundamental ideas about how science works

LEARNING OBJECTIVES

1 How do you spot when a person has an opinion that is not based on good science?
2 What is the importance of continuous, ordered and categoric variables?
3 What is meant by reliable evidence and valid evidence?
4 How can two sets of data be linked?

NEXT TIME YOU...

... read a newspaper article or watch the news on TV ask yourself if that research is valid and reliable. (See page 5.) Ask yourself if you can trust the opinion of that person.

Science is too important for us to get it wrong

Sometimes it is easy to spot when people try to use science poorly. Sometimes it can be funny. You might have seen adverts claiming to give your hair 'body' or sprays that give your feet 'lift'!

On the other hand, poor scientific practice can cost lives.

Some years ago a company sold the drug thalidomide to people as a sleeping pill. Research was carried out on animals to see if it was safe. The research did not include work on pregnant animals. The opinion of the people in charge was that the animal research showed the drug could be used safely with humans.

Then the drug was also found to help ease morning sickness in pregnant women. Unfortunately, doctors prescribed it to many women, resulting in thousands of babies being born with deformed limbs. It was far from safe.

These are very difficult decisions to make. You need to be absolutely certain of what the science is telling you.

a) Why was the opinion of the people in charge of developing thalidomide based on poor science?

Deciding on what to measure

You know that you have an independent and a dependent variable in an investigation. These variables can be one of four different types:

- A **categoric variable** is one that is best described by a label (usually a word). The colour of eyes is a categoric variable, e.g. blue or brown eyes.
- A **discrete variable** is one that you describe in whole numbers. The number of leaves on different plants is a discrete variable.
- An **ordered variable** is one where you can put the data into order, but not give it an actual number. The height of plants compared to each other is an ordered variable, e.g. the plants growing in the woodland are taller than those on the open field.
- A **continuous variable** is one that we measure. Therefore its value could be any number. Temperature (as measured by a thermometer or temperature sensor) is a continuous variable, e.g. 37.6°C, 45.2°C.

When designing your investigation you should always try to measure continuous data whenever you can. This is not always possible, so you should then try to use ordered data. If there is no other way to measure your variable then you have to use a label (categoric variable).

Figure 1 Student recording a range of temperatures

b) Imagine you were growing seedlings in different volumes of water. Would it be better to say that some were tall and some were short; or some were taller than others; or to measure the heights of all of the seedlings?

Making your investigation reliable and valid

When you are designing an investigation you must make sure that others can get the same results as you – this makes it **reliable**.

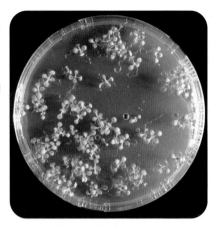

You must also make sure you are measuring the actual thing you want to measure. If you don't, your data can't be used to answer your original question. This seems very obvious but it is not always quite so easy. You need to make sure that you have *controlled* as many other variables as you can, so that no-one can say that your investigation is not **valid**. A valid investigation should be reliable *and* answer the original question.

Figure 2 Cress seedlings growing in a Petri dish

c) State one way in which you can show that your results are valid.

How might an independent variable be linked to a dependent variable?

Variables can be linked together for one of three reasons:

- It could be because one variable has caused a change in the other, e.g. the more plants there are in a pond, the more oxygen there is in the water. This is a **causal link**.
- It could be because a third variable has caused changes in the two variables you have investigated, e.g. fields that have more grass also have more dandelions in them. There is an **association** between the two variables. This is caused by a third variable – how many sheep there are in the field!
- It could be due simply to **chance**, e.g. the type of weeds growing in different parts of your garden!

d) Describe a causal link that you have seen in biology.

Figure 3 Sheep grazing in a field

SUMMARY QUESTIONS

1 Name each of the following types of variables described in a), b) and c).

   a) People were asked about how they felt inside a new shopping centre: 'warm', 'hot', 'quite warm', 'cold', 'freezing!'
   b) These people were asked as they entered the new shopping centre: 'Warmer than I did outside'; 'Colder than my shed!'
   c) These people had their body temperature measured using a clinical thermometer: 37.1°C; 37.3°C; 36.8°C; 37.0°C; 37.5°C.

2 A researcher claimed that the metal tungsten 'alters the growth of leukaemia cells' in laboratory tests. A newspaper wrote that they would 'wait until other scientists had reviewed the research before giving their opinion.' Why is this a good idea?

KEY POINTS

1 Be on the lookout for non-scientific opinions.
2 Continuous data is more powerful than other types of data.
3 Check that evidence is reliable and valid.
4 Be aware that just because two variables are related it does not mean that there is a causal link between them.

H3 Starting an investigation

Figure 1 Plant showing positive phototropism

Observation

As humans we are sensitive to the world around us. We can use our many senses to detect what is happening. As scientists we use observations to ask questions. We can only ask useful questions if we know something about the observed event. We will not have all of the answers, but we know enough to start asking the correct questions.

If we observe that the weather has been hot today, we would not ask if it was due to global warming. If the weather was hotter than normal for several years then we could ask that question. We know that global warming takes many years to show its effect.

When you are designing an investigation you have to observe carefully which variables are likely to have an effect.

a) Would it be reasonable to ask if the plant in Figure 1 is 'growing towards the glass'? Explain your answer.

A farmer noticed that her corn was much smaller at the edge of the field than in the middle (observation). She noticed that the trees were quite large on that side of the field. She came up with the following ideas that might explain why this was happening:

- The trees at the edge of the field were blocking out the light.
- The trees were taking too many nutrients out of the soil.
- The leaves from the tree had covered the young corn plants in the spring.
- The trees had taken too much water out of the soil.
- The seeds at the edge of the field were genetically small plants.
- The drill had planted fewer seeds on that side of the field.
- The fertiliser spray had not reached the side of the field.
- The wind had been too strong over winter and had moved the roots of the plants.
- The plants at the edge of the field had a disease.

b) Discuss each of these ideas and use your knowledge of science to decide which four are the most likely to have caused the poor growth of the corn.

Observations, backed up by really creative thinking and good scientific knowledge can lead to a **hypothesis**.

What is a hypothesis?

A hypothesis is a 'great idea'. Why is it so great? – well because it is a great observation that has some really good science to try to explain it.

For example, you observe that small, thinly sliced chips cook faster than large, fat chips. Your hypothesis could be that the small chips cook faster because the heat from the oil has a shorter distance to travel before it gets to the potato in the centre of the chips.

c) Check out the photograph in Figure 2 and spot anything that you find interesting. Use your knowledge and some creative thought to suggest a hypothesis based on your observations.

When making hypotheses you can be very imaginative with your ideas. However, you should have some scientific reasoning behind those ideas so that they are not totally bizarre.

Remember, your explanation might not be correct, but you think it is. The only way you can check out your hypothesis is to make it into a prediction and then test it by carrying out an investigation.

Figure 2 A frog

Observation ➕ knowledge ➡ hypothesis ➡ prediction ➡ investigation

Starting to design a valid investigation

An investigation starts with a prediction. You, as the scientist, predict that there is a relationship between two variables.

- An **independent variable** is one that is changed or selected by you, the investigator.

- A **dependent variable** is measured for each change in your independent variable.

- All other variables become **control variables**, kept constant so that your investigation is a fair test.

If your measurements are going to be accepted by other people then they must be valid. Part of this is making sure that you are really measuring the effect of changing your chosen variable. For example, if other variables aren't controlled properly, they might be affecting the data collected.

d) Look at Figure 3. When investigating his heart rate before and after exercise, Darren got his girlfriend to measure his pulse. Would Darren's investigation be valid? Explain your answer.

Figure 3 Measuring a pulse

SUMMARY QUESTIONS

1 Copy and complete using the words below:

> controlled   dependent   hypothesis   independent
> knowledge   prediction

Observations when supported by scientific …… can be used to make a …… . This can be the basis for a …… . A prediction links an …… variable to a …… variable. Other variables need to be …… .

2 Explain the difference between a hypothesis and a prediction.

KEY POINTS

1 Observation is often the starting point for an investigation.
2 Hypotheses can lead to predictions and investigations.
3 You must design investigations that produce valid results if you are to be believed.

H4 Building an investigation

Figure 1 Corn being harvested

Fair testing

A **fair test** is one in which only the independent variable affects the dependent variable. All other variables are controlled, keeping them constant if possible.

This is easy to set up in the laboratory, but almost impossible in fieldwork. Plants and animals do not live in environments that are simple and easy to control. They live complex lives with lots of variables changing constantly.

So how can we set up fieldwork investigations? The best you can do is to make sure that all of the many variables change in much the same way, except for the one you are investigating. Then at least the plants get the same weather, even if it is constantly changing.

a) Imagine you were testing how close together you could plant corn to get the most cobs. You would plant five different plots, with different numbers of plants in each plot. List some of the variables that you could not control.

If you are investigating two variables in a large population then you will need to do a survey. Again it is impossible to control all of the variables.

Imagine you were investigating the effect of diet on diabetes. You would have to choose people of the same age and same family history to test. The larger the sample size you test, the more reliable your results will be.

Control groups are used in investigations to try to make sure that you are measuring the variable that you intend to measure. When investigating the effects of a new drug, the control group will be given a placebo.

The control group think they are taking a drug but the placebo does not contain the drug. This way you can control the variable of 'thinking that the drug is working' and separate out the effect of the actual drug.

Choosing values of a variable

Trial runs will tell you a lot about how your early thoughts are going to work out.

Do you have the correct conditions?
A photosynthesis investigation that produces tiny amounts of oxygen might not have enough:

- light, • pondweed, • carbon dioxide, or
- the temperature might not be high enough.

Have you chosen a sensible range?
If there is enough oxygen produced, but the results are all very similar:

- you might not have chosen a wide enough range of light intensities.

Have you got enough readings that are close together?
If the results are very different from each other:

- you might not see a pattern if you have large gaps between readings over the important part of the range.

Accuracy

Accurate results are very close to the *true value*.

Your investigation should provide data that is accurate enough to answer your original question.

However, it is not always possible to know what that true value is.

How do you get accurate data?
- You can repeat your results and your mean is more likely to be accurate.
- Try repeating your measurements with a different instrument and see if you get the same readings.
- Use high quality instruments that measure accurately.
- The more carefully you use the measuring instruments, the more accuracy you will get.

Precision and reliability

If your repeated results are closely grouped together then you have precision and you have improved the reliability of your data.

Your investigation must provide data with sufficient precision. It's no use measuring a person's reaction time using the seconds hand on clock! If there are big differences within sets of repeat readings, you will not be able to make a valid conclusion. You won't be able to trust your data!

How do you get precise and reliable data?
- You have to use measuring instruments with sufficiently small scale divisions.
- You have to repeat your tests as often as necessary.
- You have to repeat your tests in exactly the same way each time.

A word of caution!

Be careful though – just because your results show precision does not mean your results are accurate. Look at the box opposite.

b) Draw a thermometer scale showing 4 results that are both accurate and precise.

The difference between accurate and precise results

Imagine measuring the temperature after a set time when a food is burned and used to heat a fixed volume of water. Two students repeated this experiment, four times each. Their results are marked on the thermometer scales below:

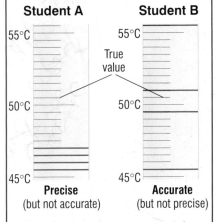

- Precise results are grouped closely together.
- Accurate results will have a mean (average) close to the true value.

SUMMARY QUESTIONS

1 Copy and complete using the following terms:

range   repeat   conditions   readings

Trial runs give you a good idea of whether you have the correct ......; whether you have chosen the correct ......; whether you have enough ......; if you need to do ...... readings.

2 Use an example to explain how results can be accurate, but not precise.

3 Briefly describe how you would go about setting up a fair test in a laboratory investigation. Give your answer as general advice.

KEY POINTS

1 Care must be taken to ensure fair testing – as far as is possible.
2 You can use a trial run to make sure that you choose the best values for your variables.
3 Careful use of the correct equipment can improve accuracy.
4 If you repeat your results carefully they are likely to become more reliable.

H5 Making measurements

LEARNING OBJECTIVES

1 Why do results always vary?
2 How do you choose instruments that will give you accurate results?
3 What do we mean by the sensitivity of an instrument?
4 How does human error affect results and what do you do with anomalies?

Figure 1 Student testing the rate at which oxygen is produced using an enzyme

Using instruments

Do not panic! You cannot expect perfect results.

Try measuring the temperature of a beaker of water using a digital thermometer. Do you always get the same result? Probably not. So can we say that any measurement is absolutely correct?

In any experiment there will be doubts about actual measurements.

a) Look at Figure 1. Suppose, like this student, you tested the rate at which oxygen was produced using an enzyme. It is unlikely that you would get two readings exactly the same. Discuss all the possible reasons why.

When you choose an instrument you need to know that it will give you the accuracy that you want. That is, it will give you a true reading.

If you have used an electric water bath, would you trust the temperature on the dial? How do you know it is the true temperature? You could use a very expensive thermometer to calibrate your water bath. The expensive thermometer is more likely to show the true temperature. But can you really be sure it is accurate?

You also need to be able to use an instrument properly.

b) In Figure 1 the student is reading the amount of gas in the measuring cylinder. Why is the student unlikely to get a true reading?

When you choose an instrument you have to decide how accurate you need it to be. Instruments that measure the same thing can have different sensitivities. The **sensitivity** of an instrument refers to the smallest change in a value that can be detected.

Choosing the wrong scale can cause you to miss important data or make silly conclusions. For example, 'The cells were all the same size – they were less than 1 millimetre.'

c) Match the following types of measuring instrument to their best use:

Used to measure	Sensitivity of measuring instrument
Size of a cell	millimetres
Human height	metres
Length of a running race to test fitness	micrometres
Growth of seedlings	centimetres

Errors

Even when an instrument is used correctly, the results can still show differences.

Results may differ because of **random error**. This is most likely to be due to a poor measurement being made. It could be due to not carrying out the method consistently.

The error might be a **systematic error**. This means that the method was carried out consistently but an error was being repeated.

Check out these two sets of data that were taken from the investigation that Mark did. He tested 5 different volumes of enzyme. The third line is the amount of oxygen that was expected from calculations:

Amount of enzyme used (cm³)	1	2	3	4	5
Oxygen produced (cm³)	3.2	8.9	9.5	12.7	15.9
	3.1	6.4	9.7	12.5	16.1
Calculated oxygen production (cm³)	4.2	8.4	12.5	16.6	20.7

d) Discuss whether there is any evidence for random error in these results.
e) Discuss whether there is any evidence for systematic error in these results.

Anomalies

Anomalous results are clearly out of line. They are not those that are due to the natural variation you get from any measurement. These should be looked at carefully. There might be a very interesting reason why they are so different. If they are simply due to a random error, then they should be discarded (rejected).

If anomalies can be identified while you are doing an investigation, then it is best to repeat that part of the investigation.

If you find anomalies after you have finished collecting data for an investigation, then they must be discarded.

SUMMARY QUESTIONS

1 Copy and complete using the words below:

accurate   discarded   random   sensitivity   systematic   use   variation

There will always be some ...... in results. You should always choose the best instruments that you can to get the most ...... results. You must know how to ...... the instrument properly. The ...... of an instrument refers to the smallest change that can be detected. There are two types of error – ...... and .... Anomalies due to random error should be ....

2 Which of the following will lead to a systematic error and which to a random error?
   a) Using a weighing machine, which has something stuck to the pan on the top.
   b) Forgetting to re-zero the weighing machine.

DID YOU KNOW?

Sir Alexander Fleming was showing his research assistant some plates on which he had grown bacteria. He noticed an anomaly. There was some mould growing on one of the plates and around it there were no bacteria. He investigated further and grew the mould, identifying it as *Penicillium rubrum*.

He persuaded an assistant to taste it and he said it tasted like Stilton cheese! He later injected the assistant with it – and he didn't die!

Only because Fleming checked out his anomaly did it lead to the discovery of penicillin. Oh, and Fleming also let his nose dribble onto one plate and he discovered lysozyme!!

KEY POINTS

1 Results will nearly always vary.
2 Better instruments give more accurate results.
3 Sensitivity of an instrument refers to the smallest change that it can detect.
4 Human error can produce random and systematic errors.
5 We examine anomalies; they might give us some interesting ideas. If they are due to a random error, we repeat the measurements. If there is no time to repeat them, we discard them.

H6 Presenting data

LEARNING OBJECTIVES

1 What do we mean by the 'range' and the 'mean' of the data?
2 How do you use tables of results?
3 How do you display your data?

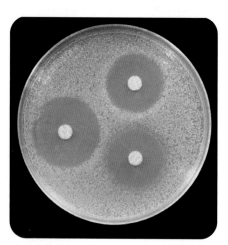

Figure 1 Petri dish with discs showing growth inhibition

For this section you will be working with data from this investigation:

Mel spread some bacteria onto a dish containing nutrient jelly. She also placed some discs onto the jelly. The discs contained different concentrations of an antibiotic. The dish was sealed and then left for a couple of days.

Then she measured the diameter of the clear part around each disc. The clear part is where the bacteria have not been able to grow. The bacteria grew all over the rest of the dish.

Tables

Tables are really good for getting your results down quickly and clearly. You should design your table **before** you start your investigation.

Your table should be constructed to fit in all the data to be collected. It should be fully labelled, including units.

In some investigations, particularly fieldwork, it is useful to have an extra column for any notes you might want to make as you work.

While filling in your table of results you should be constantly looking for anomalies.

- Check to see if a repeat is sufficiently close to the first reading.
- Check to see if the pattern you are getting as you change the independent variable is what you expected.

Remember a result that looks anomalous should be checked out to see if it really is a poor reading or if it might suggest a different hypothesis.

Planning your table

Mel knew the values for her independent variable. We always put these in the first column of a table. The dependent variable goes in the second column. Mel will find its values as she carries out the investigation.

So she could plan a table like this:

Concentration of antibiotic (μg/ml)	Size of clear zone (mm)
4	
8	
16	
32	
64	

Or like this:

Concentration of antibiotic (μg/ml)	4	8	16	32	64
Size of clear zone (mm)					

All she had to do in the investigation was to write the correct numbers in the second column to complete the top table.

Mel's results are shown in the alternative format in the table below:

Concentration of antibiotic (µg/ml)	4	8	16	32	64
Size of clear zone (mm)	4	16	22	26	28

The range of the data

Pick out the maximum and the minimum values and you have the range. You should always quote these two numbers when asked for a range. For example, the range is between .... (the lowest value) and .... (the highest value) – and don't forget to include the units!

a) What is the range for the dependent variable in Mel's set of data?

The mean of the data

Often you have to find the mean of each repeated set of measurements.

You add up the measurements in the set and divide by how many there are. Miss out any anomalies you find.

The repeat values and mean can be recorded as shown below:

Concentration of antibiotic (µg/ml)	Size of clear zone (mm)			
	1st test	2nd test	3rd test	Mean

Displaying your results

Bar charts

If you have a categoric or an ordered independent variable and a continuous dependent variable then you should use a bar chart.

Line graphs

If you have a continuous independent and a continuous dependent variable then a line graph should be used.

Scatter graphs (or scattergrams)

Scatter graphs are used in much the same way as line graphs, but you might not expect to be able to draw such a clear line of best fit. For example, if you wanted to see if lung capacity was related to how long you could hold your breath, you would draw a scatter graph with your results.

NEXT TIME YOU...

... make a table for your results remember to include:
- headings,
- units,
- a title.

... draw a line graph remember to include:
- the independent variable on the x-axis,
- the dependent variable on the y-axis,
- a line of best fit,
- labels, units and a title.

GET IT RIGHT!

Marks are often dropped in the exam by candidates plotting points incorrectly. Also use a **line of best fit** where appropriate – don't just join the points 'dot-to-dot'!

KEY POINTS

1 The range states the maximum and the minimum values.
2 The mean is the sum of the values divided by how many values there are.
3 Tables are best used during an investigation to record results.
4 Bar charts are used when you have a categoric or an ordered independent variable and a continuous dependent variable.
5 Line graphs are used to display data that are continuous.

SUMMARY QUESTIONS

1 Copy and complete using the words below:

categoric   continuous   mean   range

The maximum and minimum values show the ...... of the data. The sum of all the values divided by the total number of the values gives the .... Bar charts are used when you have a ...... independent variable and a continuous dependent variable.
Line graphs are used when you have ...... independent and dependent variables.

2 Draw a graph of Mel's results from the top of this page.

H7 | Using data to draw conclusions

Identifying patterns and relationships

Now that you have a bar chart or a graph of your results you can begin to look for patterns. You must have an open mind at this point.

Firstly, there could still be some anomalous results. You might not have picked these out earlier. How do you spot an anomaly? It must be a significant distance away from the pattern, not just within normal variation.

A line of best fit will help to identify any anomalies at this stage. Ask yourself – do the anomalies represent something important or were they just a mistake?

Secondly, remember a line of best fit can be a straight line or it can be a curve – you have to decide from your results.

The line of best fit will also lead you into thinking what the relationship is between your two variables. You need to consider whether your graph shows a **linear** relationship. This simply means can you be confident about drawing a straight line of best fit on your graph? If the answer is yes – then is this line positive or negative?

a) Say whether graphs (i) and (ii) in Figure 1 show a positive or a negative linear relationship.

Look at the graph in Figure 2. It shows a positive linear relationship. It also goes through the origin (0,0). We call this a **directly proportional** relationship.

Your results might also show a curved line of best fit. These can be predictable, complex or very complex! Look at Figure 3 below.

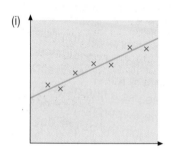

(i)

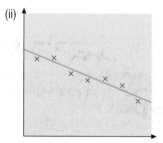

(ii)

Figure 1 Graphs showing linear relationships

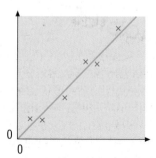

Figure 2 Graph showing a directly proportional relationship

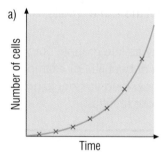

a) Number of cells / Time

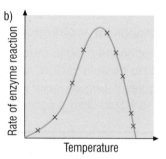

b) Rate of enzyme reaction / Temperature

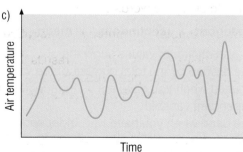

c) Air temperature / Time

Figure 3 a) Graph showing predictable results. b) Graph showing complex results. c) Graph showing very complex results.

Drawing conclusions

Your graphs are designed to show the relationship between your two chosen variables. You need to consider what the relationship means for your conclusion.

There are three possible links between variables. (See page 5.) They can be:

● causal,

● due to association, or

● due to chance.

You must decide which is the most likely. Remember a positive relationship does not always mean a causal link between the two variables.

Poor science can often happen if a wrong decision is made here. Newspapers have said that living near electricity sub-stations can cause cancer. All that scientists would say is that there is possibly an association. Getting the correct conclusion is very important.

You will have made a prediction. This could be supported by your results. It might not be supported or it could be partly supported. Your results might suggest some other hypothesis to you.

Your conclusion must go no further than the evidence that you have. A grass snake is not poisonous, but this does not mean that all snakes are safe to handle!

Evaluation

If you are still uncertain about a conclusion, it might be down to the reliability and the validity of the results. You could check these by:

● looking for other similar work on the Internet or from others in your class,

● getting somebody else to re-do your investigation, or

● trying an alternative method to see if you get the same results.

FEATURE!

HELPING YOU TO SLIM

Thermogenics is described as a new science that describes how the body changes food into heat energy rather than fat. The research claims to have found chemicals that will cause food to be used for making heat rather than fat and so help to make you thinner, by stimulating the thermogenic process.

KEY POINTS

1 Drawing lines of best fit help us to study the relationship between variables.
2 The possible relationships are linear, positive and negative; directly proportional; predictable and complex curves.
3 Conclusions must go no further than the data available.
4 The reliability and validity of data can be checked by looking at other similar work done by others, perhaps on the Internet. It can also be checked by using a different method or by others checking your method.

SUMMARY QUESTIONS

1 Copy and complete using the words below:

anomalous   complex   directly   negative   positive

Lines of best fit can be used to identify …… results. Linear relationships can be …… or …… . If a graph goes through the origin then the relationship could be …… proportional. Often a line of best fit is a curve which can be predictable or …… .

2 Nasma found a newspaper article about slimming. (See above.) Discuss the type of experiment and the data you would expect to see to support this conclusion.

H8 Scientific evidence and society

LEARNING OBJECTIVES

1 How can science encourage people to have faith in its research?
2 How might bias affect people's judgement of science?
3 Can politics influence judgements about science?
4 Do you have to be a professor to be believed?

DID YOU KNOW?

During the middle part of the last century, Stalin had an iron grip on the agriculture in what was the USSR. The effect was to starve to death up to 8 million people. Biological research was to be the scapegoat and many scientists had their work expunged. Vavilov was on a field trip, he had just discovered a new species of wheat, when he was hunted down, put into prison and died of starvation two years later.

Now you have reached a conclusion about a piece of scientific research. So what is next? If it is pure research then your fellow scientists will want to look at it very carefully. If it affects the lives of ordinary people then society will also want to examine it closely.

You can help your cause by giving a balanced account of what you have found out. It is much the same as any argument you might have. If you make ridiculous claims then nobody will believe anything you have to say.

Be open and honest. If you only tell part of the story then someone will want to know why! Equally, if somebody is only telling you part of the truth, you cannot be confident with anything they say.

a) A disinfectant claims that it kills 99.9% of germs on surfaces that you come in contact with every day. What is missing? Is it important?

You must be on the lookout for people who might be biased when representing scientific evidence. Some scientists are paid by companies to do research. When you are told that a certain product is harmless, just check out who is telling you.

b) Bottles of perfume spray contain this advice 'This finished product has not been tested on animals.' Why might you mistrust this statement?

Suppose you wanted to know about how to slim. Who would you be more likely to believe? Would it be a scientist working for 'Slimkwik', or an independent scientist? Sometimes the differences are not quite so obvious.

We also have to be very careful in reaching judgements according to who is presenting scientific evidence to us. For example, if the evidence might provoke public or political problems, then it might be played down.

Equally others might want to exaggerate the findings. They might make more of the results than the evidence suggests. Take as an example the data available on animal research. Animal liberation followers may well present the *same* evidence completely differently to pharmaceutical companies wishing to develop new drugs.

c) Check out some websites on smoking and lung cancer. Do a balanced review looking at tobacco manufacturers as well as anti-smoking lobbies such as ASH. You might also check out government websites.

The status of the experimenter may place more weight on evidence. Suppose a lawyer wants to convince a jury that a particular piece of scientific evidence is valid. The lawyer will choose the most eminent scientist in that field who is likely to support them. Cot deaths are a particularly difficult problem for the police. If the medical evidence suggests that the baby might have been murdered then the prosecution and the defence get the most eminent scientists to argue the reliability and validity of the evidence. Who does the jury believe?

EXPERT WITNESS IN COT DEATH COURT CASE MISLED THE JURY

A child abuse expert was struck off as a doctor today for giving seriously misleading evidence in a court case. The court case led to a woman being wrongly convicted of murdering her two children. **Full report – Page 6**

SUMMARY QUESTIONS

1 Copy and complete using the words below:

   status   balanced   bias   political

   Evidence from scientific investigations should be given in a …… way. It must be checked for any …… from the experimenter.
   Evidence can be given too little or too much weight if it is of …… significance.
   The …… of the experimenter is likely to influence people in their judgement of the evidence.

2 Collect some newspaper articles to show how scientific evidence is used. Discuss in groups whether these articles are honest and fair representations of the science. Consider whether they carry any bias.

3 This is the opening paragraph from a review of GM foods.

   The UK government has been promoting ... a review of the science of GM, led by Sir David King (the Government's Chief Scientific Adviser) working with Professor Howard Dalton (the Chief Scientific Adviser to the Secretary of State for the Environment, Food and Rural Affairs), with independent advice from the Food Standards Agency.

   Discuss this paragraph and decide which parts of it make you want to believe the evidence they might give. Then consider which parts make you mistrust any conclusions they might reach.

KEY POINTS

1 Scientific evidence must be presented in a balanced way that points out clearly how reliable and valid the evidence is.
2 The evidence must not contain any bias from the experimenter.
3 The evidence must be checked to appreciate whether there has been any political influence.
4 The status of the experimenter can influence the weight placed on the evidence.

H9

How is science used for everybody's benefit?

LEARNING OBJECTIVES

1 How does science link to technology?
2 How is science used and abused?
3 How are decisions made about science?
4 What are the limitations of science?

The development of oral contraceptives shows how science can be used for technological development. There were many unscientific ways in which women would attempt contraception.

In Europe, women wore the foot of a weasel around their neck to prevent them becoming pregnant. In North Africa the flower silphium was thought to be an oral contraceptive. It became very expensive to buy and eventually was used so much that it became extinct.

These ideas might have had some basis, because some plants do contain human sex hormones. Some plants such as black kohosh, are used today as relief from problems related to the menopause.

Figure 1 A yam plant

Yam was used as a pain relief. A Japanese scientist found diosgenin in yam plants. Diosgenin was used as the starting point for an investigation that led to the development of the hormone progesterone. This led to the development of the first contraceptive pill. It was some unscientific thinking that encouraged people to eat yam plants as a natural contraceptive – they aren't!

Frank Colton developed Enovid, one of the first oral contraceptives, in 1960. By 1961 the birth control pill was available to 'everyone'. Some doctors were in a dilemma for social as well as medical reasons. They would not prescribe it to unmarried women, because it 'encourages sex outside marriage'.

The UK Government said it couldn't afford the cost. They said the pill could have long-term effects. A woman's body was likened to a clock; 'Whilst it was running well it should be left alone,' said Sir Charles Dodds, a leading expert on drugs. The pill allowed women to take control of their own fertility.

Figure 2 Frank Colton

It was known that there was a risk of heart disease and stroke. However, new developments have reduced the dose and therefore this risk.

Today there are hundreds of different contraceptive pills. There are even male contraceptive pills. The morning-after pill has raised problems for some people who consider it is a form of abortion.

Some of these hormones are now in such a high concentration in river water that it affects the ability of some fish to reproduce.

Figure 3 Contraceptive pills

The contraceptive pill still raises social, ethical, economic and even environmental issues.

There are many questions left for science to answer. For example, how to develop an oral contraceptive that is 100% safe for all people. However, science cannot answer questions about whether or not we should use contraception.

SUMMARY QUESTIONS

Use the account of the development of contraceptive pills to answer these questions.

1 What scientific knowledge was available to Frank Colton that enabled him to develop Enovid?

2 How did different groups of people react to the development of Enovid?

3 a) Identify some of these issues raised by the development of contraceptive pills: i) ethical, ii) social, iii) economic, iv) environmental.
   b) Which of these issues are decided by individuals and which by society?

KEY POINTS

1 Scientific knowledge can be used to develop technologies.
2 People can exploit scientific and technological developments to suit their own purposes.
3 The uses of science and technology can raise ethical, social, economic and environmental issues.
4 These issues are decided upon by individuals and by society.
5 There are many questions left for science to answer. But science cannot answer questions that start with 'Should we .... ?'

SUMMARY QUESTIONS

1 Fit these words into order. They should be in the order in which you might use them in an investigation.

design; prediction; conclusion; method; repeat; controls; graph; results; table; improve; safety

2 a) How would you tell the difference between an opinion that was scientific and a prejudiced opinion?

b) Suppose your were describing the height of plants for some fieldwork. What type of variable would you choose and why?

c) Explain the difference between a causal link between two variables and one which is due to association.

3 You might have observed that lichens do not grow where there is air pollution. You ask the question why. You use some theory to try to answer the question.

a) Explain what you understand by a hypothesis.

b) Sulfur dioxide in the air forms acids that attack the lichens. This is a hypothesis. Develop this into a prediction.

c) Explain why a prediction is more useful than a hypothesis.

d) Suppose you have tested your prediction and have some data. What might this do for your hypothesis?

e) Suppose the data does not support the hypothesis. What should you do to the theory that gave you the hypothesis?

4 a) What do you understand by a fair test?

b) Explain why setting up a fair test in fieldwork is difficult.

c) Describe how you can make your results valid in fieldwork.

d) Suppose you were carrying out an investigation into how pulse rates vary with exercise. You would need to carry out a trial. Describe what a trial would tell you about how to plan your method.

5 Suppose you were watching a friend carry out an investigation measuring the carbon dioxide produced by yeast cells. You have to mark your friend on how accurately she is making her measurements. Make a list of points that you would be looking for.

6 a) How do you decide on the range of a set of data?

b) How do you calculate the mean?

c) When should you use a bar chart?

d) When should you use a line graph?

7 a) What should happen to anomalous results?

b) What does a line of best fit allow you to do?

c) When making a conclusion, what must you take into consideration?

d) How can you check on the reliability of your results?

8 a) Why is it important when reporting science to 'tell the truth, the whole truth and nothing but the truth'?

b) Why might some people be tempted not to be completely fair when reporting their opinions on scientific data?

9 a) 'Science can advance technology and technology can advance science.'
What do you think is meant by this statement?

b) Who answers the questions that start with 'Should we . . . '?

10 You can see from this electron micrograph below that stomata are very small holes in the leaves of plants. They allow carbon dioxide to diffuse into the leaf cells for photosynthesis. The size of the hole is controlled by guard cells. It was suggested that the size of the hole might affect the rate at which carbon dioxide diffused through the hole.

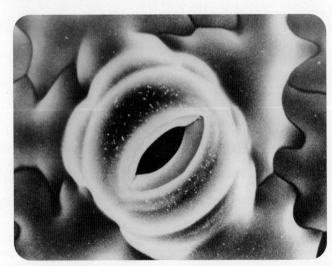

Electron micrograph of stomatal guard cells

Stomata are very small holes (when fully open they are 10–20 μm in diameter). The question was: Are small holes better than large holes? This would seem reasonable as plants have very small stomata. The hypothesis was that small holes would allow more carbon dioxide to pass through than large holes.

It was decided to use much larger holes than the stomata because it would be easier to get accurate measurements. The investigation was carried out and the results were as follows.

Diameter of hole (mm)	Volume of CO_2 diffusing per hour (cm^3)
22.7	0.24
12.1	0.10
6.0	0.06
3.2	0.04
2.0	0.02

a) What was the observation on which this investigation was based?

b) What was the original hypothesis?

c) What was the likely prediction?

d) What was the independent variable?

e) What was the dependent variable?

f) What is the range for the diameter of the hole?

g) Why was the temperature kept the same during the investigation?

h) Was this a sensible range of size of holes to use? Explain your answer.

i) How could the investigation have its reliability improved?

j) Was the sensitivity of the instrument measuring volumes of CO_2 satisfactory? Provide some evidence for your answer from the data in the table.

k) Draw a graph of the results in the table above.

l) Describe the pattern in these results.

m) What conclusion can you make?

n) Does your conclusion support the prediction?

B1a | Human biology

What you already know

Here is a quick reminder of previous work that you will find useful in this unit:

- When you reach puberty your body changes so that you can reproduce.

- Girls have a regular menstrual cycle when their body prepares for pregnancy. Boys begin making sperm which can fertilise an egg.

- Eating a balanced diet is an important part of keeping healthy.

- A balanced diet will include carbohydrates, proteins, fats, minerals, vitamins, fibre and water.

- The food you eat is used as a fuel during respiration. It provides energy for the cells of your body. Your food also gives you the raw materials you need to grow and to repair worn out cells.

- Your food needs breaking down (digesting) before it is any use in your body.

- Diseases caused by bacteria and viruses can affect your health.

- Your body can defend itself against disease-causing microorganisms. However sometimes you give it a helping hand by taking medicines or being immunised.

- Both legal and illegal drugs can damage your health if you abuse them. It is against the law to use illegal drugs.

RECAP QUESTIONS

1 What are the main food groups you need to eat to have a balanced diet?

2 How does your body use the food that you eat?

3 What are the main changes which take place in:

   a) boys and

   b) girls

   when they go through puberty?

4 How often is an egg produced in the menstrual cycle?

5 a) Which types of microorganism can cause disease?

   b) What is an *infectious* disease?

   c) Give the names of three infectious diseases.

6 How does your body defend itself against disease?

7 Explain how medicines and immunisation can help to keep you healthy.

Making connections

Becki and Sam want to have a baby but although they've been trying for over a year, Becki still isn't pregnant. She might not be making enough hormones to release an egg each month, so we are doing some blood tests to find out what is happening. If lack of hormones really is the problem, we can soon sort things out for them.

Sara makes sure baby Jaz gets his injections right on time. Years ago, lots of children died young from infectious diseases. By immunising babies like Jaz against many of the most dangerous diseases, we can help today's children grow up as healthily as possible.

Keeping well is important – if you are healthy you can enjoy life to the full at work, at school and at home. As doctors, we want to help people to be as healthy as possible, but usually we only see people when they are not feeling at their best. Meet some of our patients!

Abusing drugs – even legal ones like alcohol and tobacco – can cause all sorts of health problems, from a bad hangover to death! We run special sessions to help people control their drinking and smoking. In spite of this we have quite a few patients every year who die from diseases like lung cancer, liver disease and heart disease. These are often closely linked with their smoking and drinking habits.

Liam's got a temperature, a headache and a terribly sore throat. When we looked into his throat it was covered in yellow pus. He's got a bad case of tonsillitis. A course of antibiotics will soon have him feeling much better and ready to go back to school!

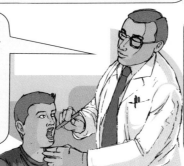

Being overweight isn't good for your joints or your heart. Some of our overweight patients find it hard to join in activities and they often get teased about being so big. We help them to lose weight and take more exercise, which makes them feel fitter – and better.

ACTIVITY

Many people think that doctors are only there to make them better when they feel ill. As you can see, most doctors do much more than this!

Design a poster to encourage people to use their GP to help them get healthy and stay healthy.

Your poster is going to be put up in all sorts of places – schools, libraries, shopping centres – so make sure it is clear, easy to read and gets the message across.

Chapters in this unit

 Co-ordination and control

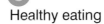 Healthy eating

 Drug abuse

Controlling infectious disease

B1a 1.1   Responding to change

You need to know what is going on in the world around you. Your *nervous system* makes this possible. It enables you to react to your surroundings and co-ordinate your behaviour.

Your nervous system carries electrical signals (*impulses*) which travel fast – from 1 to 120 metres per second. This means you can react to changes in your surroundings very quickly indeed.

Figure 1 Your body is made up of millions of cells which have to work together. Whatever you do with your body – whether it's winning a race or playing on the computer – your movements need to be co-ordinated. The conditions inside your body must also be controlled.

a) What is the main job of the nervous system?

Hormones are chemical substances. They control many of the processes going on inside your body. Special *glands* make and release (*secrete*) these hormones into your body. Then the hormones are carried around your body to their target organs in the bloodstream. They can act very quickly, but often their effects are quite slow and long lasting.

b) What type of messengers are hormones?

The nervous system

Like all living things, you need to avoid danger, find food and – eventually – find a mate! This is where your nervous system comes into its own. Your body is particularly sensitive to changes in the world around you. Any changes (known as **stimuli**) are picked up by cells called *receptors*.

These receptors are usually found clustered together in special **sense organs**, such as your eyes and your skin. You have many different types of sensory receptors (see Figure 3).

c) Where would you find receptors which would respond to i) a loud noise, ii) touching a hot oven, iii) a strong perfume?

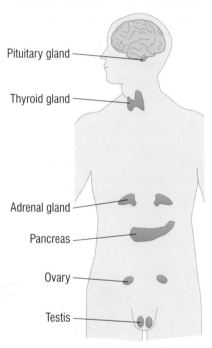

Pituitary gland
Thyroid gland
Adrenal gland
Pancreas
Ovary
Testis

Figure 2 Hormones act as chemical messengers. They are made in glands in one part of your body but having an effect somewhere else entirely

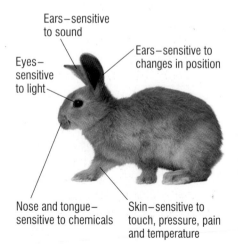

Ears – sensitive to sound
Ears – sensitive to changes in position
Eyes – sensitive to light
Nose and tongue – sensitive to chemicals
Skin – sensitive to touch, pressure, pain and temperature

Figure 3 Look at this rabbit. Being able to detect changes in the environment is important. It can often be a matter of life and death.

How your nervous system works

Once a sensory receptor detects a stimulus, the information (sent as an electrical impulse) passes along special cells called **neurones**. These are usually found in bundles of hundreds or even thousands of neurones known as *nerves*.

The impulse travels along the neurone until it reaches the **central nervous system** or **CNS**. The CNS is made up of the brain and the spinal cord. The cells which carry impulses from your sense organs to your central nervous system are called *sensory neurones*.

d) What is the difference between a neurone and a nerve?

Your brain gets huge amounts of information from all the sensory receptors in your body. It co-ordinates the information and sends impulses out along special cells. These cells carry information from the CNS to the rest of your body. The cells are called *motor neurones*. They carry impulses to make the right bits of your body – the **effector organs** – respond.

Effector organs are muscles or glands. Your muscles respond to the arrival of impulses by contracting. Your glands respond by releasing (**secreting**) chemical substances.

The way your nervous system works can be summed up as:

receptor → sensory neurone → co-ordinator → motor neurone → effector
(CNS)

e) What is the difference between a sensory neurone and a motor neurone?

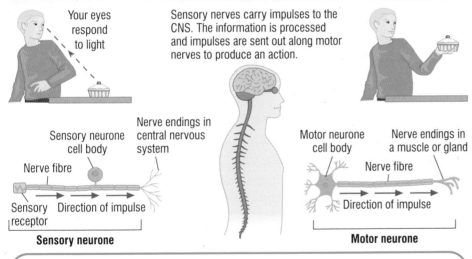

GET IT RIGHT!

Be careful to use the terms **neurone** and **nerve** correctly. Talk about **impulses** (*not* **messages**) travelling along a neurone.

Figure 4 The rapid responses of our nervous system allow us to respond to our surroundings quickly – and in the right way!

KEY POINTS

1 Hormones, secreted by special glands, are chemicals that help control and co-ordinate processes in your body.
2 The nervous system uses electrical impulses to enable you to react to your surroundings and co-ordinate what you do.
3 Cells called receptors detect stimuli (changes in the environment).
4 Impulses from receptors pass along sensory neurones to the brain. Impulses are sent from the brain to the effector organs along motor neurones.

SUMMARY QUESTIONS

1 Copy and complete using the words below:

   blood   chemical   electrical   glands   nervous

   Your …… system carries fast …… impulses. Your hormones are …… messengers secreted by special …… and carried around the body in the ……. .

2 Make a table to show the different types of sense receptors. For each one, give an example of the sort of things it responds to, e.g. touch receptors respond to an insect crawling on your skin.

3 Explain i) what happens in your nervous system when you see a piece of chocolate, pick it up and eat it, ii) the differences between hormonal and nervous control in your body.

B1a 1.2   Reflex actions

Your nervous system lets you take in information about the world around you and respond in the right way. However some of your responses are so fast that they happen without giving you time to think.

When you touch something hot, or sharp, you pull your hand back before you feel the pain. If something comes near your face, you blink. Automatic responses like these are known as **reflexes**.

What are reflexes for?

Reflexes are very important both for human beings and for other animals. They help you to avoid danger or harm because they happen so fast. There are also lots of reflexes which take care of your basic body functions. These functions include breathing and moving the food through your gut.

It would make life very difficult if you had to think consciously about those things all the time – and could be fatal if you forgot to breathe!

a) Why are reflexes important?

How do reflexes work?

Reflex actions involve just three types of neurone. These are:

- sensory neurones,
- motor neurones, and
- relay neurones which simply connect a sensory neurone and a motor neurone. We find relay neurones in the CNS, often in the spinal cord.

An electrical impulse passes from the sensory receptor along the sensory neurone to the CNS. It then passes along a relay neurone (usually in the spinal cord) and straight back along a motor neurone. From there the impulse arrives at the effector organ (usually a muscle in a reflex). We call this a *reflex arc*.

The key point in a reflex arc is that the impulse bypasses the conscious areas of your brain. The result is that the time between the stimulus and the reflex action is as short as possible. When you put your hand on something hot, you have moved your hand away before you feel the pain!

b) Why is it important that the impulses in a reflex arc do not go to the conscious brain?

How synapses work

Your nerves are not joined up directly to each other. There are junctions between them called **synapses**. The electrical impulses travelling along your neurones have to cross these synapses but they cannot leap the gap. Look at Figure 1 to see what happens next.

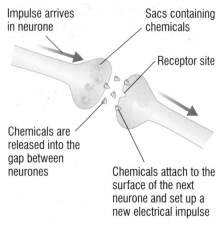

Impulse arrives in neurone

Sacs containing chemicals

Receptor site

Chemicals are released into the gap between neurones

Chemicals attach to the surface of the next neurone and set up a new electrical impulse

Figure 1 When an impulse arrives at the junction between two neurones, chemicals are released which cross the synapse and arrive at **receptor sites** on the next neurone. This starts up an electrical impulse in the next neurone.

The reflex arc in detail

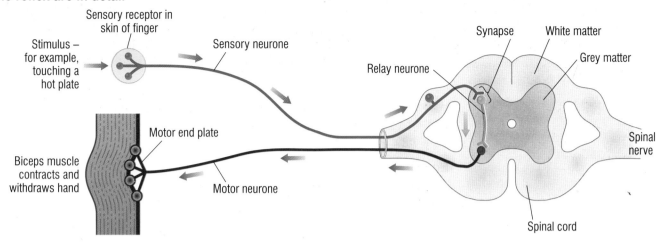

Figure 2 The reflex action which moves your hand away from something hot can save you from a nasty burn!

Look at Figure 2. It shows what would happen if someone touched a hot object.

When they touch it, a receptor in their skin is stimulated. An electrical message passes along a sensory neurone to the central nervous system – in this case the spinal cord.

When an impulse from the sensory neurone arrives in the synapse with a relay neurone, a chemical message is released. This crosses the synapse to the relay neurone and sets off an electrical impulse that travels along the relay neurone.

When the impulse reaches the synapse between the relay neurone and a motor neurone returning to the arm, another chemical message is released.

This crosses the synapse and starts an electrical impulse travelling down the motor neurone. When the impulse reaches the organ (effector), it is stimulated to respond. In this example the impulses arrive in the muscles of the arm, causing them to contract and move the hand rapidly away from the source of pain.

Most reflex actions can be shown as follows:

stimulus → receptor → co-ordinator → effector → response

This is not very different from a normal conscious action. However, in a reflex action the co-ordinator is a relay neurone either in the spinal cord or in the unconscious areas of the brain. The whole reflex is very fast indeed.

Figure 3 A baby's gripping reflex is very strong

SUMMARY QUESTIONS

1 Copy and complete using the words below:

**conscious   motor   reflex   relay   response
sensory   stimulus**

In a ...... arc the electrical impulse bypasses the ...... areas of your brain. The time between the ...... and the ...... is as short as possible. Only ...... neurones, ...neurones and ...... neurones are involved.

2 Explain why some actions, such as breathing and swallowing, are reflex actions, while others such as speaking and eating are under your conscious control.

3 Draw a flow chart to explain what happens when you step on a pin. Make sure you include an explanation of how a synapse works.

KEY POINTS

1 Some responses to stimuli are automatic and rapid and are called reflex actions.
2 Reflex actions run everyday bodily functions and help you to avoid danger.

B1a 1.3 The menstrual cycle

LEARNING OBJECTIVES

1 How is the menstrual cycle controlled?
2 When is a woman most likely to conceive?

Figure 1 The bodies of young boys and girls work in very similar ways. But once the sex hormones kick in during puberty, some big differences appear in the shape of their bodies and how they work. This shows you the power of the hormones!

Hormones control the functions of many of your body organs. They also control the activities of your individual cells. A woman's **menstrual cycle** is a good example of how this control works.

Hormones made in a woman's brain and in her ovaries control her menstrual cycle. The levels of the different hormones rise and fall in a regular pattern. This affects the way her body works.

What is the menstrual cycle?

The average length of the menstrual cycle is about 28 days. Each month the lining of the womb thickens ready to support a developing baby. At the same time an egg starts maturing in the ovary.

About 14 days after the egg starts maturing it is released from the ovary. This is known as **ovulation**. The lining of the womb stays thick for several days after the egg has been released.

If the egg is fertilised by a sperm, then pregnancy takes place. The lining of the womb provides protection and food for the developing embryo. If the egg is not fertilised, the lining of the womb and the dead egg are shed from the body. This is the monthly bleed or *period*.

All of these changes are brought about by hormones. These are made and released by the **pituitary gland** (a pea sized gland in the brain) and the **ovaries**.

a) What controls the menstrual cycle?
b) Why does the lining of the womb build up each month?

How the menstrual cycle works

Once a month, a surge of hormones from the pituitary gland in the brain starts eggs maturing in the ovaries. The hormones also stimulate the ovaries to produce the female sex hormone *oestrogen*.

- **FSH:** secreted by the pituitary gland. It makes eggs mature in the ovaries. *FSH* also stimulates the ovaries to produce *oestrogen*.

- **Oestrogen:** made and secreted by the ovaries. It stimulates the lining of the womb to build up ready for pregnancy. It also stimulates the pituitary gland to make another hormone known as *LH*.

- **LH:** secreted by the pituitary gland. It stimulates the release of a mature egg from one of the ovaries in the middle of the menstrual cycle.

c) Which hormones are made in the pituitary gland?
d) Which hormone is made by the ovary?

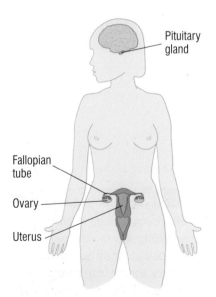

Pituitary gland

Fallopian tube

Ovary

Uterus

Figure 2 Hormones from the pituitary and the ovaries work together to control a woman's fertility

The hormones produced by the pituitary gland and the ovary act together to control what happens in the menstrual cycle. As the oestrogen levels rise they inhibit (slow down) the production of FSH and encourage the production of LH by the pituitary. When LH levels reach a peak in the middle of the cycle, they stimulate the release of a mature egg.

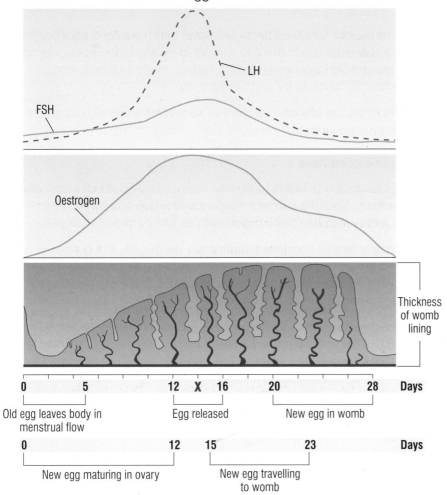

Figure 3 The changing levels of the female sex hormones control the different stages of the menstrual cycle

GET IT RIGHT!

Make sure you know the difference between eggs maturing and eggs being released.

SUMMARY QUESTIONS

1 Copy and complete using the list below:

   28   hormones   FSH   LH   menstrual   oestrogen   ovary

   During the …… cycle a mature egg is released from the …… about every …… days. The cycle is controlled by several …… including ……, …… and …… .

2 Look at Figure 3.

   a) On which day is the woman most likely to get pregnant?
   b) On which days is she having a menstrual period?
   c) On which day is the level of LH highest?
   d) Which hormone controls the build up of the lining of the womb?

3 Produce a leaflet to explain the events of the menstrual cycle to women who are hoping to start a family. You will need to explain the graphs at the top of this page and show when they are most likely to get pregnant.

KEY POINTS

1 Hormones control the release of an egg from the ovary and the build up of the lining of the womb in the menstrual cycle.

2 The main hormones involved are FSH and LH from the pituitary gland and oestrogen from the ovary.

B1a 1.4

The artificial control of fertility

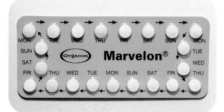

Figure 1 The contraceptive pill contains a mixture of hormones which effectively trick the body into thinking it is already pregnant, so no more eggs are released

For centuries people have tried to control when they have children. They have used substances from camel dung to vinegar to try and stop people having babies. Other people have carved fertility figures, made sacrifices and swallowed horrible herbs to try and have a child.

But it is only in the last fifty years or so that scientists have really been able to help couples control their own fertility, if they choose to do so.

Contraceptive chemicals

In the 21st century it is possible to choose when to have children – and when not to have them. One of the most important and widely used ways of controlling fertility is to use *oral contraceptives* (the *contraceptive pill*).

The pill contains female hormones, particularly oestrogen. The hormones affect your ovaries, preventing the release of any eggs. The pill inhibits (stops) the production of FSH so no eggs mature in the ovaries. Without mature eggs, you can't get pregnant.

Anyone who uses the pill as a contraceptive has to take it very regularly. If they forget to take it, the artificial hormone levels drop. Then their body's own hormones can take over very quickly. This can lead to the unexpected release of an egg – and an unexpected baby!

Fertility treatments

In the UK as many as one couple in six have problems having a family when they want one. There are many reasons for this infertility. It may be linked to a lack of female hormones. Some women want children but simply do not make enough FSH to stimulate the eggs in their ovaries. Fortunately artificial FSH can be used as a fertility drug. It stimulates the eggs in the ovary to mature and also triggers oestrogen production.

Figure 2 Most people who take fertility drugs end up with one or two babies. But the Walton family in the UK had six baby girls who all survived and are now young adults in their own right!

Fertility drugs are also used when a couple is trying to have a baby by IVF (*in vitro* fertilisation). If your fallopian tubes are damaged, eggs cannot reach your womb so you cannot get pregnant naturally.

Fortunately doctors can now help. They remove eggs from the ovary and fertilise them with sperm outside the body. Then they place the tiny developing embryos back into the uterus of the mother, bypassing the faulty tubes.

To produce as many ripe eggs as possible for IVF, the woman is given fertility drugs as part of her treatment. IVF is expensive and not always successful.

Advantages and disadvantages

The use of hormones to control fertility has been a major scientific breakthrough. But like most things there are pros and cons!

In the developed world, using the pill has helped make families much smaller than they used to be. There is less poverty because people have fewer mouths to feed.

The pill has also helped to control population growth in countries such as China, where they find it difficult to feed all their people. In many other countries of the developing world the pill is not available because of a lack of money, education and doctors.

The pill can cause health problems so a doctor always oversees its use.

The use of fertility drugs can also have some health risks for the mother and it can be expensive for society. A large multiple birth can be tragic for the parents if some or all of the babies die. It also costs hospitals a lot of money to keep very small premature babies alive.

Controlling fertility artificially also raises many ethical issues for society and individuals. For example, some religious groups think that preventing conception is denying life and ban the use of the pill.

The mature eggs produced by a woman using fertility drugs may be stored, or fertilised and stored, until she wants to get pregnant later. But what happens if the woman dies, or does not want the eggs or embryos any more?

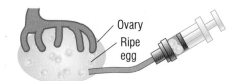

1 Fertility drugs are used to make lots of eggs mature at the same time for collection

2 The eggs are collected and placed in a special solution in a petri dish

3 A sample of semen is collected

4 The eggs and sperm are mixed in the petri dish

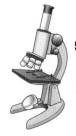

5 The eggs are checked to make sure they have been fertilised and the early embryos are developing properly

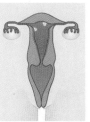

6 When the fertilised eggs have formed tiny balls of cells, 1 or 2 of the tiny embryos are placed in the uterus of the mother. Then, if all goes well, at least one baby will grow and develop successfully.

Figure 3 New reproductive technology using hormones and IVF has helped thousands of infertile couples to have babies

DID YOU KNOW?

In the early days of using fertility drugs there were big problems with the doses used. In 1971 an Italian doctor removed fifteen four-month-old fetuses (ten girls and five boys) from the womb of a 35-year-old woman after treatment with fertility drugs. Not one of them survived.

SUMMARY QUESTIONS

1 Define the following terms: oral contraceptive, fallopian tube, fertility drug, *in vitro* fertilisation.

2 Explain how artificial female hormones can be used to:

a) prevent unwanted pregnancies,
b) help people overcome infertility.

3 What, in your opinion, are the main advantages and disadvantages of using artificial hormones to control female fertility?

KEY POINTS

1 Hormones can be used to control fertility.
2 Oral contraceptives contain hormones, which stop FSH production so no eggs mature.
3 FSH can be used as a fertility drug for women, to stimulate eggs to mature in their ovaries. These eggs may be used in IVF treatments.

B1a 1.5   **Controlling conditions**

LEARNING OBJECTIVES

1 How are conditions inside your body controlled?
2 Why is it so important to control your internal environment?

The conditions inside your body are known as its *internal environment*. Your organs cannot work properly if this keeps changing. Many of the processes which go on inside your body aim to keep everything as constant as possible. This balancing act is called **homeostasis**.

It involves your nervous system, your hormone systems and many of your body organs.

a) Why is homeostasis important?

Controlling water and ions

Water can move in and out of your body cells. How much it moves depends on the concentration of mineral ions (like salt) and the amount of water in your body. If too much water moves into or out of your cells, they can be damaged or destroyed.

You take water and minerals into your body as you eat and drink. You lose water as you breathe out, and in your sweat. You lose salt in your sweat as well. You also lose water and salt in your *urine*, which is made in your *kidneys*.

Your kidneys can change the amount of salt and water lost in your urine, depending on your body conditions. They play an important part in controlling the balance of water and mineral ions in your body. The concentration of the urine produced by your kidneys is controlled by a combination of nerves and hormones.

b) What do your kidneys control?

Figure 1 Running a marathon affects your internal environment

PRACTICAL

Helping your body out

When you do a lot of exercise you lose a lot of salt and water from your body as you sweat. It is important to keep your cells hydrated so your body can work properly.

There are lots of special 'sports drinks' you can buy. They claim to rehydrate your body fast, supply you with energy and replace the salt you have lost. Some people think that plain water is just as good! Your kidneys control your internal environment very effectively unless you are exercising really hard for a long time. However, the manufacturers of sports drinks have scientific evidence to back up their claims. You can investigate these claims, and discover just what the drinks contain. See if they help you to perform better!

● How will you carry out your investigation?

Figure 2 A real help in sport – or a good way of making money? Sports drinks are becoming more and more popular, but do most of us really need them?

Controlling temperature

It is vital that your deep core body temperature is kept at 37°C. At this temperature your **enzymes** work best. At only a few degrees above or below normal body temperature the reactions in your cells stop and you will die.

Your body controls your temperature in several ways. For example, you can sweat to cool down and shiver to warm up. You can also change your clothing or turn up the heating! Your nervous system is very important in coordinating the way your body responds to changes in temperature.

c) What is the ideal body temperature?

Controlling blood sugar

When you digest a meal, lots of sugar (glucose) passes into your blood. Left alone, the blood glucose levels would keep changing. The levels would be very high straight after a meal, but very low again a few hours later. This would cause chaos in your body. However, the concentration of glucose in your blood is kept constant by hormones made in your *pancreas*.

d) What would happen to your blood sugar level if you ate a packet of sweets?

Figure 3 Sweets like this are almost all sugar. When you eat them your body has to deal with the effect on your blood.

SUMMARY QUESTIONS

1 Copy and complete using the words below:

> **body   constant   homeostasis   hormones**
> **internal environment   nervous system**

Your ...... cannot work properly if your .... keeps changing. ...... describes the processes which keep everything as ...... as possible. This control of conditions involves your .... and your ....

2 Why is it important to control:

   a) water levels in the body
   b) the body temperature
   c) sugar (glucose) levels in the blood?

3 Look at the marathon runners in Figure 1 on page 32. List the ways in which the running is affecting their:

   a) water balance,   b) ion balance,   c) temperature.

   d) It is much harder to run a marathon in a costume than in running clothes. Explain why this is.

4 Outline in a flow chart how you could evaluate the claims of the manufacturers of sports drinks.

KEY POINTS

1 Humans need to maintain a constant internal environment, controlling levels of water, ions, blood sugar and temperature.

2 Homeostasis is the result of the coordination of your nervous system, your hormones and your body organs

B1a 1.6   Controlling fertility

The contraceptive question

About 3.5 million women in the UK rely on the contraceptive pill to help them plan their families. This month's post-bag is full of queries about this widely-used form of contraception. We're letting our experts get to grips with your FAQs!

Question:

'My daughter is getting married soon and she has gone on the pill. How does it work, and can she really rely on it?' — **Mary**

Our expert's reply:

The pill contains female hormones, which are very similar to the ones in your daughter's body that control her menstrual cycle. The pill stops the production of a hormone called FSH made in your daughter's brain. FSH stimulates the eggs in your daughter's ovaries to mature each month. No FSH means no mature eggs — and no pregnancy.

Your daughter and her husband can certainly rely on the pill to stop her getting pregnant — but only if your daughter remembers to take her pill regularly! If the artificial hormone levels drop, her own body hormones can take over and an egg can be released — which can lead to an unplanned pregnancy!

A tiny pill containing a mixture of female hormones can trick your body and prevent it from releasing any eggs — which makes sure you can't get pregnant!

Question:

'What are the risks of using the pill?' — **Geeta**

Our expert's reply:

The big advantage of the pill is that it prevents pregnancy — and half a million women die each year around the world giving birth. But no medical treatment is risk-free.

The pill can have some mild side effects like headaches and slight weight gain. But in a few people it can cause blood clots, high blood pressure, heart attacks and strokes.

For most people the benefits of the pill outweigh the risks — but only a doctor can prescribe the pill. This ensures your health is looked after carefully and the risks kept as small as possible.

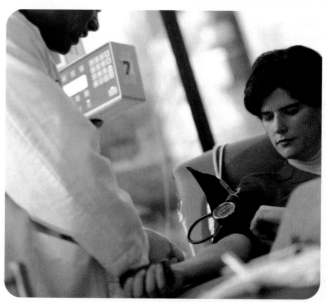

Anyone in the UK who takes the contraceptive pill will have a regular check-up of their blood pressure and general health to make sure they get the benefits of the pill with as little risk as possible

Question:

'If the pill is so good, why doesn't everyone use it so we can really control the world population?' — **Sarah**

Our expert's reply:

The pill is playing an important part in helping to control population growth in many countries, including China. It has also helped people out of the poverty trap in parts of the developing world — smaller families mean more food to go around.

But in some countries the pill is not available because there isn't enough money, education or doctors for it to be used properly. Also, there are some religious groups who feel that preventing conception is denying life, and so they ban the use of the pill for their believers. You're right — it's a difficult issue!

New reproductive technology using hormones and IVF has made it possible for women in their 50s and 60s to have babies of their own – but is it a good idea?

I married late – I was 40 – and we wanted a family, but my periods stopped when I was 41. Now we have a chance again. I haven't got any eggs so doctors will use FSH as a fertility drug to help them take lots of eggs from my donor (a younger woman). We want this child so much!

We've got three lovely children. I decided to donate some of my eggs to help couples who aren't as lucky as we are. I don't mind the age of the woman who gets my eggs as long as she manages to have a baby and loves it!

I think it is disgraceful and un-natural for women to have babies at this age. We are interfering with nature and with God's will and no good will come of it. The mother might die before the child is an adult!

I can't see anything wrong with older women having babies as long as they are fit and well. I know some people object to it, but some women have babies in their fifties naturally – and lots of men father children in their 60s and even their 70s and no-one objects to that, do they?

All our evidence shows that infertility treatment is just as successful in older women as it is in younger ones. We have to use artificial hormones to get the womb ready but once the women are pregnant their own hormones take over.

There is a lot of debate about the issues explored on these two pages. Use what you have learned in this chapter to help you write a 2–3 minute report for your school radio. It will go out in a regular slot called *Science Issues*. Choose one of these for your report:

- The contraceptive pill – good or bad?
- Older mothers – should science help?

SUMMARY QUESTIONS

1 Match up the following parts of sentences:

a) Many processes in the body	A ...... effector organs.
b) The nervous system allows you	B ...... secreted by glands.
c) The cells which are sensitive to light	C ...... to react to your surroundings and co-ordinate your behaviour.
d) Hormones are chemical substances	D ...... are found in the eyes.
e) Muscles and glands are known as	E ...... are known as nerves.
f) Bundles of neurones	F ...... are controlled by hormones.

2 a) What is the job of your nervous system?

b) Where in your body would you find nervous receptors which respond to:

i) light?

ii) sound?

iii) heat?

iv) touch?

c) Draw a simple diagram of a reflex arc. Explain carefully how a reflex arc works and why it allows you to respond quickly to danger.

3 a) What is the menstrual cycle?

b) What is the role of:

i) FSH

ii) LH

iii) oestrogen

in the menstrual cycle?

4 a) Explain carefully the difference between nervous and hormone control of your body.

b) What is a synapse and why are they important in your nervous system?

c) How can hormones be used to control the fertility of a woman?

5 It is very important to keep the conditions inside the body stable.

a) Taking part in school sports on a hot day without a drink bottle for the afternoon would be difficult for your body. Explain how your body would keep the internal environment as stable as possible.

b) Plan an investigation to see whether sports drinks or water are most effective in helping you perform well when you are exercising.

EXAM-STYLE QUESTIONS

1 Oral contraceptives can stop someone becoming pregnant by . . .

A preventing the ovaries from releasing an egg.

B killing eggs that have been released from the ovaries.

C killing sperms before they can reach an egg.

D preventing a fertilised egg from implanting in the uterus lining. (1)

2 A person puts their foot on a sharp object. They automatically lift their foot. The structures involved in this reflex action are shown below:

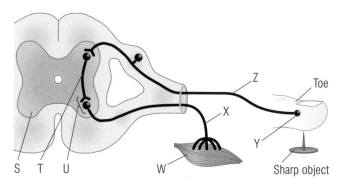

(a) The relay neurone is labelled with the letter

A T	**B** U
C X	**D** Z

(b) The motor neurone is labelled with the letter

A T	**B** X
C Y	**D** Z

(c) A synapse is labelled with the letter

A S	**B** U
C W	**D** Y

(d) The structure labelled W is known as

A a synapse	**B** a receptor
C a coordinator	**D** an effector organ

(e) In this reflex action the correct path taken by an impulse is . . .

A sensory neurone → receptor → coordinator → motor neurone → effector.

B effector → coordinator → receptor → sensory neurone → motor neurone.

C receptor → sensory neurone → coordinator → motor neurone → effector.

D coordinator → receptor → sensory neurone → effector → motor neurone. (5)

3 The graph below shows the concentrations of three hormones involved in the menstrual cycle:

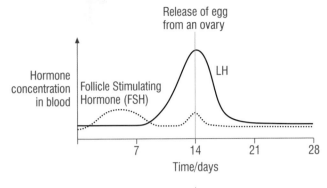

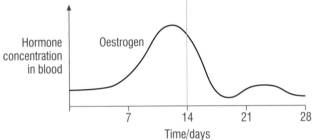

(a) Both FSH and LH are produced in the same gland. What is the name of this gland? (1)

(b) Where in the body is the hormone oestrogen produced? (1)

(c) What is the name given to the release of an egg from the ovary? (1)

(d) The lining of the uterus thickens from around day 5 until some days after the egg has been released. Suggest two purposes for this thickened lining? (2)

(e) Use the information in the graph as well as your knowledge to explain how the concentration of oestrogen affects and controls the release of an egg during the menstrual cycle. (4)

4 Some women are unable to have children naturally. The hormone FSH can sometimes be used to help these women have children.

(a) (i) Harriet does not produce eggs at ovulation. Explain how FSH could be the cause of the problem. (1)

(ii) Explain how FSH could help Harriet to produce and release an egg. (4)

(b) Sharon has had an infection of her Fallopian tubes that has left them blocked. Although she still produces eggs, they are unable to pass down the Fallopian tubes. Describe a method by which Sharon and her partner John could still have children. Include in your account the role of FSH. (4)

HOW SCIENCE WORKS QUESTIONS

The girls in the class set a challenge to the boys. The girls suggested that they had much better control over their nervous reactions than the boys did. The boys accepted the challenge and agreed to the investigation.

The equipment was set up as you can see in this picture.

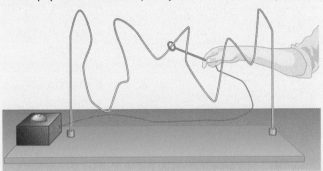

Five girls and five boys took it in turns to move the metal ring along the wire. If anyone touched the ring onto the wire the circuit would be completed and the bell would ring. The teacher counted the number of times the bell rang for each student.

The results are in this table:

Girls' names	Number of touches	Boys' names	Number of touches
Alexandra	6	Arthur	7
Farzana	0	Barnaby	2
Kerry	4	Zahir	4
Summer	3	Jameel	1
Annabel	8	Terry	5

a) In this investigation, which was the dependent variable? (1)

b) Suggest one variable that had not been controlled. (1)

c) Why did the group decide to use the teacher to record the results? (1)

d) Calculate the average for:
i) the boys   ii) the girls. (2)

e) Which of the following words would you use to describe the independent variable?
i) continuous   ii) categoric
iii) ordered   iv) discrete. (1)

f) How might you present these results?
i) bar chart   ii) line graph
iii) scattergraph   iv) use a line of best fit. (1)

g) Do you think that the girls' prediction is supported by the data collected? Explain your answer. (2)

B1a 2.1   Diet and exercise

Figure 1 Everyone needs a source of energy to survive – and your energy source is your food. Whatever food you eat – whether you prefer sushi, dahl, or roast chicken – most people eat a varied diet that includes everything you need to keep your body healthy.

What makes a healthy diet?

A healthy diet contains:

- carbohydrates,
- proteins,
- fats,
- vitamins,
- minerals,
- fibre and
- water

and the energy you need to live, all in the right amounts!

If your diet isn't balanced, you will end up **malnourished**. If you don't take in enough vitamins and minerals, you will end up with deficiency diseases like scurvy. (Scurvy is caused by a lack of vitamin C.)

Fortunately, in countries like the UK, most of us take in all the minerals and vitamins we need from the food we eat. However, our diet can easily be less well balanced in terms of the energy we take in. If we take in too much energy we get fat – but if we don't eat enough we get too thin.

It isn't always easy to get it right because different people need different amounts of energy.

a) Why do you need to eat food?

How much energy do you need?

The amount of energy you need to live depends on lots of different things. Some of these things you can change and some you can't.

If you are male, you will need to take in more energy than a female of the same age – unless she is pregnant.

If you are a teenager, you will need more energy than if you are in your 70s – and there isn't much you can do about it!

b) Why does a pregnant woman need more energy than a woman who isn't pregnant?

The amount of exercise you do affects the amount of energy you use up. If you do very little exercise, then you don't need much food. The more you exercise, the more food you need to take in. Your food supplies energy to your muscles as they work.

People who exercise regularly are usually much fitter than people who take little exercise. They make bigger muscles – and muscle tissue burns up much more energy than fat. But exercise doesn't always mean time spent training or 'working out' in the gym. Walking to school, running around the house and garden looking after small children or having a physically active job all count as exercise too.

c) Why do athletes need to eat more food than the average person?

Figure 2 Athletes who spend a lot of time training and playing a sport will have a great deal of muscle tissue on their bodies – up to 40% of their body mass. So they have to eat a lot of food to supply the energy they need.

The temperature where you live affects your energy needs as well. The warmer it is, the less energy you need. This is because you have to use less energy keeping your body temperature at a steady level. So you need to take in less food!

Figure 3 If you live somewhere really cold, you need lots of high-energy fats in your diet. You need the energy to keep warm!

The metabolic rate

Imagine two people who are very similar in age, sex and size. However, they may still need quite different amounts of energy in their diet. This is because the rate at which all the chemical reactions in the cells of the body take place (the **metabolic rate**) varies from person to person.

The proportion of muscle to fat in your body affects your metabolic rate. Men generally have a higher proportion of muscle to fat than women, so they have a higher metabolic rate. You can change the proportion of muscle to fat in your body by exercising and building up more muscle.

Your metabolic rate is also affected by the amount of activity you do. Exercise increases your metabolic rate for a time even after you stop exercising.

Finally, scientists think that your basic metabolic rate may be affected by factors you inherit from your parents.

GET IT RIGHT!

Metabolic rate is not the same as heart rate or breathing rate – make sure you know the difference.

SUMMARY QUESTIONS

1 What do we mean by 'a balanced diet'?

2 a) Why does an old person need less energy in their diet than a teenager?
  b) Why does a top footballer need more energy in their diet than you do? Where does the energy in the diet come from?

3 a) What is meant by the 'metabolic rate'?
  b) Explain why some people put on weight more easily than others.

KEY POINTS

1 Most people eat a varied diet, which includes everything needed to keep the body healthy.
2 Different people need different amounts of energy.
3 The metabolic rate varies from person to person.
4 The less exercise you take, the less food you need.

B1a 2.2 Weight problems

1 What health problems are linked to being overweight?
2 Why is it unhealthy to be too thin?

Figure 1 In spite of some of the media hype, most people are not obese – but the amount of weight people carry certainly varies a great deal!

Human beings come in all sorts of shapes and sizes. Most people look about right but there will always be extremes. Some people are very overweight and others appear unnaturally thin. Scientists and doctors don't just measure what you weigh. They look at your **body/mass index** or **BMI**. This compares your weight to your height in a simple formula:

$$\text{BMI} = \frac{\text{weight}}{(\text{height})^2}$$

Most people have a BMI in the range 20–30. But if you have a BMI of below 18.5, or above 35, then you may have some real health problems.

a) What does your body/mass index measure?

Obesity

If you take in more energy than you use, the excess is stored as fat. You need some body fat to cushion your internal organs. Your fat also acts as an energy store for when you don't feel like eating. But if someone eats a lot more food than they need, over a long period of time, they could end up **obese**.

Carrying too much weight is often inconvenient and uncomfortable. Far worse, it can lead to serious health problems. Obese people are more likely to suffer from **arthritis** (worn joints), **diabetes** (high blood sugar levels which are hard to control), **high blood pressure** and **heart disease**. They are more likely to die young than slimmer people.

b) What health problems are linked to obesity?

Losing weight

Many people want to lose weight. This might be for their health or just to look better. You gain fat by taking in more energy than you need, so there are three main ways you can lose it.

● You can reduce the amount of energy you take in by cutting back the amount of food you eat – particularly energy-rich foods like biscuits, crisps and chips.
● You can increase the amount of energy you use up by taking more exercise.
● And the best way to lose weight is to do both – reduce your energy intake and exercise more!

Many people find it easier to lose weight by attending slimming groups. At these weekly meetings they get lots of advice, plus support from other slimmers. All the different slimming programmes involve eating less food and/or taking more exercise!

Increasing your exercise levels can be an important part of losing weight and getting fitter. However, you need to take care. If you suddenly start working out hard in the gym, or taking other vigorous exercise, you can cause other health problems.

Different slimming programmes approach weight loss in different ways. Many simply give advice on healthy living. They advise lots of fruit and vegetables, not too much fat or too many calories and plenty of exercise. Some are more extreme suggesting you cut out almost all of the fat or the carbohydrates from your diet.

Others claim that 'slimming teas' or 'herbal pills' will enable you to eat what you like and still lose weight. What sort of evidence would you look for to decide which approaches worked best?

c) What must you do to lose weight?

Starvation

In some parts of the world obesity is rare, because the biggest problem is lack of food. Civil wars, droughts and pests can destroy local crops so people cannot get enough to eat. Starvation leads to a number of symptoms including:

- You become very thin and your muscles waste away.

- Your immune system can't work properly so you pick up infections.

- If you are female, your periods will become irregular or stop altogether.

These symptoms are also sometimes seen in the developed world in people suffering from the mental disorder called **anorexia** (loss of appetite) **nervosa**.

d) What are the main symptoms of starvation?

Figure 2 Hundreds of thousands of people around the world suffer the symptoms of malnutrition and starvation. There is simply not enough food for them to eat.

GET IT RIGHT!

Make sure you can give specific examples of the problems caused by obesity and starvation.

SUMMARY QUESTIONS

1 Copy and complete using the words below:

energy   fat   less   more   obese

If you take in more ...... than you use, the excess is stored as .... If you eat too much over a long period of time, you will eventually become .... To lose weight you need to eat ...... and exercise ....

2 Plan a simple information sheet about the dangers of being overweight and how to lose weight sensibly.

3 Research the claims of two slimming programmes. Compare and evaluate the claims they make.

KEY POINTS

1 If you take in more energy than you use, you will store the excess as fat.

2 Obese people have more health problems than people of normal weight.

3 People who do not have enough to eat can develop serious health problems.

B1a 2.3   Fast food

1 What is cholesterol?
2 Why do your cholesterol levels matter?
3 Is too much salt bad for us?

Figure 1 Fast food tastes good. But it has had lots of things added to make it easy to cook and eat. These often include fat and salt.

People eat fast processed food because it is quick and easy and fits in with their busy lives. But it often contains a lot of fat and salt. These make the food taste good. However, there are some real concerns about the effect that too much fat and salt in your diet can have on your health.

a) What substances do you often find in fast foods?

Cholesterol

Fat is an energy-rich food. So too much fat in your diet can easily make you overweight. But that isn't the only problem with fatty food. The amount and type of fat you eat also seems to affect the levels of **cholesterol** in your blood.

Cholesterol is a substance which you make in your liver. It gets carried around your body in your blood. You need it to make the membranes of your body cells, your sex hormones and the hormones that help your body deal with stress. Without cholesterol, you wouldn't survive. Yet people often talk about cholesterol as if it is a bad thing. Why?

High levels of cholesterol in your blood seem to increase your risk of getting heart disease or diseased blood vessels. The cholesterol builds up in your blood vessels and can even block them. Heart disease is one of the main causes of death in the UK and USA, so no wonder doctors are worried.

b) Why do you need cholesterol in your body?

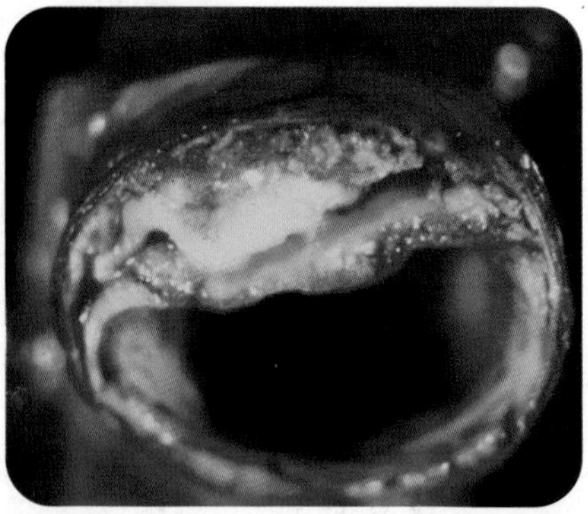

Controlling cholesterol

The amount of cholesterol you have in your blood depends on two things:

● The way your liver works, which is something you inherit from your parents and cannot change.
● The amount of fat in your diet.

Some people have livers that can deal with almost any amount of fat. Their blood cholesterol seems to stay within healthy levels. But for many people the level of cholesterol in their blood is linked to the amount and type of fat they eat.

It isn't just the overall level of cholesterol in your blood which affects your risk of developing heart disease. Cholesterol is carried around your body by two types of *lipoproteins*:

● **Low-density lipoproteins (LDLs)** are known as 'bad' cholesterol. Raised levels of LDLs increase your risk of heart problems.
● **High-density lipoproteins (HDLs)** are known as 'good' cholesterol and they reduce your risk of heart disease.

The balance of LDLs and HDLs in your blood is very important for a healthy heart.

Figure 2 When you get cholesterol building up in the wrong place – like the arteries leading to your heart – it can be very serious indeed

There are three main types of fats in the food you eat and they seem to have different effects on your cholesterol:

- **Saturated fats** increase (raise) blood cholesterol levels. You find them in animal fats like meat, butter and cheese.
- **Mono-unsaturated fats** seem to have two useful effects. They may reduce your overall blood cholesterol levels and improve the balance between LDLs and HDLs in your blood. You find them in foods like olive oil, olives, peanuts and lots of margarines.

- **Polyunsaturated fats** seem to be even better at reducing your blood cholesterol levels and balancing LDLs and HDLs than mono-unsaturates. You find them in foods including corn oil, sunflower oil, many margarines and oily fish.

c) What is the big difference between saturated fats and the other types of fats?
d) Why are raised blood cholesterol levels a worry?

What about salt?

Like fat, salt is vital in your diet. Without it, your nervous system would not work and the chemistry of all your cells would be in chaos. But for about a third of you (30% of the population), too much salt in your diet can lead to high blood pressure. This can damage your heart and kidneys and increase your risk of a stroke.

Many people eat too much salt each day without knowing it. That's because many processed, 'fast' foods contain large amounts of salt. But you can control your salt intake by doing your own cooking – or by reading the labels very carefully when you buy ready-made food!

NEXT TIME YOU...

... eat a burger and fries, think about all the fat you are taking on board. Will your liver be able to deal with it, or are your blood cholesterol levels about to go up?

Oil/fat	% saturated fat	% polyunsaturated fat	% mono-unsaturated fat
Butter	66	4	30
Corn oil	13	62	25
Olive oil	14	12	74
Sunflower oil	11	69	20

Figure 3 Once you start to look at the different fats in the food you are buying, shopping can get very complicated!

SUMMARY QUESTIONS

1 Copy and complete using the words below:

    salt   heart   salt   blood pressure   fat   cholesterol

Fast food can contain too much ...... and .... Raised ...... in the blood can lead to ...... disease, while too much ...... can give some people high ....

2 Look at Figure 3 and use it to help you answer these questions:

  a) Which fat or oil has the highest percentage of mono-unsaturates?
  b) Which fat or oil has the highest percentage of polyunsaturates?
  c) Which fat or oil has the highest percentage of saturated fats?
  d) Decide which of these fats or oils would be the best to use for a healthy heart, and which would be the worst. Explain your answer carefully, including the balance of LDLs and HDLs in your blood.

3 Many people want to lower the amount of salt in fast foods.

  a) Explain why salt is important in your diet.
  b) Why are people worried about high salt levels in food?
  c) Would lowering the salt levels in processed foods make everyone healthier? Explain your answer.

KEY POINTS

1 Fast food often contains high proportions of fat and/or salt.
2 Cholesterol is made in the liver and found in the blood. High cholesterol levels have been linked to heart disease.
3 The level and type of cholesterol in your blood is influenced by the type of fat you eat.
4 Too much salt in the diet can lead to raised blood pressure in about a third of the population.

B1a 2.4 Health issues

The Statin Revolution

ACTIVITY

Write a short report on statins for the health page of your local paper.

Doctors have an amazing new weapon against high cholesterol levels and the problems they can bring. They can use a group of drugs called **statins**. Statins stop the liver producing so much cholesterol. Patients need to keep to a relatively low fat diet as well for the best effects.

Here are some different opinions about these exciting new drugs:

Some people just can't get their cholesterol balance right by changing their diet. It doesn't matter how hard they try. I've been very pleased with the results using statins. Almost all my patients have now got healthy cholesterol levels. What's more, we have lost far fewer people to strokes and heart attacks since we started using the drugs.

We are delighted with the results we are getting with statins. We have got data from several really large, powerful research trials involving over 30,000 patients. The trials all show similar results. Using a statin drug can lower your chances of having a heart attack or stroke by 25 to 40% – and we didn't find too many side effects.

The great thing about these new statins that the doctor's given me is that they control my cholesterol for me. It's back to the cream cakes and chips for me – and I won't have to worry about my heart!

I'm so pleased with my new medicine – the pills have brought my cholesterol levels right down and I'm feeling really well

I'm very worried about possible side effects with these new tablets – the leaflet said they can cause liver damage. I know my cholesterol levels were very high without the tablets, but I think I'm going to stop taking them. I don't want my liver to rot!

Scientists wear blinkers – and we pay the price!

For many years now scientists and doctors have been telling us that we are at risk from heart disease because we eat too much animal fat and our blood cholesterol is too high. But a lot of people still die of heart disease. Now it seems that vitamins might be just as important to our hearts as fat. What's more, this idea was first discovered years ago – so why didn't we find out sooner?

Thirty years ago, Kilmer McCully was a young researcher at Harvard University in the USA. He discovered a possible link between an amino acid called homocysteine and changes in the blood supply to the heart which can lead to heart attacks. High levels of homocysteine are linked to low levels of B vitamins in the diet – and these B vitamins are often missing in processed foods!

Changing your diet or taking a cheap supplement of B vitamins lets your body remove the homocysteine and prevents the damage to your heart.

Unfortunately McCully did his research at the same time as many top scientists were supporting the link between fats and heart disease. Time and money had been spent developing anti-cholesterol drugs and low-fat foods. No-one wanted to hear about McCully's cheap and simple solution. He lost his funding at Harvard and his ideas were quashed.

Thirty years on – and in spite of the fact that we have all cut back on our fat levels and taken our anti-cholesterol medicines, deaths from heart disease are still high. Kilmer McCully's work is finally being taken seriously. Major trials on B vitamins and heart disease are taking place around the world. It seems increasingly likely that McCully really has found one of the pieces in the jigsaw which explains heart disease. It is just a pity that no-one would use it for so long! Perhaps scientists need to take the blinkers off and realise that there can be more than one solution to a problem!

ACTIVITY

Write a letter:
Either from the young Kilmer McCully to a friend explaining what you have discovered about a link between B vitamins and heart disease and what it might mean for patients;

Or from a senior scientist who has been working on treatments for high cholesterol and heart disease to one of his colleagues about McCully's work and how you feel about it.

Menu 1

Turkey twizzlers

Chicken nuggets

Pizza

Chips

Spaghetti hoops

Iced bun

Doughnut

Menu 2

Char-grilled chicken

Spaghetti Bolognese

Fish with pesto topping

Baked potato

Fresh fruit

Yoghurt

ACTIVITY

Plan an assembly for the year 7 pupils in your school on the importance of a healthy diet. It should include help with the food they should choose in the school canteen for lunch.

SUMMARY QUESTIONS

1 a) Define the following:
   i) Balanced diet.
   ii) Metabolic rate.

   b) A top athlete needs to eat a lot of food each day. This includes protein and carbohydrate. Explain how they can eat so much without putting on weight.

2 a) What is obesity?

   b) Why is obesity a threat to your health?

   c) Suggest some ways in which an obese person might lose weight.

   d) How do following a slimming diet and suffering from starvation differ?

3 a) What is the link between a high-fat diet, raised blood cholesterol and the LDL/HDL balance in your blood?

   b) Why are doctors concerned if a patient has raised cholesterol levels or a high ratio of LDLs to HDLs?

   c) Fast food is often linked to an unhealthy lifestyle. Explain the problems with fast foods – and why people still eat them.

   d) Recently there has been a lot of media interest in school dinners. People think they contain far too many 'fast foods' and not enough fresh produce, fruit and vegetables.

   Plan a short report for your local radio station on why healthy school meals are important for the future health of the children who eat them.

4 Here are two young people who have written to a lifestyle magazine problem page for advice about their diet and lifestyle. Produce an 'answer page' for the next edition of the magazine.

   a) Melanie: *I'm 16 and I worry about my weight a lot. I'm not really overweight but I want to be thinner. I've tried to diet but I just feel so tired when I do – and then I buy chocolate bars on the way home from school when my friends can't see me! What can I do?*

   b) Jaz: *I'm nearly 17 and I've grown so fast in the last year that I look like a stick! So my clothes look pretty silly. I'm also really good at football, but I don't seem as strong as I was and my legs get really tired by the end of a match. I want to build up a bit more muscle and stamina – but I don't just want to eat so much I end up getting really heavy. What can I do about it?*

EXAM-STYLE QUESTIONS

1 The table is about some conditions that affect the body as a result of certain diets.
Match descriptions **A**, **B**, **C** and **D** with the words 1 to 4 in the table.

A an imbalance of nutrients in the diet

B a severe shortage of food in the diet

C a psychological disorder leading to a dangerously low body mass

D a body/mass index above 30   (4)

	Condition
1	Obesity
2	Anorexia
3	Malnutrition
4	Starvation

2 The body/mass index (BMI) compares body mass to height. The BMI is calculated using the following formula:

$$BMI = \frac{\text{body mass in kg}}{(\text{height in metres})^2}$$

(a) A woman has a body mass of 60 kg and a height of 1.6 metres. Her BMI is equal to . . .

A 18.7   **B** 23.4   **C** 29.2   **D** 37.5   (1)

The graph shows the percentage of a population of people in different BMI groups who suffer from a form of arthritis.

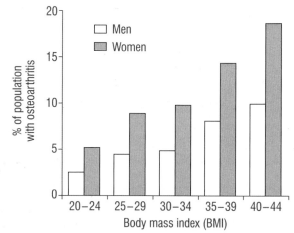

(b) From these data, which group of people have the highest percentage of osteoarthritis?

A Men with a BMI of 30–34.

B Men with a BMI of 35–39.

C Women with a BMI of 35–39.

D Women with a BMI of 40–44.   (1)

(c) The data were collected by measuring the BMI of a large sample of people. What is the main advantage of using a large sample of people?

A The data obtained are more reliable.

B The mean BMI can be calculated.

C The results obtained are fairer.

D People of many ages are included.   (1)

(d) What conclusion can be drawn from the data?

    A Women with a BMI of 35–39 are the group most likely to suffer from osteoarthritis.

    B The higher the BMI in both men and women the greater the risk of suffering from osteoarthritis.

    C The lower the BMI in men, the more likely they are to suffer from osteoarthritis.

    D Slimming will prevent osteoarthritis. (1)

3 (a) State the seven components that make up a healthy diet. (2)

An investigation was carried out over four different periods to find the energy intake of 14- and 15-year-old girls and boys. The results are shown in the table below.

Period of time	Average energy intake/kJ per day	
	Boys	**Girls**
1930s	12 873	11 088
1960s	11 739	9 534
1970s	10 962	8 484
1980s	10 478	8 316

(b) Calculate the percentage decrease in energy intake for girls between the 1930s and the 1980s. Show your working. (1)

(c) Explain why the intake of energy for both boys and girls decreased between the 1930s and the 1980s. (2)

(d) If the same study had been carried out with groups of 70-year-olds, how might the results have been different? (1)

(e) Suggest a reason why girls need to take in less energy than boys of the same age. (2)

(f) What would be the best way to display the data from the table above? Explain your choice. (2)

(g) Calculate the mean of the average energy intake of boys from the 1930s to the 1980s. (1)

(h) Who has the larger range of average energy intake between the 1930s and 1980s – boys or girls? Show your working out. (1)

4 What a person eats can affect their health. Explain how doing each of the following might help to keep a person healthy:

(a) Reducing the amount of saturated fat that is eaten. (3)

(b) Eating less salt. (2)

HOW SCIENCE WORKS QUESTIONS

A class of students were asked to test some fruit juices for their vitamin C content. They were given the apparatus set up as in the diagram below. They had to put the sample of fruit juice into the test tube and add the dye (DCPIP) from the burette drop by drop until the mixture retained the blue colour of the dye.

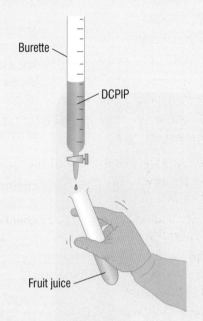

a) Name a control variable they would have to use. (1)

b) Describe how they should use the burette to get accurate results. (5)

The class results were as follows:

Juice	Volume of DCPIP added/cm³				
Orange	21.1	26.2	24.8	25.5	25.7
Apple	20.9	19.7	21.3	20.5	21.0

c) Calculate the mean for the amount of DCPIP added by the class to the apple juice. (1)

d) It was suggested that the first result for the orange juice was an anomaly.
Why was this thought to be an anomaly? (1)

e) Calculate the average for the amount of DCPIP added to the orange juice. (1)

f) What conclusion can you make from these results? (1)

g) One student commented that when carrying out the titration with the orange juice it was quite difficult to tell when the DCPIP stayed blue. She thought this was due to the colour of the orange juice. How does this idea affect your conclusion? (2)

B1a 3.1   Drugs

Figure 1 Millions of pounds worth of illegal drugs are brought into the UK every year. It is a constant battle for the police to find and destroy drugs like these.

A drug is a substance that alters the way in which your body works. It can affect your mind, your body or both. In every society there are certain drugs which people use for medicine, and other drugs which they use for pleasure.

Many of the drugs that are used both for medicine and for pleasure come originally from natural substances, often plants. Many of them have been known to and used by indigenous peoples for many years. Usually some of the drugs that are used for pleasure are socially acceptable, while others are illegal.

a) What do we mean by 'indigenous peoples'?

Drugs are everywhere in our society. People drink coffee and tea, smoke cigarettes and have a beer, an alcopop or a glass of wine. They think nothing of it. Yet all of these things contain drugs – caffeine, nicotine and alcohol (the chemical ethanol). These drugs are all legal.

Other drugs, such as cocaine, ecstasy and heroin are illegal. Which drugs are legal and which are not varies from country to country. Alcohol is legal in the UK as long as you are over 18, but it is illegal in many Arab states. Heroin is illegal almost everywhere.

b) Give an example of one drug which is legal and one which is illegal in the UK.

Because drugs affect the chemistry of your body, they can cause great harm. This is even true of drugs we use as medicines. However, because medical drugs make you better, it is usually worth taking the risk.

But legal drugs, such as alcohol and tobacco, and illegal substances, such as solvents, cannabis and cocaine, can cause terrible damage to your body. Yet they offer no long-term benefits to you at all.

What is addiction?

Some drugs change the chemical processes in your body so that you may become addicted to them. You can become dependent on them. If you are addicted to a drug, you cannot manage properly without it.

Once addicted, you generally need more and more of the drug to keep you feeling normal. When addicts try to stop using drugs they usually feel very unwell. They often have aches and pains, sweating, shaking, headaches and cravings for their drug. We call these **withdrawal symptoms**.

c) What do we mean by 'addiction'?

The problems of drug abuse

People take drugs for a reason. Drugs can make you feel very good about yourself. They can make you feel happy and they can make you feel as if your problems no longer matter. Unfortunately, because most recreational drugs are addictive, they can soon become a problem themselves.

No drugs are without a risk. Cannabis is often thought of as a relatively 'soft' – and therefore safe – drug. But evidence is growing which shows that it can cause serious psychological problems to develop in some people.

Hard drugs, such as cocaine and heroin, are extremely addictive. Using them often leads to very severe health problems. Some of these come from the drugs themselves, and some come from the lifestyle which often goes with drugs. Because they are illegal, they are expensive. Young people often end up turning to crime to pay for their drug habit. They don't eat properly or look after themselves. They can also end up with serious illnesses, such as hepatitis, STDs and HIV/AIDS.

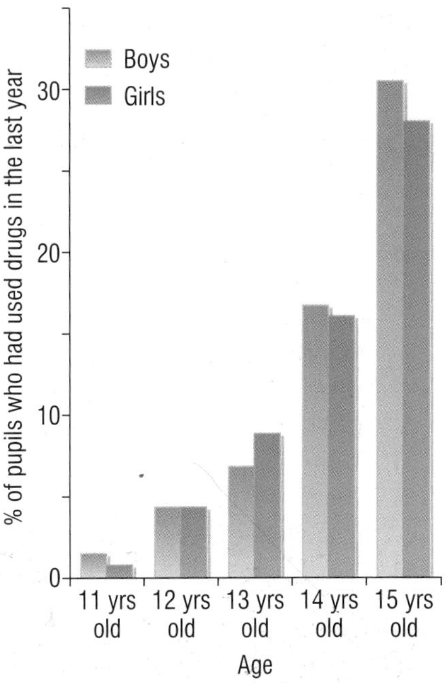

DON'T LET DRUG DEALERS CHANGE THE FACE OF YOUR NEIGHBOURHOOD.
Call Crimestoppers anonymously on 0800 555 111.

Figure 2 Most of the young people who have used drugs have smoked cannabis – but the number of 15-year-old pupils who have tried drugs is causing a lot of concern

Figure 3 Drugs can seem appealing, exciting and fun when you first take them. Many people use them for a while and then leave them behind. But the risks of addiction are high, and no-one can predict who drugs will affect most.

SUMMARY QUESTIONS

1 Copy and complete using the words below:

mind   cocaine   ecstasy   legal   alcohol   drug   body

A …… alters the way in which your body works. It can affect the ……the …… or both. Some drugs are …… e.g. caffeine and …… . Other drugs, such as ……, …… and heroin are illegal.

2 a) Why do people often need more and more of a drug?
   b) What happens if you stop taking a drug when you are addicted to it?

3 a) Look at Figure 2. Explain what this tells you about drug-taking in young people.
   b) Why do people take drugs?
   c) Explain some of the problems linked with using cannabis, cocaine and heroin.
   d) Why do you think young people continue to take these drugs when they are well aware of the dangers?
   e) Why does the impact of drug use vary from person to person?

KEY POINTS

1 Drugs change the chemical processes in your body, so you may become addicted to them.
2 Alcohol, tobacco and illegal drugs may harm your body.
3 Smoking cannabis may cause psychological problems.
4 Hard drugs, such as cocaine and heroin, are very addictive and can cause serious health problems.

B1a 3.2

Legal and illegal drugs

Figure 1 Even everyday drugs, like the caffeine we take for granted, have an effect on your nervous system and brain

Figure 2 NASA scientists have shown that common house spiders spin their webs very differently when given some of the commonly used drugs shown in the table on page 51. The effect of caffeine on the nervous system of a spider is particularly dramatic!

What is the most widely used drug in the world? It is probably one that most of you will have used at least once today, yet no-one really thinks about. The caffeine in your cup of tea, mug of coffee or can of cola is a drug.

Many people find it hard to get going in the morning without their first mug of coffee – they are probably addicted to the drug! Caffeine is a mild stimulant. It stimulates your brain and increases your heart rate and blood pressure.

a) What makes a can of cola a drug?

How do drugs affect you?

Many of the drugs used for medical treatments have little or no effect on your nervous system. However, all of the drugs which people use for pleasure (see table on page 51) affect the way your brain and nervous system work. It is these changes which people enjoy when they use the drugs. The same changes can cause addiction, so your body doesn't work properly without the drug.

Some drugs like caffeine, nicotine and cocaine speed up the activity of your brain. They make you feel more alert and energetic.

Others like alcohol and cannabis slow down the responses of your brain, making you feel calm and better able to cope. Heroin actually stops impulses travelling in your nervous system, so you don't feel any pain or discomfort. Other drugs like cannabis produce vivid waking dreams. You see or hear things which are not really there.

Why do people use drugs?

People use these drugs for a variety of reasons. For example, people feel that caffeine, nicotine and alcohol help them cope with everyday life. Few people who use these drugs would think of themselves as addicts, yet the chemicals can have a big impact on your brain (see Figure 2).

As for the other recreational drugs – people who try them may be looking for excitement or escape. They might want to be part of the crowd or just want to see what happens. Yet because many of these drugs are addictive, you don't have to try them many times before your body starts to demand a regular fix!

Marijuana   Benzedrine   Caffeine   Chloral hydrate

Some of these recreational drugs are more harmful than others. Most media reports on the dangers of drugs use focus on illegal drugs (see table below). But in fact the impact of legal drugs on health is much greater than the impact of illegal drugs. That's because far more people take them. Millions of people in the UK smoke – but only a few thousand take heroin.

b) Why do legal drugs cause many more health problems than illegal drugs?

Legal recreational drugs	Illegal recreational drugs
Ethanol (alcoholic drinks)	cannabis
Nicotine (cigarette smoke)	cocaine
Caffeine (coffee, tea, cola etc)	heroin
	ecstasy
	LSD

Drugs in sport

The world of sport has a major problem with the illegal use of drugs. In theory competition in sport is to find the best natural athlete. The only difference between the competitors should be their natural ability and the amount they train. However, there are many drugs which can enhance your performance in sport – and sadly some athletes use them and cheat. Drugs can build up your muscle mass, make you body produce more blood, make you more alert and speed up your reactions.

The sports authorities produce new tests for drugs and run random drugs tests to try and identify the cheats. Athletes are banned from competing if they are discovered using illegal drugs. But competitors are always looking for new ways to get ahead, so the illegal use of drugs in sport continues.

Figure 3 At the 2000 Summer Olympic Games in Sydney, Australia, the Romanian gymnast Andrea Raducan won a gold medal. It was taken away when she tested positive for a banned stimulant.

SUMMARY QUESTIONS

1 Copy and complete using the words below:

> **brain   health   illegal   legal   recreation**

> Drugs which people use for …… all affect the …… and nervous system. Some of these drugs are legal but some of them are …… More people suffer …… problems caused by the …… drugs than illegal ones.

2 a) What are the main reasons for using illegal drugs?
   b) Plan a TV advert against the use of illegal substances in sport to be shown in the run-up to the 2012 Olympics in the UK.

3 Compare the impact of legal and illegal drugs on individuals and on society.

KEY POINTS

1 Many recreational drugs affect the brain and nervous system.
2 Some recreational drugs are legal and others are illegal.
3 The overall impact of legal drugs on health is much greater than illegal drugs because more people use them.

B1a 3.3

Alcohol – the acceptable drug?

LEARNING OBJECTIVES

1 How does alcohol affect your body?
2 What is binge drinking?

Figure 1 You can buy alcoholic drinks like these legally in the UK once you are 18. But you have to be over 21 in many places in the USA and they are completely illegal in some other countries.

For many people **alcohol** is part of their social life. They like to share a drink with friends or enjoy a glass of wine with a meal. They probably don't think of themselves as drug users.

In small amounts, alcohol makes people feel relaxed and cheerful. It makes you less inhibited. So shy people can feel more confident when they've had an alcoholic drink.

But alcohol has a powerful effect on your body. It is very addictive and it is also very poisonous. Just imagine if alcohol was discovered today. It would almost certainly be illegal and thought of as a very dangerous drug.

In fact, alcohol is one of the most widely used drugs in the UK. Although some religions ban the use of alcohol, it is accepted all over the world. Perhaps this is because alcohol has been around for thousands of years. We also see that many important and famous people like a drink!

a) Why is alcohol described as a drug?

How does alcohol affect your body?

Alcohol is poisonous. However, your liver can usually break it down. Your liver gets rid of the alcohol before it causes permanent damage and death.

When you have an alcoholic drink, the alcohol passes through the wall of your gut and goes into your bloodstream. From your blood, the alcohol passes easily into nearly every tissue of your body. It gets into your nervous system and brain. This slows down your reactions. It can make you lose your self-control.

When you have had too much to drink, you lack judgement. You can end up making stupid or dangerous decisions. Some people end up making mistakes they regret for the rest of their lives.

If you drink large amounts of alcohol, like a whole bottle of spirits, your liver simply cannot cope. You would suffer from alcohol poisoning. This can quickly lead to unconsciousness, coma and death.

b) Give an example of a poor decision that someone under the influence of alcohol might make.

Some people drink heavily for many years, becoming **alcoholics**. They are addicted to the drug. Their liver and brain suffer long-term damage and eventually the drink may kill them.

They may develop **cirrhosis of the liver**. This disease destroys your liver tissue. They can also get *liver cancer*, which spreads quickly and can be fatal. In some heavy drinkers their brain is so damaged (it becomes soft and pulpy) that it can't work any longer. This causes death.

Short bouts of very heavy drinking can cause the same symptoms to develop quite quickly.

c) Which organs are most affected by heavy drinking?

DID YOU KNOW?

It takes your liver about one hour to break down the alcohol in a glass of wine or half a pint of beer (one unit of alcohol).

The effects of drinking on society

Alcohol can also put you at risk because of the way you behave under the influence of the drug. Because alcohol slows down your reactions, you are much more likely to have an accident. This is very dangerous if you drive after drinking. Alcohol is a factor in about 20% of all fatal road accidents in the UK.

Alcohol abuse affects personal lives as well. Domestic violence is often linked to patterns of heavy drinking. Many crimes take place when people are under the influence of alcohol, often mixed with other drugs.

Binge drinking is a recent problem. This often involves young people. They go out and get very drunk several nights a week. They become violent and abusive, damage property and put their own health at risk.

Alcohol related crime in the UK costs us around 20 billion pounds a year and causes great unhappiness. Add to this the medical costs of alcohol abuse and you see that we all pay a high price for this socially acceptable drug!

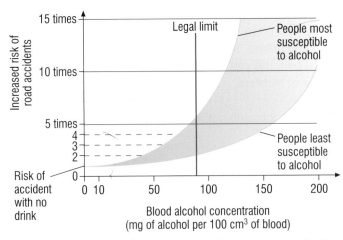

Figure 3 Alcohol affects your brain and your reactions – it is not surprising that if you drink alcohol and then drive a car you are much more likely to have an accident

SUMMARY QUESTIONS

1 Copy and complete using the words below:

drug   alcohol   brain   alcoholics   liver

...... is a poisonous .... It is broken down in your .... It can have a big effect on your ...... and your liver.

...... are people who are addicted to alcohol.

2 a) How does alcohol reach your brain after you have had a drink?

   b) What effect does alcohol have on your brain?

3 Look at Figure 3.

   a) What is the approximate legal limit of alcohol that you are allowed in your blood before you drive a car?

   b) Young people are often easily affected by alcohol. If a young person drinks enough to have 125 mg of alcohol per 100 cm³ of their blood and then drives a car, how will this affect their risk of having an accident?

   c) Police advise you not to drink alcohol at all if you are going to drive a car. Based on the evidence of the graph, explain why this is good advice?

   d) Summarise the effects of alcohol on society and discuss banning its use.

A healthy liver

Diseased liver from a heavy drinker with cirrhosis

Figure 2 Your liver deals with all the poisons you put into your body. But if you drink too much alcohol, your liver may not be able to cope. The difference between the healthy liver and the liver with cirrhosis shows just why people are warned against heavy drinking!

KEY POINTS

1 Alcohol affects your nervous system by slowing down your reactions.

2 Alcohol can lead to loss of self-control, unconsciousness, coma and death.

3 Alcohol can cause damage to your liver and brain.

B1a 3.4

Smoking and health

LEARNING OBJECTIVES

1 How does smoking tobacco affect your health?
2 How do pregnant women put their babies at risk by smoking?

Figure 1 Cigarette smoking increases your risk of developing many serious and fatal diseases. Every packet of cigarettes sold in the UK has to carry a clear health warning. Yet people still buy them in their millions!

Smoking is big business. There are 1.1 billion smokers worldwide, smoking around 6000 billion cigarettes each year!

As a cigarette burns it produces a smoke cocktail of around 4000 chemicals. If you smoke, you breathe this cocktail straight into your lungs. You absorb some of the chemicals into your blood stream. This carries them around your body to your brain.

a) How many chemicals do you inhale in cigarette smoke?

Nicotine is the addictive drug in *tobacco smoke*. It makes people feel calm, contented and able to cope. But you gradually need more and more of the drug to get the same effect. So the number of cigarettes you smoke each day tends to increase over the years. What's more, you feel awful if you don't get your regular dose of nicotine.

b) What is the drug in tobacco smoke?

Smoking related diseases

If you are a non-smoker, small hairs in your breathing system are constantly moving mucus away from your lungs. The mucus traps dirt, dust and bacteria from the air you breathe in. The hairs make sure you get rid of it all.

If you smoke, each cigarette anaesthetises these hairs. They stop working for a time, allowing dirt down into your lungs. This makes you much more likely to suffer from colds and other infections. The mucus also builds up and causes coughing (smoker's cough!).

Tar is a sticky black chemical in tobacco smoke that builds up in your lungs, turning them from pink to grey. It makes smokers much more likely to develop bronchitis. The build-up of tar in your lungs can also lead to the delicate air sacs in the lungs breaking down. We call this *emphysema*. It makes the lungs much less efficient. Your breathing becomes difficult and you can't get enough oxygen.

c) What colour are the lungs of i) a non-smoker? ii) a smoker?
d) What causes the difference in colour in c)ii)?

Tar is also a major **carcinogen** (a cancer causing substance). A build-up of tar can cause lung cancer. We can cure this cancer if it is caught early enough. However, it often grows in the lungs with no obvious symptoms. By the time doctors diagnose it, it has spread to other parts of the body and has become fatal. There are many other carcinogens in tobacco smoke as well.

Many of the chemicals in cigarette smoke are carried right round your body in the blood. Some of them affect your heart and blood vessels. Smoking raises your blood pressure and makes it more likely that your blood vessels will become blocked. This can cause heart attacks, strokes and thrombosis.

NEXT TIME YOU...

... see someone lighting up a cigarette, think of the 4000 chemicals they are drawing down into their lungs – and the damage that some of them can do!

FOUL FACTS

Smoking 20 to 60 cigarettes a day will coat your lungs in 1 to 1.5 pounds of tar every year!

Many people want to give up smoking but because nicotine is so addictive, it isn't easy. There are many different ways of giving up. Some are more effective than others. Some are much more expensive than others. The most important thing is always how much you want to give up. Each smoker has to find the method that suits them best and helps them to become a non-smoker!

Way of stopping smoking	% effectiveness of method
Hypnosis	30
Smoke aversion therapy	25
Group withdrawal clinics	24
Acupuncture	24
Education	18
Nicotine gum alone	10
Doctor's advice	1

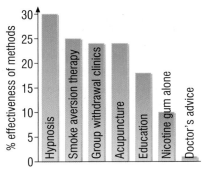

Figure 2 Many people try to give up smoking, but because nicotine is so addictive it isn't easy. There are lots of different methods which can help you – some seem to be more effective than others!

Smoking and pregnancy

Carbon monoxide is a very poisonous gas found in cigarette smoke. It is picked up by your red blood cells. This reduces the amount of oxygen carried in your blood. After smoking a cigarette, up to 10% of a smoker's blood will be carrying carbon monoxide rather than oxygen! This is one reason why smokers often get breathless going up the stairs!

Carbon monoxide in cigarette smoke affects pregnant women in particular. During pregnancy a woman needs oxygen, not just for her own cells but for her developing fetus as well.

If she smokes, the amount of oxygen in her blood will be lower than normal. This means her fetus will be deprived of oxygen. Then it may not grow as well as it should.

Mums who smoke when they are pregnant have a much higher risk of having:

- a premature birth (baby born too early so may struggle to survive),
- a baby with a low birth mass (so it is more at risk of developing problems),
- a stillbirth (where the baby is born dead).

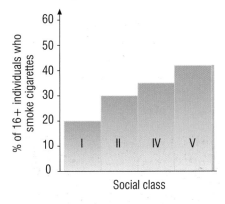

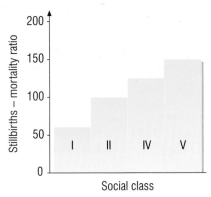

Figure 3 Data like this show that the people who smoke more are more likely to have stillborn babies. The same pattern is there for babies who are born weighing less than they should.

SUMMARY QUESTIONS

1 Define the following words: a) **nicotine**; b) **tar**; c) **carbon monoxide**; d) **carcinogenic**

2 Smokers are more likely to get infections of their breathing tubes and lungs than non-smokers. Explain why this is the case.

3 Explain how cigarette smoking is linked to an increased risk of getting:

   a) lung cancer   b) emphysema   c) heart disease.

4 a) Look at Figure 3. What is the evidence that smoking during pregnancy can be dangerous for the developing baby?
   b) Explain how smoking causes these problems during pregnancy.
   c) How do scientists try to make sure that the correlations they spot in their data are as reliable as possible?
   d) Carry out some research to find two ways of helping smokers kick the habit. Compare and contrast each method.

KEY POINTS

1 Tobacco smoke contains substances which can help to cause lung cancer, lung diseases such as emphysema and bronchitis, and diseases of the heart and blood vessels.

2 Tobacco smoke contains carbon monoxide, which reduces the oxygen-carrying capacity of the blood. In pregnant women this can deprive a fetus of oxygen and lead to a low birth mass or death.

B1a 3.5   Lung cancer and smoking

A timeline of evidence

At several stages in its history, smoking was thought to be really good for you. We now think up to 90% of all lung cancers are due to smoking! Here is a brief history of how the evidence against smoking began to build up:

1908 The sale of cigarettes to children under the age of sixteen was banned.

1912 Dr Isaac Adler suggested that there was a strong link between lung cancer and smoking based on what he saw among his patients.

1914–18 More people smoked than ever before during the First World War.

1925 The cigarette manufacturers set out to persuade women to smoke.

1930s Britain had the highest lung cancer rate in the world.

1930s Scientists in Germany found a strong statistical link between people who smoked and people who got lung cancer.

1939–45 Another war – and smoking continued to rise.

1951 Dr Richard Doll and Professor Austin Hill interviewed 5000 patients in British hospitals. They found that out of 1357 men with lung cancer, 99.5% of them were smokers. This was a very strong statistical link which connected smoking with a high risk of disease.

1953 Dr Ernst Wynder painted cigarette tar on the backs of mice and they developed cancers, showing a biological link between the chemicals in cigarettes and cancer.

1962 The Royal College of Physicians published a report suggesting the restriction of smoking, tax on tobacco products and warning of the dangers of smoking. For the first time in many years, cigarette sales fell.

1964 Doll and Hill published the results of a ten-year study into death rates in relation to smoking. They found a dramatic fall in lung cancer cases in people who had given up smoking compared to those who still smoked.

1997 Hackshaw and colleagues analysed 37 different studies. They found the risk of developing lung cancer in life-long non-smokers who lived with a smoker was 24% higher than if they lived with another non-smoker. They also found carcinogens from tobacco smoke in the blood of the non-smokers. People who lived with heavy smokers were more at risk than people living with light smokers. They concluded that '*The epidemiological and biochemical evidence . . . provides compelling confirmation that breathing other people's tobacco smoke is a cause of lung cancer.*'

> I really like smoking – and I'm really fit. I don't think it does you any harm at all.

> It is a free country – if people want to smoke they have every right to do so.

> My gran smoked all her life – and she lived until she was ninety!

> We're going to sue the tobacco company. Dad couldn't stop smoking, and it killed him. Yet they knew the risks all along!

> We need to keep them hooked. We can do this by increasing the amount of nicotine. This won't affect the tar figures, so we can still advertise them as low tar.

> We found out years ago that nicotine is addictive. However, we didn't publish our results in any scientific journals.

A change of image?

Tobacco arrived in Britain in the 16th century, when it was seen as a new and exciting thing. Since then it has been seen as a foul habit, the height of fashion and a calming influence during the wars. Now in the developed world, scientific evidence shows that smoking is a serious health risk. But in the developing world – and for many young people even in the UK – smoking is still seen as a glamorous and desirable habit . . .

ACTIVITY

Discuss the evidence shown here in small groups.

Deaths from lung cancer and smoking

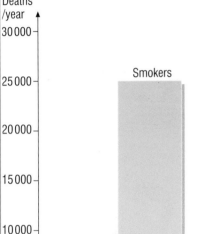

Cigarette consumption and risk of lung cancer death

Number of cigarettes smoked per day	Annual death rate per 100 000	Relative risk
0	14	–
1–14	105	8
15–24	208	15
25+	355	25

Death rates from lung cancer in men by age group: England and Wales 1974–92

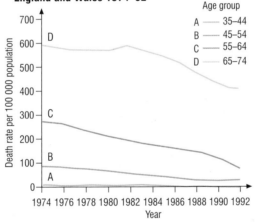

Age group
A —— 35–44
B —— 45–54
C —— 55–64
D —— 65–74

Age-adjusted death rates for lung cancer and breast cancer among women, United States, 1930–1997

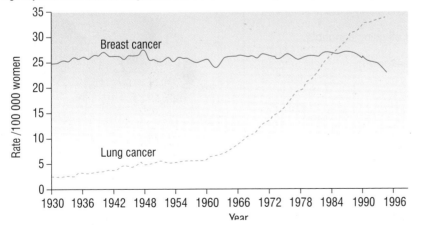

Evidence like this gradually builds up a compelling picture showing that smoking is a major risk factor in the development of lung cancer. Similar evidence can be put together linking smoking with cancers of the throat and tongue and with heart disease.

ACTIVITY

In 21st-century Britain most people accept that there is a link between smoking and lung cancer, heart disease and other health problems – but it doesn't stop them smoking! There has also been a lot in the press about *passive smoking* – breathing in other people's cigarette smoke. In some countries smoking has been banned in almost all public places.

Look at the evidence published by Hackshaw and colleagues in the *British Medical Journal* in 1997 along with all the other evidence here. Use it to carry out one of these two tasks:

Either: Plan an anti-smoking campaign in your school. Target teachers and pupils alike. Plan an article for the school magazine and a presentation which can be used in Citizenship lessons with year 9 pupils.

Or: Plan a campaign to have smoking banned in all public places in your local community – shops, pubs, cafes and restaurants, bars and bowling alleys. Think carefully about how to get the issues across to the general public. You might need posters, leaflets and/or a speech to make at a public meeting.

Whichever task you chose, use plenty of scientific evidence to help make people take notice!

B1a 3.6   Does cannabis lead to hard drugs?

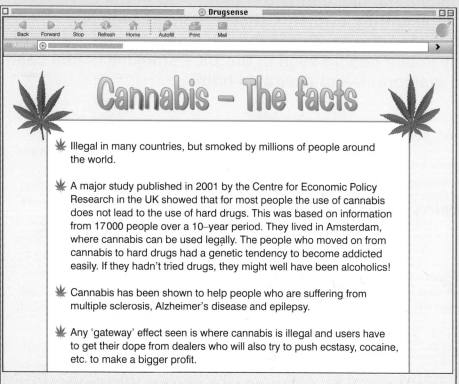

Drugsense

Cannabis – The facts

- Illegal in many countries, but smoked by millions of people around the world.

- A major study published in 2001 by the Centre for Economic Policy Research in the UK showed that for most people the use of cannabis does not lead to the use of hard drugs. This was based on information from 17 000 people over a 10–year period. They lived in Amsterdam, where cannabis can be used legally. The people who moved on from cannabis to hard drugs had a genetic tendency to become addicted easily. If they hadn't tried drugs, they might well have been alcoholics!

- Cannabis has been shown to help people who are suffering from multiple sclerosis, Alzheimer's disease and epilepsy.

- Any 'gateway' effect seen is where cannabis is illegal and users have to get their dope from dealers who will also try to push ecstasy, cocaine, etc. to make a bigger profit.

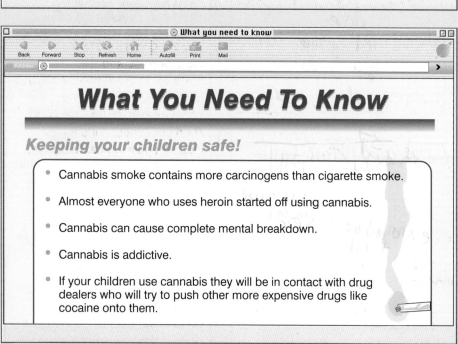

What you need to know

What You Need To Know

Keeping your children safe!

- Cannabis smoke contains more carcinogens than cigarette smoke.

- Almost everyone who uses heroin started off using cannabis.

- Cannabis can cause complete mental breakdown.

- Cannabis is addictive.

- If your children use cannabis they will be in contact with drug dealers who will try to push other more expensive drugs like cocaine onto them.

In the minds of many people – parents, teachers and politicians – cannabis is a 'gateway' drug. It opens the door to the use of other much harder drugs, such as cocaine and heroin. Your health – and indeed your life itself – is at risk. How accurate is this picture? Let's look at what the scientists say . . .

Scientist A: In some people cannabis use can trigger mental illness, which can be very serious and permanent. The people who are affected have usually been ill or have a family history of mental illness. In fact it seems as if these people are particularly attracted to drug use.

There have been some major studies in New Zealand involving many thousands of young people. These show a definite link between using cannabis and both depression and schizophrenia. In fact one research team calculated that if we stopped people using cannabis in the UK, the number of people with schizophrenia would drop by 13%.

Scientist B: There is very little evidence based on serious scientific research to say that cannabis use affects the long- or short-term memory of users or that cannabis has many of the physically damaging effects often reported in the popular press.

Scientist C: Almost all heroin users were originally cannabis users. This sounds as if cannabis use leads to heroin use. But think again! This is not a case of 'cause and effect' as we call it in science. Almost all cannabis users were originally smokers – but we don't claim that smoking cigarettes leads to cannabis use! In fact the vast majority of smokers do not go on to use cannabis – and the vast majority of cannabis users do not move on to hard drugs like heroin.

Scientist D: The number of people who are damaged or killed as a result of the use of illegal drugs each year is a tiny fraction of the people affected by the legal drugs alcohol and tobacco.

A lot of scientific research has been done into the effects of cannabis on our health, and on the links between cannabis use and addiction to hard drugs.

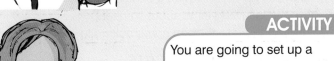

Unfortunately many of the studies have been quite small. They have not used large sample sizes, so the evidence is not reliable.

Most of the bigger studies show that the effects of cannabis use are not as serious as was thought. They also show that cannabis acts as a 'gateway' to other drugs. That's **not** because it makes people want a stronger drug but because it puts them in touch with illegal dealers.

The UK Government downgraded cannabis in 2004, although it is still an illegal drug. More states in the USA are looking at decriminalising cannabis use as the evidence grows steadily that it is less dangerous than alcohol. Some scientists support the moves, while others feel that cannabis should remain illegal.

ACTIVITY

You are going to set up a classroom debate. The subject is: *'We believe that cannabis should not be made a legal drug.'*

You are going to prepare **two** short speeches – one **for** the idea of legalising cannabis and one **against**.

You can use the information on these pages and also look elsewhere for information – in books and leaflets from PSHE, in the media and on the Internet.

In both of your speeches you must base your arguments on scientific evidence as well as considering the social, moral and ethical implications of any change in the law. You have to be prepared to argue your case (both 'for' and 'against') and answer any questions – so do your research well!

SUMMARY QUESTIONS

1 a) What is a drug?
   b) Why can drugs be so harmful?
   c) Give two examples of legal drugs and two examples of illegal drugs.
   d) What is addiction to a drug?

2 a) How does alcohol affect your body?
   b) Why do so many people use alcohol?
   c) Alcohol can cause serious damage to your body, both if you take a single big overdose or if you drink too much over many years.
   Explain what happens to your body in both cases.
   d) Alcohol costs our society millions of pounds in health care and in sorting out the social problems that it causes. It is a very dangerous drug.
   Explain why you think it is still legal and easy to get hold of.

3 a) Copy and complete:
   Every cigarette contains ...... leaves which burn to produce around ...... chemicals which are breathed into your lungs. Some of those chemicals are ...... into the blood stream to be carried around your body and to your brain. ...... is the......drug found in tobacco smoke. It is absorbed into your .... On the other hand ......stays in your ...... where it can cause cancer.
   b) Explain the following facts about smokers.
   i) They are more likely than non-smokers to cough.
   ii) They are more likely than non-smokers to suffer from lung diseases like emphysema.
   iii) They are more likely than non-smokers to suffer from lung cancer.
   iv) They are more likely than non-smokers to have a baby which is born dead or has a low birth mass.

4 Use the data on page 57 to help you answer the following questions:
   a) What percentage of the people who die of lung cancer are smokers?
   b) Draw a bar chart to show the effect of the number of cigarettes you smoke on your relative risk of dying.
   c) i) What happened to the death rates from cancer from 1974–1992 in men in England and Wales?
   ii) What do you think happened to the numbers of men smoking over the same time period? Explain your answer.
   d) The numbers of women smoking increased steadily from the 1950s. How does the evidence suggest that smoking is linked to lung cancer but not to breast cancer.

EXAM-STYLE QUESTIONS

1 The table below contains statements about certain drugs. Match the words **A**, **B**, **C** and **D** with the statements **1** to **4** in the table.

   A Alcohol   **B** Nicotine
   C Cannabis   **D** Heroin

	Statement
1	May cause cirrhosis of the liver.
2	May cause psychological problems.
3	Is an illegal drug.
4	Acts as a stimulant.

   (4)

2 Which of the following substances in tobacco smoke is addictive?
   A Tar   **B** Nicotine
   C Carbon monoxide   **D** Carbon dioxide   (1)

3 Which of the following is **not** a disease that can be caused by smoking cigarettes?
   A Liver damage   **B** Bronchitis
   C Emphysema   **D** Lung cancer   (1)

4 Which of the options **A**, **B**, **C** or **D** in the following statement is **not** correct?
   Women who smoke during pregnancy have a higher risk of having a:
   A longer pregnancy.
   B premature birth.
   C baby of lower than average birth mass.
   D baby that is dead at birth.   (1)

5 The graph shows the number of cigarettes smoked per male per year in the UK from 1910–1980. The dotted line shows the number of male deaths from lung cancer over the same period.

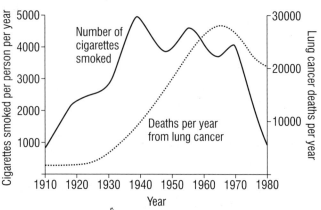

(a) Name the independent variable shown on the graph. (1)

(b) Most graphs display data collected on one dependent variable.
   (i) Why is this graph unusual? (1)
   (ii) Explain why the data has been presented like this. (1)

(c) Use the information in the graph to show that there is a link between cigarette smoking and lung cancer. (2)

(d) Assuming smoking does cause lung cancer, give two reasons why there is a time lag of around 30 years between a high level of cigarette smoking and a high rate of death from lung cancer. (2)

(e) What is the effect on the body of the carbon monoxide in tobacco smoke? (1)

6 The table shows the results of a survey carried out in 1995 to find out how much alcohol youngsters drink in a week.

Amount drunk* (units of alcohol)	Aged 11 (%)	Aged 12 (%)	Aged 13 (%)	Aged 14 (%)	Aged 15 (%)
None	75.7	64.1	56.7	44.7	31.4
1–6	20.1	29.0	29.7	34.2	38.8
7–10	1.7	3.5	5.4	8.8	13.3
11–14	1.6	1.4	3.7	5.2	5.4
15–20	0.7	1.0	2.1	3.8	4.6
21+	0.3	1.0	2.4	3.3	7.5

(* One unit = half a pint of beer/one glass of wine/one measure of spirit).

(a) Calculate the percentage of 14-year-olds who drink more than six units of alcohol in a week. (1)

(b) In a school with 240 pupils aged 15, how many of these pupils does the data suggest drink more than 21 units per week? (1)

(c) Calculate the mean percentage of 11 to 15 year olds who drink 1–6 units of alcohol in a week. (1)

(d) What is the range of the proportion of youngsters aged 11–15 who drink 15–20 units of alcohol per week? (1)

(e) Explain why driving a car under the influence of alcohol can be dangerous. (3)

(f) Suggest two other effects on society of alcohol abuse apart from drink-driving. (2)

HOW SCIENCE WORKS QUESTIONS

Some students decided to test whether drinking coffee could affect heart rate. They asked the class to help them with their investigation. They divided the class into two groups. Both groups had their pulses taken. They gave one group a drink of coffee. They waited for 10 minutes and then took their pulses again. They then followed the same procedure with the second group.

a) What do you think the second group were asked to drink? (1)

b) State a control variable that should have been used. (1)

c) Explain why it would have been a good idea not to tell the two groups exactly what they were drinking. (1)

d) Which of the following best describes the type of dependent variable being measured?
   i) continuous
   ii) discrete
   iii) categoric
   iv) ordered. (1)

e) Study this table of results that they produced.

Group	Increase in pulse rates (beats per min.)
With caffeine	12, 15, 13, 10, 15, 16, 10, 15, 16, 21, 14, 13, 16
Without caffeine	4, 3, 4, 5, 7, 5, 7, 4, 2, 6, 5, 4, 7

Can you detect any evidence for systematic error in these results? If so, describe this evidence. (2)

f) Is there any evidence for a random error in these results? If so, describe this evidence. (2)

g) What is the range for the increase in pulse rates without caffeine? (1)

h) What is the mean (or average) increase in pulse rate:
   i) with caffeine
   ii) without caffeine. (2)

i) What conclusion, if any, does the data collected suggest? (1)

j) How could you make your data more reliable? (1)

.a.E.
3/09/08

B1a 4.1   Pathogens

LEARNING OBJECTIVES

1 What are the differences between bacteria and viruses?
2 How do pathogens cause disease?
3 How did Ignaz Semmelweiss change the way we look at disease?

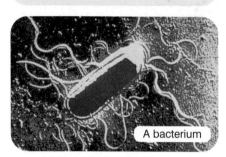

A bacterium

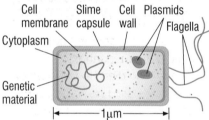

Cell membrane   Slime capsule   Cell wall   Plasmids
Cytoplasm
Genetic material
Flagella

|←——1μm——→|

Figure 1 Bacteria come in a variety of shapes and sizes, which help us to identify them under the microscope, but they all have the same basic structure

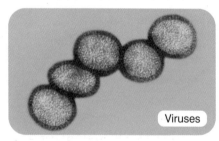

Viruses

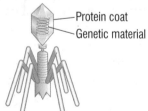

Protein coat
Genetic material

Figure 2 Viruses are really tiny with a very simple structure. Scientists are still arguing about whether they are actually living organisms or not.

Infectious diseases are found all over the world, in every country. The diseases range from relatively mild ones, such as the common cold and tonsillitis, through to known killers, such as tetanus, influenza and HIV/AIDS.

An infectious disease is caused by a **microorganism** entering and attacking your body. People can pass these microorganisms from one person to another. This is what we mean by *infectious*.

Microorganisms which cause disease are called **pathogens**. Common pathogens are bacteria and viruses.

a) What causes infectious diseases?

The differences between bacteria and viruses

Bacteria are single celled living organisms that are much smaller than animal and plant cells.

A bacterium is a single cell. It is made up of cytoplasm surrounded by a membrane and a cell wall. Inside the bacterial cell is the genetic material. Unlike animal and plant cells, this genetic material is not contained in a nucleus.

Although some bacteria cause disease, many are harmless and some are really useful to us. We use them to make food like yoghurt and cheese, in sewage treatment and to make medicines.

b) How are bacteria different from animal and plant cells?

Viruses are even smaller than bacteria. They usually have regular shapes. A virus is made up of a protein coat surrounding simple genetic material. They do not carry out any of the functions of normal living organisms except reproduction. But they can only reproduce by taking over another living cell. As far as we know, all naturally occurring viruses cause disease.

c) Give one way in which viruses differ from bacteria?

How pathogens cause disease

Bacteria and viruses cause disease because once they are inside the body they reproduce rapidly. Bacteria simply split in two. They often produce toxins (poisons) which affect your body. Sometimes they directly damage your cells. Viruses take over the cells of your body as they reproduce, damaging and destroying them. They very rarely produce toxins.

Common disease symptoms are a high temperature, headaches and rashes. These are caused by the damage and toxins produced by the pathogens. The symptoms also appear as a result of the way your body responds to the damage and toxins.

d) How do pathogens make you feel ill?

The work of Ignaz Semmelweiss

When Ignaz Phillipp Semmelweiss was a doctor in the mid-1850s, many women who gave birth in hospital died a few days later. They died from childbed fever, but no-one knew what caused it.

Semmelweiss realised that his medical students were going straight from dissecting a dead body to delivering a baby without washing their hands. He wondered if they were carrying the cause of disease from the corpses to their patients.

Then another doctor cut himself while working on a body and died from symptoms which were identical to childbed fever. Now Semmelweiss was sure the fever was caused by an infectious agent.

He insisted that his medical students wash their hands before delivering babies. Immediately, fewer mothers died.

Getting his ideas accepted

Semmelweiss presented his findings to other doctors. He thought his evidence would prove to them that childbed fever was spread by doctors. But his ideas were mocked.

Many doctors thought that childbed fever was God's punishment to women. They didn't want to accept the idea that the disease was caused by something invisible passed from patient to patient. Also it was hard for doctors to admit that they might have spread the disease and killed their patients instead of curing them.

What's more, hand-washing seemed a strange idea at the time. There was no indoor plumbing, the water was cold, and the chemicals used eventually damaged the skin of your hands. It is difficult for us to imagine just how difficult hand-washing must have seemed in the 19th century! It took years for Semmelweiss's ideas to be accepted.

In hospitals today, bacteria such as MRSA, which are resistant to antibiotics, are causing lots of problems (see page 68). Getting doctors, nurses and visitors to wash their hands more often is seen as part of the solution – just as it was in Semmelweiss's time!

Make sure you know the differences between bacteria and viruses.

Figure 3 Ignaz Semmelweiss – his battle to persuade medical staff to wash their hands to prevent infections is still going on today!

DID YOU KNOW?

Semmelweiss couldn't bear to think of the thousands of women who died because other doctors ignored his findings. By the 1860s he suffered a major breakdown and in 1868, aged only 47, he died – from an infection picked up from a patient during an operation!

SUMMARY QUESTIONS

1 Copy and complete using the words below:

**toxins   viruses   microorganisms   reproduce   pathogens
damage   symptoms   bacteria**

The ...... which cause infectious diseases are known as .... Once ...... and ...... get inside your body they ...... rapidly. They ...... your tissues and may produce ...... which cause the ...... of disease.

2 Bacteria and viruses can both cause disease. Make a table which shows how bacteria and viruses are different, and how they are similar.

3 Give five examples of the way we now accept the germ theory of disease in our everyday lives, e.g. washing your hands after using the toilet.

4 Write a letter by Ignaz Semmelweiss to a friend explaining how you formed your ideas and the struggle to get them accepted.

KEY POINTS

1 Infectious diseases are caused by microorganisms such as bacteria and viruses.
2 Microorganisms which cause disease are called pathogens.
3 Bacteria and viruses reproduce rapidly inside your body. They may produce toxins which make you feel ill.
4 Viruses use and damage your cells as they reproduce. This can also make you feel ill.

B1a 4.2

Defence mechanisms

LEARNING OBJECTIVES

1 How does your body stop pathogens getting in?
2 How do white blood cells protect us from disease?

Figure 1 Droplets carrying millions of pathogens fly out of your mouth and nose at up to 100 miles an hour when you sneeze!

There are a number of ways in which we can spread pathogens from one person to another. The more pathogens that get into your body, the more likely it is that you will get an infectious disease.

Droplet infection: When you cough, sneeze or talk you expel tiny droplets full of pathogens from your breathing system. Other people breathe in the droplets, along with the pathogens they contain. So they pick up the infection, e.g. 'flu (influenza), tuberculosis or the common cold. (See Figure 1.)

Direct contact: Some diseases are spread by direct contact of the skin, e.g. impetigo and some sexually transmitted diseases like genital herpes.

Contaminated food and drink: Eating raw or undercooked food, or drinking water containing sewage can spread disease, e.g. diarrhoea or salmonellosis. You get these by taking large numbers of microorganisms straight into your gut.

Through a break in your skin: Pathogens can enter your body through cuts, scratches and needle punctures, e.g. HIV/AIDS or hepatitis.

When people live in crowded conditions, with no sewage treatment, infectious diseases can spread very rapidly.

a) What are the four main ways in which infectious diseases are spread?

Preventing microbes getting into your body

Each day you come across millions of disease-causing microorganisms. Fortunately your body has a number of ways of stopping these pathogens getting inside.

Your skin covers your body and acts as a barrier. It prevents bacteria and viruses from reaching the vulnerable tissues underneath.

If you damage or cut your skin in any way you bleed. Platelets in your blood quickly help to form a clot which dries into a scab. The scab forms a seal over the cut, stopping pathogens getting in through the wound. (See Figure 2.)

Your breathing system could be a weak link in your body defences. That's because every time you breathe you draw air loaded with pathogens right inside your body. However, your breathing organs produce a sticky liquid, called mucus, which covers the lining of your lungs and tubes. It traps the pathogens. The mucus is then moved out of your body or swallowed down into your gut. Then the acid in your stomach destroys the microorganisms. In the same way, the stomach acid destroys most of the pathogens you take in through your mouth.

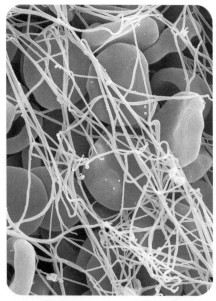

Figure 2 When you get a cut, the platelets in your blood set up a chain of events to form a clot which dries to a scab. This stops pathogens from getting into your body. It also stops you bleeding to death as well!

b) What are the three main ways in which your body prevents pathogens from getting in?

How white blood cells protect you from disease

In spite of your various defence mechanisms, some pathogens still manage to get inside your body. Once there, they will meet your second line of defence – the *white blood cells* of your **immune system**. The white blood cells help to defend your body against pathogens in several ways:

Role of white blood cell	How it protects you against disease
Ingesting microorganisms 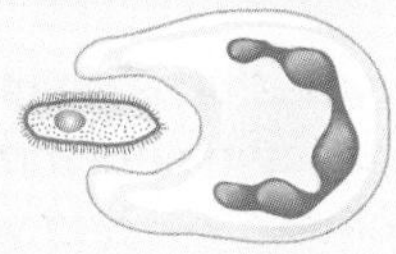	Some white blood cells ingest (take in) pathogens, destroying them and preventing them from causing disease.
Producing antibodies 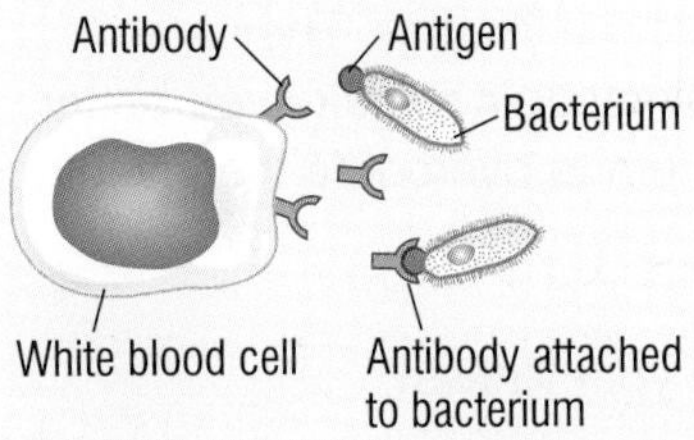 Antibody Antigen Bacterium White blood cell Antibody attached to bacterium	Some white blood cells produce special chemicals called **antibodies**. These target particular bacteria or viruses and destroy them. You need a unique antibody for each type of bacterium or virus. Once your white blood cells have produced antibodies against a particular pathogen, they can produce them again very rapidly if that pathogen invades again.
Producing antitoxins 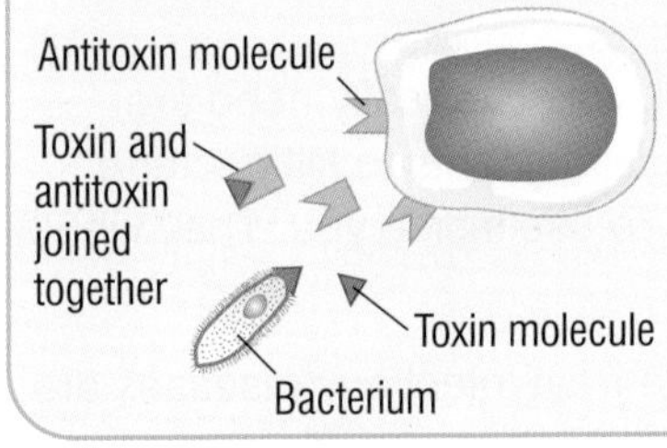 Antitoxin molecule Toxin and antitoxin joined together Toxin molecule Bacterium	Some white blood cells produce antitoxins. These counteract the toxins (poisons) released by pathogens.

Figure 3 Ways in which your white blood cells destroy pathogens and protect you against disease

GET IT RIGHT!

Avoid using words like 'battle' and 'fight' when you explain how antibodies work. Such words suggest that the white blood cells think about what they are doing.

SUMMARY QUESTIONS

1 Explain how diseases are spread by:

   a) droplet infection
   b) direct contact
   c) contaminated food and drink
   d) through a cut in the skin.

2 Certain diseases mean you cannot fight infections very well. Explain why the following symptoms would make you less able to cope with pathogens.

   a) Your blood won't clot properly.
   b) The number of white cells in your blood falls.
   c) Your skin is damaged to expose a large area of raw tissue underneath.

3 Here are four common things we do. Explain carefully how each one helps to prevent the spread of disease.

   a) Washing your hands before preparing a salad.
   b) Throwing away tissues after you have blown your nose.
   c) Making sure that sewage does not get into drinking water.
   d) Putting your hand in front of your mouth when you cough or sneeze.

4 Explain in detail how the white blood cells in your body work.

KEY POINTS

1 Your body has several methods of defending itself against the entry of pathogens using the skin, the mucus of the breathing system and the clotting of the blood.
2 Your white blood cells help to defend you against pathogens by ingesting them, making antibodies and making antitoxins.

B1a 4.3   Using drugs to treat disease

When you have an infectious disease, you often take medicines which contain useful drugs. Often the medicine has no effect at all on the pathogen that is causing the problems. It just eases the symptoms and makes you feel better.

For example, drugs like aspirin and paracetamol are very useful as painkillers. When you have a cold they will help relieve your headache and sore throat. On the other hand, they will have no effect on the virus which has invaded your tissues and made you feel ill!

Many of the medicines you can buy at a chemist's are like this. They are symptom relievers rather than pathogen killers. They do not make you better any faster. You have to wait for your immune system to overcome the invading microorganisms.

a) Why don't medicines like aspirin actually cure your illness?

Antibiotics

While drugs which make us feel better are useful, what we really need are drugs that can *cure* us. We use antiseptics and disinfectants to kill bacteria outside the body. But they are far too poisonous to use inside you. They would kill you and your pathogens at the same time!

The drugs which have really changed the way we treat infectious diseases are **antibiotics**. These are medicines which can kill disease-causing bacteria inside your body.

b) What is an antibiotic?

Alexander Fleming was a scientist who studied bacteria and was keen to find ways of killing them. In 1928, he was growing lots of bacteria on agar plates. Alexander was a rather sloppy scientist, and his lab was quite untidy. So he often left the lids off his plates for a long time. He also often forgot about cultures he had set up!

When he came back from a holiday, Fleming noticed that lots of his culture plates had mould growing on them. Just before he put the plates in the washing up bowl, he noticed a clear ring in the jelly around some of the spots of mould. Something had killed the bacteria covering the jelly.

There and then Fleming saw how important this was. He worked hard on the mould, extracting a 'juice' which he called **penicillin**. But he couldn't get much penicillin from the mould. He couldn't make it keep, even in the fridge. So he couldn't prove it would actually kill bacteria and make people better. By 1934 he gave up on penicillin and went on to do different work!

About 10 years after penicillin was first discovered, Ernst Chain and Howard Florey set about trying to use it on people. Eventually they managed to make penicillin on an industrial scale. The process was able to produce enough penicillin to supply the demands of the Second World War. We have used it as a medicine ever since.

Figure 1 Taking paracetamol will make this child feel better, but she will not actually get well any faster as a result.

DID YOU KNOW?

A patient called Albert Alexander was dying of a blood infection when Florey and Chain gave him some of their experimental penicillin for five days. The effect was almost miraculous and Albert recovered. But then the penicillin ran out. Florey and Chain even tried to collect spare penicillin from Albert's urine, but it was no good. The infection came back and sadly Albert died.

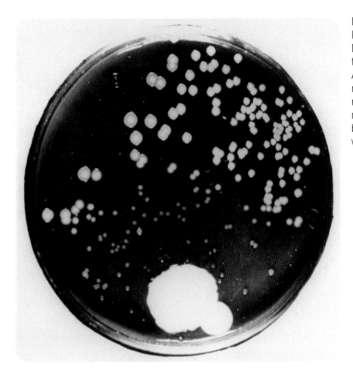

Figure 2 Alexander Fleming was on the lookout for something that would kill bacteria. As a result of him noticing the effect of this mould on his cultures, millions of lives have been saved around the world.

c) Who was the first person to discover penicillin?

How antibiotics work

Antibiotics, such as penicillin, work by killing the bacteria which cause disease while they are inside your body. They damage the bacterial cells without harming your own cells. They have had an enormous effect, because we can now cure bacterial diseases such as plague and TB. These same diseases killed millions of people in years gone by.

Unfortunately antibiotics have not been a complete answer to the problem of infectious diseases. Antibiotics have no effect on diseases caused by viruses. What's more, developing drugs which do have an effect on viral diseases is proving very difficult indeed.

The problem with viral pathogens is that they reproduce inside the cells of your body. It is extremely difficult to develop drugs which kill the viruses without damaging the cells and tissues of your body at the same time.

d) How do antibiotics work?

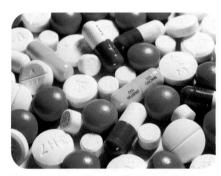

Figure 3 Penicillin was the first antibiotic. Now we have many different ones which kill different types of bacteria. In spite of this, scientists are always on the look out for new antibiotics to keep us ahead in the battle against the pathogens.

SUMMARY QUESTIONS

1 What is the main difference between drugs like paracetamol and drugs such as penicillin?

2 a) How did Alexander Fleming discover penicillin?
  b) Why was it so difficult to make a medicine out of penicillin?
  c) Who developed the industrial process which made it possible to mass-produce penicillin?

3 Explain why it is so much more difficult to develop medicines against viruses than it has been to develop anti-bacterial drugs.

KEY POINTS

1 Some medicines relieve the symptoms of disease but do not kill the pathogens which cause it.

2 Antibiotics cure bacterial diseases by killing the bacteria inside your body.

3 Antibiotics do not destroy viruses because viruses reproduce inside the cells. It is difficult to develop drugs that can destroy viruses without damaging your body cells.

B1a 4.4 Changing pathogens

1 What is antibiotic resistance?
2 Why is mutation such a problem?

If you are given an antibiotic and use it properly, the bacteria which have made you ill are almost all killed. The ones that are left are the ones which have a natural **mutation** which means they are not affected by the antibiotic. They are resistant to it – but your body finishes them off!

Antibiotic-resistant bacteria

If antibiotics are used too often, or you don't take the full course of medicine prescribed by your doctor, more of the resistant bacteria survive. If they go on to make someone else ill, they will not be killed by the original antibiotic. They are *resistant* to that antibiotic. This resistance is the result of a process of **natural selection**. (See Figure 1.)

As more types of bacteria become resistant to more antibiotics, so diseases caused by bacteria are becoming harder and harder to treat.

To prevent more resistant strains of bacteria appearing it is important not to over-use antibiotics. It's best to only use them when you really need them. It is also very important that people finish their course of medicine every time.

a) Why is it important not to use antibiotics too frequently?

The MRSA story

Hospitals treat many patients with infectious diseases. They use large amounts of antibiotics. As a result of natural selection, hospitals often contain a number of bacteria which are not affected by most of the commonly used antibiotics. This is what has happened with *MRSA* (the bacterium *methicillin resistant Staphylococcus aureus*).

In hospitals, where doctors and nurses move from patient to patient, these antibiotic-resistant bacteria are spread easily. MRSA alone now causes around a thousand deaths every year in hospital patients who, although ill, might otherwise have recovered.

How can we control the spread of these antibiotic-resistant bacteria in hospitals? There are a number of simple steps which can have a big effect on the spread of microorganisms such as MRSA. We have known some of them since the time of Semmelweiss, but they sometimes get forgotten!

- Doctors, nurses and other medical staff wash their hands between patients.
- Visitors wash their hands as they come into and leave the hospital.
- Look after patients infected with the bacteria in isolation from other patients.
- Keep hospitals clean – high standards of hygiene.
- Medical staff wear either disposable clothing or clothing which is regularly sterilised.

b) Is MRSA a bacterium or a virus?

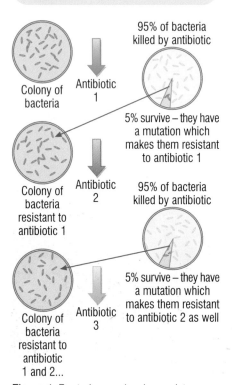

Figure 1 Bacteria can develop resistance to many different antibiotics in a process of natural selection

95% of bacteria killed by antibiotic
Colony of bacteria
Antibiotic 1
5% survive – they have a mutation which makes them resistant to antibiotic 1
Colony of bacteria resistant to antibiotic 1
Antibiotic 2
95% of bacteria killed by antibiotic
5% survive – they have a mutation which makes them resistant to antibiotic 2 as well
Colony of bacteria resistant to antibiotic 1 and 2...
Antibiotic 3

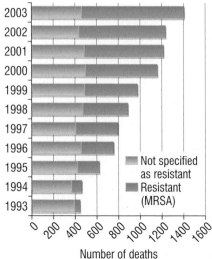

Not specified as resistant
Resistant (MRSA)

Number of deaths

Source: National Statistics Office

Figure 2 The growing impact of MRSA in our hospitals can be seen from this data

Mutation and pandemics

Another problem caused by the mutation of bacteria and particularly viruses is that new forms of diseases can appear. Because no-one is immune to them, they can cause widespread illness. A good example of this in action is influenza, commonly known as 'flu.

'Flu is caused by a virus which mutates easily, so every year new strains appear which can fool your immune system. These new strains are usually quite similar to the old 'flu but just occasionally a very different form of 'flu virus appears.

These usually cause a 'flu *epidemic* (in one country) or even a *pandemic* (across several countries). In 1918–19 a new strain of 'flu emerged which spread rapidly around the world and killed between 20 and 40 million people.

Animals such as chickens and pigs get 'flu-like diseases too. Sometimes an animal virus will mutate so it can infect people. For example, the 1918–19 'flu pandemic may have been linked to bird 'flu.

Many scientists think that a new and serious form of human 'flu is likely to be linked to one of the bird influenzas which keep appearing in China, Thailand and Asia.

We are trying to prevent a new pandemic using research to discover what makes certain types of 'flu so dangerous. The World Health Organisation is monitoring all 'flu outbreaks. Scientists are also trying to produce different *vaccines* to protect us against new and dangerous forms of the disease.

There is no 'antibiotic for viruses' but there are several drugs which can reduce the length of time you suffer with 'flu. This lessens the chance of it spreading.

Finally, many countries have plans to restrict travel and even put people in isolation if there is a 'flu outbreak. This worked very well in 2003 when SARS, a 'flu-like illness, appeared for the first time in China.

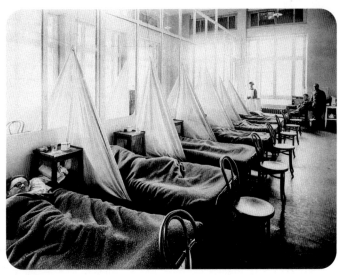

Figure 3 In the early 20th century a 'flu pandemic killed millions. In the 21st century a new form of 'flu could sweep the world even more quickly because of the way we all travel around on aircraft for business and holidays.

GET IT RIGHT!

Remember bacteria don't 'want' to develop resistance – they just do!
Make sure you understand how bacteria become resistant to antibiotics through natural selection.

SUMMARY QUESTIONS

1 Copy and complete using the words below:

**antibiotics   bacterium   better   disease
mutation   mutate   resistant   virus**

If bacteria change or …… they may become …… to …… This means the medicine no longer makes you …… . A …… in a …… or ……can also lead to a new form of …… .

2 Make a flow chart to show how bacteria develop resistance to antibiotics.

3 A new strain of bird 'flu has been discovered in Asia and it is in the national news. Write a piece for the science pages of your local paper. Explain what people can do to protect themselves and what problems they might face preventing the new disease spreading.

KEY POINTS

1 Many types of bacteria have developed antibiotic resistance as a result of natural selection. To prevent the problem getting worse we must not over-use antibiotics.

2 If bacteria or viruses mutate, new strains of a disease can appear which spread rapidly to cause epidemics and pandemics.

B1a 4.5   Developing new medicines

LEARNING OBJECTIVES

1 What are the stages in testing and trialling a new drug?
2 Why is testing new drugs so important?

We are developing new medicines all the time, as scientists and doctors try to find ways of curing more diseases. Every new medical treatment has to be extensively tested and trialled. This process makes sure that it works well and is as safe as possible.

A good medicine is:

● **Effective** – it must prevent or cure the disease it is aimed at, or at least make you feel better.
● **Safe** – the drug must not be toxic (poisonous) and there must be no unacceptable side effects.
● **Stable** – you need to be able to use the medicine under normal conditions and store it for some time.
● **Successfully taken into and removed from your body** – a medicine is no use unless it can reach its target in your body. Then your body must be able to remove the medicine once it has done its work.

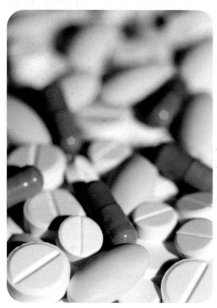

Figure 1 No matter how many medicines we have, there is always room for more as we tackle new diseases!

Developing and testing a new drug

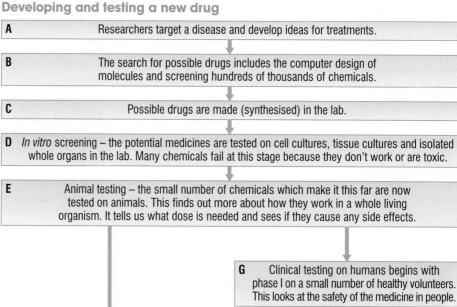

A	Researchers target a disease and develop ideas for treatments.
B	The search for possible drugs includes the computer design of molecules and screening hundreds of thousands of chemicals.
C	Possible drugs are made (synthesised) in the lab.
D	*In vitro* screening – the potential medicines are tested on cell cultures, tissue cultures and isolated whole organs in the lab. Many chemicals fail at this stage because they don't work or are toxic.
E	Animal testing – the small number of chemicals which make it this far are now tested on animals. This finds out more about how they work in a whole living organism. It tells us what dose is needed and sees if they cause any side effects.

F	Animal testing continues, looking at the effect of longer term use of the medicine.

G	Clinical testing on humans begins with phase I on a small number of healthy volunteers. This looks at the safety of the medicine in people.
H	Human phase II trials run with a small number of the patients suffering from the target disease. This is where scientists can really begin to see if the drug will be safe and effective.
I	Human phase III trials continue with a larger number of patients.

J	When the medicine has passed all the tests set down in law, it will be granted a licence. Now your doctor can use the new medicine to treat your illness.
K	Once the medicine is in use, phase IV trials continue. The medicine will be monitored for as long as patients use it. This makes sure it works and is as safe as possible.

GET IT RIGHT!

You won't have to recall all the steps in testing a new drug. However, you will need to know that they are extensively tested and trialled before the public can use them.

Figure 2 It takes a long time and a lot of work to develop a successful new drug

When scientists research a new medicine they have to make sure that all these conditions are met. This is why it takes a very long time. It can take up to 12 years to bring a new medicine into your doctor's surgery. It can also cost a lot of money, up to about £350 million!

a) What are the important properties of a good new medicine?

Testing drugs

We test new medicines in the laboratory. This is to find out if they are toxic and if they seem to do their job. We also trial new medicines on human volunteers. This is to discover if they have any side effects.

Take a look at all the stages a new drug has to go through – no wonder it is a slow process! (See Figure 2.)

Why do we test new medicines so thoroughly?

Thalidomide is a medicine which was developed in the 1950s as a sleeping pill. This was before we had agreed standards for studying the effects of new medicines. In particular, the specific animal tests on pregnant animals which are now known to be essential were not carried out.

Then it was discovered that thalidomide stopped sickness in pregnancy. Because thalidomide seemed very safe for adults, it was assumed that it was also safe for unborn children. Doctors gave it to pregnant women to relieve their morning sickness.

Tragically, thalidomide was *not* safe for developing fetuses. It affected many of the women who took the drug in the early stages of pregnancy. They went on to give birth to babies with severe limb deformities.

The thalidomide tragedy led to a new law which set standards for the testing of all new medicines. Since the Medicines Act 1968, new medicines *must* be tested on animals to see if they have an effect on developing fetuses.

There is another twist in the thalidomide story. Although thalidomide is never given to anyone who is or might become pregnant, doctors are finding more and more uses for the drug. They can use it to treat leprosy and autoimmune diseases (where the body attacks itself). There have also been some very exciting results using thalidomide to treat certain types of cancer.

b) Why was thalidomide prescribed to pregnant women?

Figure 3 This man has limb deformities because his mother took thalidomide during her pregnancy. He was just one of thousands of people affected by the thalidomide tragedy, many of whom have gone on to live full and active lives.

KEY POINTS

1 When we develop new medicines they have to be tested and trialled extensively before we can use them.
2 Drugs are tested to see if they work well. We also make sure they are not toxic and have no unacceptable side effects.
3 Thalidomide was developed as a sleeping pill and was found to prevent morning sickness in early pregnancy. It had not been fully tested and it caused birth defects. Thalidomide is now used to treat leprosy and other diseases.

SUMMARY QUESTIONS

1 Copy and complete using the words below:

effective   trialled   safe   medicine   stable   tested

Every new …… has to be extensively …… and …… before you can use it to make sure that it works well. A good medicine can be taken into and removed from your body, and it is ……, …… and …… .

2 a) Testing a new medicine costs a lot of money and can take up to 12 years. Explain the main stages in testing new drugs.
  b) What were the flaws in the original development of Thalidomide?
  c) Comment on the benefits and drawbacks of using Thalidomide to treat leprosy and cancer.

B1a 4.6 Immunity

LEARNING OBJECTIVES

1 How does your immune system work?
2 How does vaccination protect you against disease?

Figure 1 No-one likes having a vaccination very much – but they save millions of lives!

Every cell has unique proteins on its surface called **antigens**. The antigens on the microorganisms which get into your body are different to the ones on your own cells. Your immune system recognises they are different.

Your white blood cells then make antibodies to attack the antigens. This destroys the pathogens. (See page 65.)

Your white blood cells seem to 'remember' the right antibody needed to tackle a particular pathogen. If you meet that pathogen again, they can make the same antibody very quickly. So you become *immune* to that disease.

The first time you meet a new pathogen you get ill. That's because there is a delay while your body sorts out the right antibody needed. The next time, you completely destroy the invaders before they have time to make you feel unwell.

a) How does your immune system work?

Vaccination

Some pathogens can make you seriously ill very quickly. In fact you can die before your body manages to make the right antibodies. Fortunately, you can be protected against many of these diseases by **immunisation** (also known as *vaccination*).

Immunisation involves giving you a *vaccine*. A vaccine is usually made of a dead or weakened form of the disease-causing microorganism. It works by triggering your natural immune response to invading pathogens.

A small amount of dead or inactive pathogen is introduced into your body. This gives your white blood cells the chance to develop the right antibodies against the pathogen **without** you getting ill.

Then if you meet the live pathogens, your white blood cells can respond rapidly. They can make the right antibodies just as if you had already had the disease. This is how vaccination protects you against disease.

b) How do vaccines work?

Figure 2 This is how vaccines protect you against dangerous infectious diseases

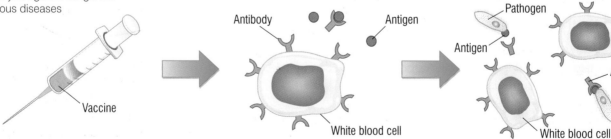

Small amounts of dead or inactive pathogen are put into your body, often by injection.

The antigens in the vaccine stimulate your white blood cells into making antibodies. The antibodies destroy the antigens without any risk of you getting the disease.

You are immune to future infections by the pathogen. That's because your body can respond rapidly and make the correct antibody as if you had already had the disease.

We use vaccines to protect us against both bacterial diseases (e.g. tetanus and diphtheria) and viral diseases (e.g. polio, measles, mumps and rubella). Vaccines have saved millions of lives around the world. One disease – smallpox – has been completely wiped out by vaccinations. It also looks as if polio will disappear as well in the next few years.

c) Give an example of one bacterial and one viral disease which you can be immunised against.

The vaccine debate

No medicines are completely risk free. Very rarely, a child will react badly to a vaccine with tragic results. Making the decision to have your baby immunised can be difficult.

Society as a whole needs as many people as possible to be immunised against as many diseases as possible. This keeps the pool of infection in the population as low as we can get it. On the other hand, by taking your healthy child along for a vaccination, you know there is a remote chance that something will go wrong.

Because vaccines are so successful, we never see the terrible diseases they protect us against. We forget that 60 years ago in the UK thousands of children died every year from infectious diseases. Many more were left permanently damaged. So parents today are often aware of the very small risks from vaccination – but sometimes forget about the terrible dangers of the diseases we vaccinate against.

If you are a parent it can be difficult to find unbiased advice to help you make a decision. The media emphasise scare stories which make good headlines. The pharmaceutical companies want to sell vaccines. Doctors and health visitors can weigh up all the information, but they have vaccination targets set by the government.

Most people agree that vaccination is a good thing both for you and for society. The great majority of parents still choose to protect their children – but it is not always an easy choice. (See pages 74 and 75.)

(See pages 74 and 75.)

DID YOU KNOW...

... in the first 10 years of the 20th century, nearly 50% of *all* deaths in people aged up to 44 years old were caused by infectious diseases? The development of antibiotics and vaccines means that now only 0.5% of all deaths in the same age group are due to infectious disease!

GET IT RIGHT!

High levels of antibodies do not stay in your blood forever – immunity is the ability of your white blood cells to produce them quickly if you are re-infected by a disease.

KEY POINTS

1 You can be immunised against a disease by introducing small amounts of dead or inactive pathogens into your body.
2 Your white blood cells produce antibodies to destroy the pathogens. Then your body will respond rapidly to future infections by the same pathogen, by making the correct antibody. You become immune to the disease.
3 We can use vaccination to protect against both bacterial and viral pathogens.

SUMMARY QUESTIONS

1 Copy and complete using the words below:

> **antibodies**   **pathogen**   **immunised**   **dead**   **immune**
> **inactive**   **white blood**

People can be ...... against a disease by introducing small quantities of ...... or ...... forms of a ...... into your body. They stimulate the .... cells to produce ...... to destroy the pathogen. This makes you ...... to the disease in future.

2 Explain carefully, using diagrams if they help you:

a) how the immune system of your body works,
b) how vaccines use your natural immune system to protect you against serious diseases.

3 Make a table to show the advantages and disadvantages of giving your child the MMR vaccine. What would you choose to do?

B1a 4.7   How *should* we deal with disease?

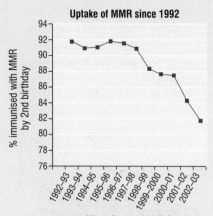

Uptake of MMR since 1992

(Source: *MMR Decision Making Study*, Durham University)

Media reporting of the 'controversial MMR vaccine' helped to feed people's fears about the safety of their children. In fact, as far as the great majority of scientists and doctors were concerned, there was *no* controversy. All the other evidence before and since has shown no link between MMR and autism.

Find out as much as you can about the MMR vaccine and the way the media reported the situation (www.thevaccinesite.org is a good starting point).
Either Design a web site for parents which answers the sort of questions they might ask about MMR. Make it user-friendly – the sort of thing health workers could use to reassure couple [B] about the importance of the MMR vaccine.
Or Produce a PowerPoint® presentation on the importance of responsible media reporting of science and medicine, using the MMR case as your main example.

The MMR dilemma

MMR is a vaccine which protects you against measles, mumps and rubella. Measles and mumps can both cause serious problems such as brain damage and death. Rubella can damage unborn babies. Once the MMR vaccine was introduced into the UK, cases of these diseases fell rapidly until no more children died of measles or mumps.

In February 1998, Dr Andrew Wakefield and his team published a paper in *The Lancet* (a medical journal). His research suggested that the MMR vaccine might be linked to the development of autism in some children. (Autistic children tend to be very withdrawn from people and cannot cope easily with normal life.)

The story got a lot of media coverage. People became very worried and uptake of the MMR vaccine started to fall. By 2000, the number of measles cases in Ireland had gone up hugely. For the first time in years, two babies died of the disease. By 2001, uptake of the MMR vaccine in England had also dropped – from 92% to 75%. These levels are not enough to keep people safe.

It turned out that the research had been done on a tiny group of twelve children. The scientist had been paid £55 000 by the parents of some of the children to prepare evidence against the vaccine for a court case. What's more, Dr Wakefield had developed some measles treatments which would not have been used if parents were confident in MMR. No-one knew this when he published his results.

This research has now been completely discredited. Many studies on thousands of children have shown no link between MMR and autism. Sadly, children have been damaged and some have died from measles because of the poor research and irresponsible media reporting that managed to scare so many parents unnecessarily.

I'm sure doctors wouldn't recommend the vaccine if they didn't think it was safe. We trust our GP, and we don't want to risk Kirsty having any of these dreadful diseases so she's going to have the MMR.

There's no smoke without fire. I'm sure these scientists couldn't publish their research if it wasn't true. We just daren't risk Cameron becoming autistic so we're not giving him the MMR.

I did things my way . . .

Each of these scientists made a great breakthrough in the treatment of disease – but the methods they used would not be acceptable today!

Edward Jenner (1794): I was sure it was the pus from cowpox which was protecting the milkmaids. I knew from smallpox! I proved my point by scratching the pus from some cowpox spots into the skin of young James Phipps, who soon went down with cowpox. Two months later I scratched lots of pus from a smallpox blister into his arm – and the young lad showed no signs of illness at all!

Lady Mary Wortley Montagu (1718): When I was in Turkey, I saw people giving themselves small doses of pus from smallpox blisters. Most of them seemed to get the disease mildly – and then to be protected for life. I had my children done straight away!
When I got back to England I wanted to show people how this protection worked. I offered several prisoners who had been condemned to death the chance to be free – if they let me try my experiments on them. They all agreed, we infected them with smallpox pus, they all survived – and they all went free.

Louis Pasteur (1885): I had been working on my vaccine against rabies for years when I came up with a new idea, using the spinal cords of rabbits which had died of rabies. I had not had the chance to test my vaccine properly when a little boy was brought to me. He'd been badly bitten by a dog with rabies two days earlier. Two doctors agreed he was sure to die so – with the permission of his family – I gave him thirteen injections of my vaccine. He went home – and he never developed rabies. It was a triumph but I told no-one except my dear wife. I needed more evidence!

ACTIVITY

It worked – but was it right?

Some of the early workers on vaccines had some amazing successes – but the methods they used would not be acceptable today. Compare their methods with what you found out about modern medicines testing on pages 70 and 71.

a) Take each scientist in turn and explain how they carried out their work and why it would not be allowed today.

b) Do you think that the methods they used can be justified because of the great advances in medicine they brought about?

c) Draw up a schedule for how a new vaccine might be developed today.

SUMMARY QUESTIONS

1 a) Define the following terms:
   i) infectious disease
   ii) microorganism
   iii) pathogen
   iv) toxin.

   b) What are the main differences between bacteria and viruses?

   c) How do tiny organisms like bacteria and viruses make a large person like you ill?

2 There is going to be a campaign to try and stop the spread of colds in Year 7 of your school. There is going to be a poster and a simple PowerPoint® presentation.
   Make a list of all the important things that the Year 7 children need to know about how diseases are spread. Also cover how to reduce the spread of infectious diseases from one person to another.

3 a) What is the difference between a medicine which makes you feel better and a medicine which actually makes you better?

   b) What is an antibiotic?

   c) What are the limitations of antibiotics?

   d) Where have antibiotic-resistant bacteria (like MRSA found in hospitals) actually come from?

   e) What can we do to prevent the problem of antibiotic resistance getting worse?

4 a) Why do new medicines need to be tested and trialled before doctors can use them to treat their patients?

   b) Why is the development of a new medicine so expensive?

   c) Do you think it would ever be acceptable to use a new medicine before all the trials had been completed?

5 a) Draw a labelled flow diagram to show how your immune system works.

   b) Explain why thousands of children in the UK no longer die of diphtheria or become paralysed by polio.

   c) Make a table to show the risks and benefits of having children vaccinated against serious diseases.

   d) The media like a good story. Many people read the papers and watch television.
   Explain why it is important that stories about medical issues like vaccination should be reported very carefully in the media.

EXAM-STYLE QUESTIONS

1 The table below contains statements about chemicals that act against pathogens.

Match the list of chemicals **A**, **B**, **C** and **D** with the statements **1** to **4** in the table.

A Antibodies   **B** Antitoxins

C Antibiotics   **D** Antiseptics

	Statement
1	Counteract poisons released by pathogens
2	Produced by the white blood cells to destroy particular bacteria or viruses
3	Kill infective bacteria outside the body
4	Drugs that kill infective bacteria inside the body

(4)

2 Which one of the following is **not** used by white blood cells to protect us against pathogens?

   A Production of antibodies to destroy bacteria

   B Sealing of wounds to prevent infection

   C Production of antitoxins to counteract poisons released by pathogens

   D Ingestion of pathogens.   (1)

3 MRSA is a bacterium that kills around a thousand hospital patients each year. Which one of the following has led to these MRSA infections?

   A Ineffective vaccines

   B Overcrowding in hospitals

   C Antibiotic resistance

   D Shortage of hospital equipment   (1)

4 Which of the following would **not** help control the spread of MRSA?

   A Medical staff having regular health checks

   B Medical staff washing their hands between seeing patients

   C Visitors washing their hands as they enter and leave the hospital

   D Cleaning hospitals with antiseptics   (1)

5 New drugs are thoroughly trialled and tested to ensure they have certain features that make them good medicines. When tested on human volunteers, it is not possible to keep every control variable constant. So how can researchers make their data reliable?

   A Use only animals instead of humans in tests.

   B Only test people over the age of 65.

C Use as large a sample of people as possible.

D Calculate the mean age of the people tested.　(1)

6 Measles is an extremely infectious disease that is caused by a virus. Measles can cause brain damage and death. To try to prevent epidemics of measles, the MMR vaccine was developed.

(a) Which other two diseases does the MMR vaccine protect against?　(2)

(b) The MMR vaccine contains the virus that causes measles. Why does this virus not cause measles when the MMR vaccine is injected into a child?　(1)

(c) Name the type of chemical released by certain white blood cells to destroy viruses.　(1)

(d) Explain how vaccinating a child with the MMR vaccine makes them immune to measles.　(5)

(e) A child that has not been immunised with the MMR vaccine develops measles. Suggest a reason why antibiotics will not cure the child of measles.　(1)

(f) In the UK, by 2001, the uptake of the MMR vaccine had fallen from 92% to 75%. Suggest one reason for this decrease.　(1)

7 (a) A health centre gives the following advice about antibiotics to its patients. In each case suggest **one** reason why the advice is given.

(i) Do not take an antibiotic for a viral infection like a cold or the flu.

(ii) Do not take an antibiotic that has been prescribed for someone else.

(iii) Ask whether an antibiotic is likely to help your illness or whether there are alternatives.

(iv) Do not stop taking the antibiotic once you feel better – always complete the course of drugs.
　(4)

(b) Describe how antibiotic resistance arises.　(5)

HOW SCIENCE WORKS QUESTIONS

Look at the newspaper advert below and answer the following questions:

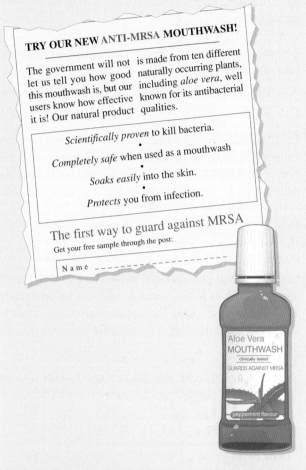

TRY OUR NEW ANTI-MRSA MOUTHWASH!

The government will not let us tell you how good this mouthwash is, but our users know how effective it is! Our natural product is made from ten different naturally occurring plants, including *aloe vera*, well known for its antibacterial qualities.

Scientifically proven to kill bacteria.

Completely safe when used as a mouthwash

Soaks easily into the skin.

Protects you from infection.

The first way to guard against MRSA

Get your free sample through the post:

Name _____

a) Why does the government want to prevent medical claims for this mouthwash?　(1)

b) Describe the tests that you think should have been carried out to make the claim that it is 'Scientifically proven to kill bacteria'.　(4)

c) Why might you mistrust any claims made by this company?　(2)

d) Mr Skeptic commented that 'it must be crazy to think that a plant could kill something as difficult as MRSA'. What do you think?　(1)

e) Describe the sort of investigation needed to be able to say that the mouthwash is 'completely safe'.　(3)

EXAMINATION-STYLE QUESTIONS

Biology A

In question **1** match the letters with the numbers.
Use **each** answer only **once**.

1   The menstrual cycle in women is controlled by hormones.
     Match words **A**, **B**, **C** and **D** with the spaces in **1** to **4** in the sentences.

     A   LH   **B**   Pituitary gland
     C   FSH   **D**   Oestrogen

     Each month, the hormone ......**1**...... is produced from the ......**2**....... and
     causes an egg to mature in a woman's ovaries. The ovaries, in turn, produce a
     hormone called ......**3**...... that stimulates the production of another hormone
     called ......**4**...... which causes the mature egg to be released.   *(4 marks)*

Question 2
In **each** part choose only **one** answer.

2   Cigarette smoking has been shown to increase the chances of early death due to
     diseases such as lung cancer, bronchitis and emphysema.
     The table below shows the percentage of the adult UK population by age group
     who were smokers. The figures were taken at five different times over a 25-year
     period.

	Age 16–19	Age 20–24	Age 25–34	Age 35–49	Age 50–59	Age 60+
1978	34	44	45	45	45	30
1988	28	37	36	36	33	23
1998	31	40	35	30	27	16
2000	29	35	35	29	27	16
2003	26	36	34	30	25	16

Figures show the percentage UK population who were smokers.

(a)   Which of the following statements is the only one supported by the data in
      the table?

      A   Fewer individuals smoked in 2003 that in 1978.

      B   A greater percentage of the population smoked in 1978 than in 2003.

      C   In 1978 a greater percentage of 20–24-year olds smoked than
              25–34-year olds.

      D   From 16 years old, the proportion of the population who smoke
              declines steadily with age.   *(1 mark)*

(b)   What would be the most reliable way of collecting the data in the table?

      A   Samples of people were asked about their smoking habits.

      B   Samples of people had their blood tested for nicotine.

      C   Tobacconists were asked about the people who buy cigarettes.

      D   Researchers recorded the number of people smoking in various places.
              (1 mark)

See pages 28–9

GET IT RIGHT!

Start with what you know.
In question 1 you have to match four letters with four numbers. Read the **whole** question first and then match up the pairs that you are certain about. If you can match all four, all well and good. If however, you are only sure about three of them, do not worry as the last one can be paired by a process of elimination. Not the best examination technique (far better to know all four) but preferable to leaving blank spaces.

See page 5

Biology B

1   In 2003 a health survey was carried out on over 8 000 men and women over 16 to find out their blood cholesterol levels. From this survey a table of results was made. It shows . . .

1   The mean level of cholesterol for each group.

2   The percentage of the group with a cholesterol level above the recommended healthy limit of 5.0 mmol/dm³.

	Age 25–34 %	Age 35–44 %	Age 45–54 %	Age 55–64 %	Age 65–74 %
Men					
Group size	718	789	675	585	401
Mean (mmol/dm³)	5.3	5.8	5.9	5.8	5.5
% 5.0 and above	59.8	76.9	81.0	79.7	67.4
Women					
Group size	717	794	674	603	455
Mean (mmol/dm³)	5.0	5.4	5.8	6.3	6.2
% 5.0 and above	54.9	69.3	79.3	83.7	77.1

(a)   The sample groups were of different sizes. Suggest which one provides the most reliable result and why.   *(2 marks)*

(b)   To remain healthy a maximum level of 5.0 mmol/dm³ for blood cholesterol is often recommended.

   (i)   Which group is most at risk from ill-health due to their cholesterol levels?   *(1 mark)*

   (ii)   Which group is the second most at risk from ill-health due to their cholesterol levels?   *(1 mark)*

   (iii)   What in particular is the health risk from having a high blood cholesterol level?   *(2 marks)*

(c)   In which organ of the body is cholesterol made?   *(1 mark)*

(d)   What a person eats affects how much cholesterol the body makes. What other factor also affects how much is made?   *(1 mark)*

(e)   The amount and type of fat in the diet affects the level of cholesterol in the blood. For each of the following, state whether their presence in the diet increases, decreases or has little effect on the blood cholesterol level.

   (i)   saturated fats

   (ii)   mono-unsaturated fats

   (iii)   polyunsaturated fats   *(3 marks)*

GET IT RIGHT!

Check the mark allowance. The number of marks allowed for a question gives you valuable information on the extent and detail of an answer. For example, in question 1 part (b) (iii), there are 2 marks compared to only 1 mark for the other sections of part (b). This suggests that two health risks are needed. (Think about which parts of the body are affected.)

See page 8

See pages 42–3

B1b | Evolution and environment

What you already know

Here is a quick reminder of previous work that you will find useful in this unit:

Our environment is precious and needs protecting

- In sexual reproduction, fertilisation happens when a male and a female sex cell join together.

- There are variations between members of the same species. Because of these variations, some individuals are more successful than others. In harsh conditions, the fittest survive.

- In sexual reproduction, information from two parents is mixed to make a new plan for the offspring. This leads to variation between members of a species.

- Variation between organisms of the same species has *environmental* as well as *inherited causes*.

- You can produce animals and plants with the features you want by a process called **selective breeding**.

- There are ways in which we can protect living organisms and the environment they live in.

- Sustainable development (when we replace the plants and animals we use) is becoming more and more important.

- Competition for the resources available is one thing which affects the size of populations of animals and plants.

- Toxic (poisonous) materials that we produce can build up in food chains and cause big problems.

RECAP QUESTIONS

1 a) How does sexual reproduction lead to variation within a species?

   b) How can the environment cause variation in a species?

2 What do we mean by 'the fittest survive'?

3 Imagine that you want to produce a type of dog with a very curly tail using selective breeding. Explain how you would do this. You can use diagrams if you think they will be helpful.

4 Think of as many ways as you can in which we might protect the environment and the living things in it. Explain why it is important to do this.

5 a) What do we mean by a 'population' of animals or plants?

   b) What sort of things do animals compete for? List as many things as you can which might affect the numbers of animals in a population.

   c) There is competition in a culture of bacteria. What sort of things might bacteria compete for?

6 a) Give two examples of food chains.

   b) Give an example of the way toxins can build up in a food chain.

   c) Explain how it is the carnivores at the end of the chain which are most likely to be affected.

Making connections

A female (top) and a male (bottom) fig wasp

There are about 700 different species of fig trees. Each one has its own species of pollinating wasps, without which it will die! The fig flowers of the trees are specially adapted so that they attract the right wasps.

Male fig wasps vary. Some species can fly but others are adapted to live in a fig fruit all their life. If they are lucky, a female wasp will arrive in the flower. Then the male will fertilise her. After this, he digs an escape tunnel for the female through the fruit and dies himself! The male wasp has special adaptations, such as a loss of his wings and very small eyes. These adaptations help him move around inside the fig fruit to find a female.

Dr James Cook

If a fig tree cannot attract the right species of wasp, it will never be able to reproduce. In fact in some areas the trees are in danger of extinction because the wasp populations are being wiped out. Dr James Cook and his team at Imperial College, London are looking at the adaptations of the different wasps and their genetic material. They are trying to work out the relationships between all the different species.

Female fig wasps have specially shaped heads for getting into fig flowers. They also have ovipositors. These allow them to place their eggs deep in the flowers of the fig tree.

ACTIVITY

Fig wasps are very strange animals. They have lots of adaptations which help them to reproduce sexually in just one species of tree.

Make a list of as many different types of animals or plants as you can that have strange ways of reproducing. Use a large piece of paper or a white board to record your ideas. You can include animals or plants which depend on just one other type of organism to be successful in life!

Chapters in this unit

Adaptation for survival   Variation   Evolution   How people affect the planet

B1b 5.1   Adaptation in animals

The variety of conditions on the surface of the Earth is huge. If you are a living organism, you could find yourself living in the dry heat of a desert or in wastelands of ice and snow. Fortunately, living organisms have special features (known as **adaptations**). These make it possible for them to survive in their particular habitat – however extreme the conditions might be!

Animals in cold climates

To survive in a cold environment you must be able to keep yourself warm. Arctic animals are adapted to reduce the heat they lose from their bodies as much as possible. You lose body heat through your body surface (mainly your skin). The amount of heat you lose is closely linked to your surface area : volume (SA/V) ratio.

Look at Figure 2. This explains why so many Arctic mammals, such as seals, walruses, whales, and polar bears, are relatively large. It keeps their surface area : volume ratio as small as possible and so helps them hold on to their body heat.

a) Why are so many Arctic animals large?

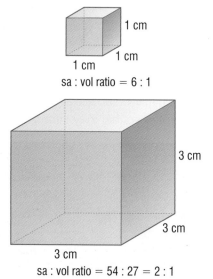

sa : vol ratio = 6 : 1

sa : vol ratio = 54 : 27 = 2 : 1

Figure 2 The ratio of surface area to volume falls as objects get bigger. You can see this clearly in the diagram. This is very important when you look at the adaptations of animals which live in cold climates.

Figure 1 The Arctic is a cold and bleak environment. However, the animals which live there are well adapted for survival. Notice the large size, small ears, thick coat and white camouflage of this polar bear.

Animals in very cold climates often have other adaptations too. The surface area of the thin skinned areas of their bodies – like their ears – is usually very small. This reduces their heat loss – look at the ears of the polar bear in Figure 1.

Many Arctic mammals have plenty of insulation, both inside and out. Blubber – a thick layer of fat that builds up under the skin – and a thick fur coat on the outside will insulate an animal very effectively. They really reduce the amount of heat lost through their skin.

The fat layer also provides a food supply. Animals often build up their blubber in the summer. Then they can live off their body fat through the winter when there is almost no food.

b) List three ways in which Arctic animals keep warm in winter.

Camouflage is important both to predators (so their prey doesn't see them coming) and to prey (so they can't be seen). The colours which would camouflage an Arctic animal in summer against plants would stand out against the snow in winter. Many Arctic animals, including the Arctic fox, the Arctic hare and the stoat, exchange the greys and browns of their summer coats for pure white in the winter.

Surviving in dry climates

Dry climates are often also hot climates – like deserts! Deserts are very difficult places for animals to live. There is scorching heat during the day, followed by bitter cold at night, while water is in short supply.

The biggest challenges if you live in a desert are:

- coping with the lack of water, and
- stopping your body temperature from getting too high.

Many desert animals are adapted to need little or no drink. They get the water they need from the food they eat.

Mammals keep their body temperature the same all the time. So as the environment gets hotter, they have to find ways of keeping cool. Most mammals sweat to help them cool down. But this means they lose water, which is not easy to replace in the desert.

c) Why do mammals try to lose heat without sweating in hot, dry conditions?

Desert animals have other adaptations for cooling down. They are often most active in the early morning and late evening, when the temperature is comfortable. During the cold nights and the heat of the day they rest. You find them in burrows well below the surface, where the temperature doesn't change much.

Many desert animals are quite small, so their surface area is large compared to their volume. This helps them to lose heat through their skin. They often have large, thin ears as well to increase their surface area for losing heat.

Another adaptation of many desert animals is that they don't have much fur. Any fur they do have is fine and silky. They also have relatively little body fat stored under the skin. Both of these features make it easier for them to lose heat through the surface of the skin. The animals keep warm during the cold nights by retreating into their burrows.

Figure 3 Animals like this fennec fox have many adaptations to help them cope with the hot dry conditions. How many can you spot?

Figure 4 An elephant is pretty big but it lives in hot, dry climates. Its huge wrinkled skin would cover an animal which was much bigger still. The wrinkles increase the surface area to aid heat loss.

FOUL FACTS

Animals from the deep oceans are adapted to cope with enormous pressure, no light and very cold water. But if these deep-water organisms are brought up to the surface too quickly, they explode because of the rapid change in pressure.

KEY POINTS

1 All living things have adaptations which help them to survive in the conditions where they live.
2 Animals which are adapted for cold environments are often large, with a small surface area : volume (SA/V) ratio. They have thick insulating layers of fat and fur.
3 Changing coat colour in the different seasons gives animals year-round camouflage.
4 Adaptations for hot, dry environments include a large SA/V ratio, thin fur, little body fat and behaviour patterns that avoid the heat of the day.

SUMMARY QUESTIONS

1 a) List the main problems which face animals living in cold conditions like the Arctic.
   b) List the main problems which face animals living in the desert.
2 Give three ways in which animals that stay in the Arctic throughout the winter keep warm. Explain how these adaptations work.
3 Give three ways in which animals which live in a desert manage to keep cool without sweating so they don't lose water.
4 Explain why being quite large helps many Arctic animals to keep warm.

B1b 5.2 Adaptation in plants

1 How are plants adapted to live in dry conditions?
2 How do plants store water?

There are some places where plants simply cannot grow. In deep oceans no light penetrates, and no plants can grow. In the icy wastes of the Antarctic, no plants grow.

Almost everywhere else, including the hot, dry areas of the world, we can find plants growing. Without them there would be no food for the animals. But plants need water both for photosynthesis and to keep their tissues upright. If a plant does not get the water it needs, it wilts and eventually dies.

a) Why do plants need water?

Plants take in water through their roots in the soil. It moves up through the plant and is lost through the leaves in the **transpiration stream**. Plants lose water all the time through their leaves. There are small openings called **stomata** in the leaves of a plant. These open to allow gases in and out for photosynthesis and respiration. But at the same time water is lost by evaporation.

The rate at which a plant loses water is linked to the conditions it is growing in. When it is hot and dry, photosynthesis and respiration take place quickly. As a result, plants also lose water very fast. So how do plants that live in dry conditions cope? Most of them either reduce their surface area so they lose less water or they store water in their tissues. Some do both!

b) How do plants lose water from their leaves?

Changing surface area

When it comes to stopping water loss through the leaves, the surface area : volume ratio (see page 82) is very important to plants. There are a few desert plants which have broad leaves with a large surface area. These leaves collect the dew that forms in the cold evenings. They then funnel the water towards their shallow roots

Some plants in dry environments have curled leaves. This reduces the surface area of the leaf. It also traps a layer of moist air around the leaf which really cuts back the amount of water they lose by evaporation.

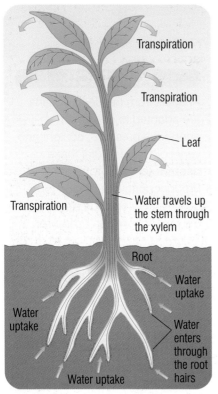

Figure 1 The transpiration stream means plants are losing water all the time by evaporation from their leaves. When the conditions are hot and dry, they lose water very quickly.

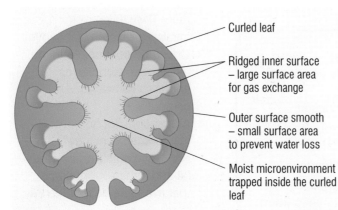

Curled leaf

Ridged inner surface – large surface area for gas exchange

Outer surface smooth – small surface area to prevent water loss

Moist microenvironment trapped inside the curled leaf

Figure 2 Plants that live on sand dunes near the sea have to survive very dry conditions. This marram grass, which you can find all around the British coast, has tightly curled leaves, which reduce the surface area available for water loss.

However, most plants which live in dry conditions have reduced the surface area of their leaves. This cuts down the area from which water can be lost. Some desert plants have small fleshy leaves with a thick cuticle to keep water loss down. The cuticle is a waxy covering on the leaf that stops water evaporating away.

The best-known desert plants are the cacti. Their leaves have been reduced to spines with a very small surface area indeed. This means the cactus only loses a tiny amount of water – and the spines put animals off eating the cactus as well!

c) Why do plants often reduce the surface area of their leaves to help them prevent water loss?

Storing water

Plants can also cope with dry conditions by storing water in their tissues. When there is plenty of water available after a period of rain, the plant stores it. Plants which store water in their fleshy leaves, stems or roots are known as **succulents**.

Cacti don't just rely on their spiny leaves to help them survive in dry conditions. They are succulents as well. The fat green body of a cactus is its stem, which is full of water-storing tissue. All these adaptations make cacti the most successful plants in a hot dry climate.

d) In which parts can a plant store its water?

Keep away!

One of the biggest problems for plants is being eaten by animals. Plants have a wide variety of adaptations designed to deal with this. Vicious thorns, unpleasant tastes and poisonous chemicals can all put animals off!

We have made use of some of these adaptations. For example, we use the bitter chemical in the bark of the cinchona tree to make quinine. This helps relieve the symptoms of malaria. What's more, the poison digitalis from foxgloves is used as a very effective heart medicine.

GET IT RIGHT!

Remember that plants need their stomata open for photosynthesis and respiration. This is why they lose water by evaporation from their leaves.

DID YOU KNOW?

An apple tree in the UK can lose a whole bath of water from its leaves every day. A large saguaro cactus in the desert loses less than one glass of water in the same amount of time!

Figure 3 Cacti are well adapted to survive in desert conditions

SUMMARY QUESTIONS

1 Copy and complete using the words below:

   adaptations   desert   plants   spiny   stem   water

   Cacti are …… which live in the …… . They have two main …… to help them survive. Their leaves have become …… and they store …… in their …… .

2 a) Explain why plants lose water through their leaves all the time.
   b) Why does this make living in a dry place such a problem?

3 Explain three adaptations which help plants living in dry conditions to reduce water loss from their leaves.

KEY POINTS

1 Plants lose water all the time by evaporation from their leaves.
2 Plants which live in dry places have adaptations which help to reduce water loss. These adaptations may often include reduced surface area of their leaves and/or water-storage tissues.

B1b 5.3   Competition in animals

1 What is competition?
2 What makes an animal a good competitor?

Figure 1 Some herbivores, like these silk worms eating their mulberry leaves and the panda with its bamboo, only feed on one particular plant. They are very open to competition from other animals or to a disease that damages their food plant.

Animals and plants grow alongside lots of other living things, some from the same species and others completely different. In any area there will only be a limited amount of food, water and space, and a limited number of mates. As a result, living organisms spend their time competing for the things they need.

The best adapted organisms are most likely to be the winners of the *competition* for resources. They will be most likely to survive and produce healthy offspring.

a) Why do living organisms compete?

What do animals compete for?

Animals compete for many things, including:

● water ● territory ● mates

Competition for food is very common. Herbivores (animals which eat only plants) sometimes feed on many types of plant, and sometimes on only one or two different sorts. Many different species of herbivores will all eat the same plants. Just think how many types of animals eat grass!

The animals which eat a wide range of plants are most likely to be successful. If you are a picky eater, you risk dying out if anything happens to your only food source. An animal with wider tastes will just eat something else for a while!

Competition is common among carnivores (animals which eat only meat). They compete for prey. Small mammals like mice are eaten by animals like foxes, owls, hawks and domestic cats. The different types of animals all hunt the same mice. So the animals which are best adapted to the area will be most successful.

Carnivores have to compete with other members of their own species for their prey as well as with members of different species. Successful predators are adapted to have long legs for running fast and sharp eyes to spot prey. These features will be passed on to their offspring.

Animals often avoid direct competition with members of other species when they can. It is the competition between members of the same species which is most intense!

Prey animals compete with each other too – to be the one which *isn't* caught! Adaptations like camouflage colouring, so you don't get seen, and good hearing, so you pick up a predator approaching, are important for success.

b) Give one adaptation which would be useful to a plant-eater and one which would be helpful to a carnivore.

Competition for mates can be fierce. In many species the male animal puts a lot of effort into impressing the females. The males compete in different ways to win the privilege of mating with her.

DID YOU KNOW?

Some animals have warning colours which let predators know they are poisonous or taste nasty. Poison arrow frogs – which give us the curare used in medicine – have very bright warning colours. Other frogs which aren't poisonous at all mimic the warning colours so that predators leave them alone.

In some species – like deer and lions – the males fight between themselves. Then the winner gets the females.

Many male animals display to the female to get her attention. Some birds have spectacular adaptations to help them stand out. Male peacocks have the most amazing tail feathers. They use them for displaying to other males (to warn them off) and to females (to attract them).

What makes a successful competitor?

A successful competitor is an animal which is adapted to be better at finding food or a mate than the other members of its own species. It also needs to be better at finding food and water than the members of other local species. What is more, it must also breed successfully.

Many animals are successful because they avoid competition with other species as much as possible. They feed in a way that no other local animals do, or they eat a type of food that other animals avoid. For example, one plant can feed many animals without direct competition. While caterpillars eat the leaves, greenfly drink the sap, butterflies suck nectar from the flowers and beetles feed on pollen.

It is much harder to avoid competition within the same species, but many animals try to do just that. They may set up and defend a **territory** – an area where they live and feed. This is a common way of making sure that they will be able to find enough food for themselves and for their young when they breed.

Figure 2 The spectacular display of a male peacock certainly attracts the attention of the females. And unlike animals which fight for their mates, the peacock doesn't risk getting hurt when he tries to win over the females.

SUMMARY QUESTIONS

1 Match the following words to their definitions:

a) Competition	A An animal which eats plants.
b) Carnivore	B An area where an animal lives and feeds.
c) Herbivore	C An animal which eats meat.
d) Territory	D The way animals compete with each other for food, water, space and mates.

2 a) Give an example of animals competing with members of other species for food.
   b) Give an example of animals competing with members of the same species for food.
   c) Why can animals which rely on a single type of food be killed off so easily?

3 a) Give two ways in which animals compete for mates.
   b) What sort of adaptations would be needed to be successful in the two types of competitions in a)?

4 Explain the adaptations would you expect to find in:

   a) an animal which hunts mice?
   b) an animal which eats grass?
   c) a fish which feeds on other fish?
   d) an animal which feeds on the tender leaves at the top of trees?

FOUL FACTS

Different types of African dung beetles avoid competition with each other by attacking the same pile of dung at different times of day and in different ways. The most active beetles work in the heat of the day and make balls of dung which they roll away. The quieter tunnellers and the beetles that actually live in the dung heaps work as dusk is falling.

GET IT RIGHT!

Learn to look at an animal and spot the adaptations which make it a successful competitor!

KEY POINTS

1 Animals often compete with each other for food and territories.
2 Animals compete for mates.
3 Animals have adaptations which make them good competitors.

B1b 5.4 | Competition in plants

EXPERIMENTAL DATA

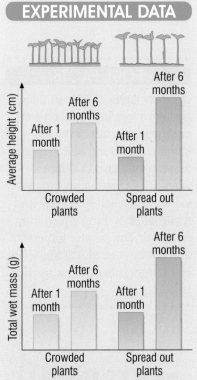

Figure 1 Experiments like this can be carried out to show the effect of competition on plants. All the conditions – light level, the amount of water and minerals available and the temperature were kept exactly the same for both sets of plants. The differences in their growth were the result of overcrowding and competition for resources in one of the groups.

Plants might look like peaceful organisms, growing silently in your local park. But the world of plants is full of cut-throat competition. They compete with each other for light, for water and for nutrients (minerals) from the soil.

They need light for photosynthesis, to make food using energy from the Sun. They need water for photosynthesis and to keep their tissues rigid and supported. And plants need minerals so they can make all the chemicals they need in their cells.

a) What do plants compete with each other for?

Why do plants compete?

Just like animals, plants are in competition both with other species of plants and with their own species. Big, tall plants like trees take up a lot of water and minerals from the soil and prevent light from reaching the plants beneath them. So the plants around them need adaptations to help them to survive.

If a plant sheds its seeds and they land nearby, the parent plant will be in direct competition with its own seedlings. Because the parent plant is large and settled, it will take most of the water, minerals and light. So the plant will deprive its own offspring of everything they need to grow successfully!

If the seeds from a plant all land close together – even if they are a long way from their parent – they will then compete with each other as they grow. So many plants have special adaptations which help them to spread their seeds over a wide area.

b) Why is it important that seeds are spread as far as possible from the parent plant?

Coping with competition

When plants are growing close to other species they often have adaptations which help them to avoid competition.

Small plants found in woodlands often grow and flower very early in the year. Although it is cold, plenty of light gets through the bare branches of the trees. The dormant trees take very little water out of the soil. The leaves shed the previous autumn have rotted down to provide minerals in the soil.

Plants like snowdrops, anemones and bluebells are all adapted to take advantage of these things. They flower, set seeds and die back again before the trees are in full leaf.

Another way plants compete successfully is by having different types of roots. Some plants have shallow roots taking water and minerals from near the surface of the soil. Others have long, deep roots, which go far underground. Both compete successfully for what they need without affecting the other.

If one plant is growing in the shade of another, it may grow taller to reach the light. It may also grow leaves with a bigger surface area to take advantage of all the light it does get.

c) How can short roots help a plant to compete successfully?

Spreading the seeds

To compete successfully, a plant has to avoid competition with its own seedlings. Usually, the most important adaptation for success is the way they shed their seeds.

Many plants use the wind to help them. Some produce seeds which are so small that they are carried easily by air currents. Many others produce fruits with special adaptations for flight to carry their seeds as far away as possible. Examples include the parachutes of the dandelion 'clock' and the winged seeds of the sycamore.

d) How do the fluffy parachutes of dandelion seeds help the seeds spread out?

Some plants use mini-explosions to spread their seeds. The pods dry out, twist and pop, flinging the seeds out and away.

Juicy berries, fruits and nuts are produced by plants to tempt animals to eat them. Once the fruit gets into the animal's gut, the tough seeds travel right through. They are deposited with the waste material in their own little pile of fertiliser, often miles from where they were eaten!

Fruits which are sticky or covered in hooks get caught up in the fur or feathers of a passing animal. They are carried around until they fall off hours or even days later.

Sometimes the seeds of different plants land on the soil and start to grow together. The plants which grow fastest will compete successfully against the slower-growing plants. For example:

- the plants which get their roots into the soil first will get most of the available water and minerals;
- the plants which open their leaves fastest will be able to photosynthesise and grow faster still, depriving the competition of light.

Plants compete at all levels, from spreading their seeds to the height they grow and how early they flower each year. The winners of the competitions are the ones we see. The losers just don't make it!

Figure 2 Coconuts will float for weeks or even months on ocean currents which can carry them hundreds of miles from their parents – and any other coconuts!

SUMMARY QUESTIONS

1 a) Give two ways in which plants can overcome the problems of growing in the shade of another plant.
   b) Explain how a primrose plant manages to grow and flower successfully in spite of living under a large oak tree.

2 a) Why is it so important that plants spread their seeds successfully?
   b) Give three examples of successful adaptations for spreading seeds.

3 The dandelion is a successful weed. Carry out some research and evaluate the adaptations that make it a better competitor than other plants on a school field.

KEY POINTS

1 Plants often compete with each other for light, for water and for nutrients (minerals) from the soil.

2 Plants have many adaptations, which make them good competitors.

B1b 5.5 How do *you* survive?

The most amazing plants in the world?

Most plants die without water – but not the resurrection plants. They can survive massive water losses. They don't prevent water loss or store water – they have adapted to cope with water loss when it happens. These amazing plants can lose up to 95% of their water content without suffering permanent damage.

When conditions get dry, the plants lose more and more water until all that is left are the small, shrivelled remains. The plant looks dead. It can last like this for weeks – but within about 24 hours of watering the tissues fill up with water again (rehydrate). The plant looks as good as new!

Dr Peter Scott and his team at the University of Sussex are trying to find out just how this survival mechanism works, because resurrection plants aren't just a fascinating fact. All over the world crops fail every year because conditions are too dry, and so millions of people don't get enough food. If scientists can find a way to produce 'resurrection crops', then starvation might become a thing of the past.

The difference 24 hours and some water can make to a resurrection plant!

ACTIVITIES

a) How might 'resurrection crops' prevent starvation in the world?

b) You want to get money for some research into the adaptations of a very unusual animal or plant (you can make one up!). Write a brief application for funding for your project. Using the example of the resurrection plants, explain how information about an unusual adaptation might lead to great benefits for people. Use this to back up your claim for money!

The fastest predator in the world?

It takes you about 650 milliseconds to react to a crisis. But the star-nosed mole takes only 230 milliseconds from the moment it first touches its prey to gulping it down. That's faster than the human eye can see!

What makes this even more amazing is that star-nosed moles live underground and are almost totally blind. Their main sense organ is a crown of fleshy tendrils around the nose – incredibly sensitive to touch and smell but very odd to look at!

It seems likely that they have adapted to react so quickly because they can't see what is going on. They need to grab their prey as soon as possible after they touch it. If they don't it might move away or try to avoid them. Then they wouldn't know where it had gone.

When you've got ultra-sensitive tendrils which can try out 13 possible targets every second, who needs eyes?!

The star-nosed mole

Death by infection

The Komodo dragon – largest reptile in the world

The Komodo dragon is the largest reptile in the world. A big male can be over three metres long! They live in Indonesia and their colour varies depending on which island they make their home. They have long, forked tongues which give them an excellent sense of smell. They can smell rotting meat five miles away!

The Komodo dragon eats carrion (dead animals) but it is also a predator. But the dragons are reptiles. They cannot run fast for long or pounce on their prey, yet they can kill a huge water buffalo. How do they do it?

The dragons have 52 sharp teeth, but it is not the sharpness which makes them deadly, it is the bacteria which grow on them. A dragon will lie in wait for a water buffalo, and then rush out and grab one of the hind legs. It tears the leg, and the 15 different species of bacteria growing on its teeth get straight into the buffalo's blood stream. Within a couple of days, the buffalo will be dead from the lethal infection. The dragon just follows after its prey until the buffalo dies.

The dragon then eats almost 80% of its own weight in buffalo meat before resting quietly for a very long time!

A carnivorous plant

The Venus flytrap is a plant that grows on bogs. Bogs are wet and their peaty soil has very few minerals in it. This makes them a difficult place for plants to live.

Venus flytraps have special 'traps' which contain a sweet-smelling nectar. They sit wide open, displaying their red insides. Insects are attracted to the colour and the smell.

Inside the trap are many small, sensitive hairs. As the insect moves about to find the nectar, it will brush against these hairs. Once the hairs have been touched, the trap is triggered. It shuts, trapping the insect inside. Special enzymes then digest the insect inside the trap.

The Venus flytrap – an insect-eating plant!

The Venus flytrap uses the minerals from the digested bodies of its victims in place of the minerals it can't get from the bog soil. After the insect has been digested, the trap reopens ready for its next victim.

ACTIVITIES

c) Look carefully at the information you have been given about resurrection plants, star-nosed moles, Komodo dragons and Venus flytraps. For each one make a list of the adaptations which help them to survive.

d) There are so many different living organisms, each with their own adaptations. Choose three organisms that you know something about – or find out about three organisms which interest you. Make your own fact file on their adaptations and how these adaptations help them to compete successfully. Choose one organism that has an adaptation which has been used by people in some way. Try to include at least one plant!

SUMMARY QUESTIONS

1 Cold-blooded animals like reptiles and snakes take their body temperature from their surroundings and cannot move until they are warm.

   a) Why do you think that there are no reptiles and snakes in the Arctic?

   b) What problems do reptiles face in desert conditions and how do they cope with them?

   c) Most desert animals are quite small. How does this help them survive in the heat?

2 a) What are the main problems for plants living in a hot, dry climate?

   b) Why does reducing the surface area of their leaves help plants to reduce water loss?

   c) Describe two ways in which plants can reduce the surface area of their leaves.

   d) How else are some plants adapted to cope with hot, dry conditions?

   e) Why are cacti such perfect desert plants?

3 a) How does marking out and defending a territory help an animal to compete successfully?

   b) Bamboo plants all tend to flower and die at the same time. Why is this such bad news for pandas, but doesn't affect most other animals?

4 Why is competition between animals of the same species so much more intense than the competition between different species?

5 Use Figure 1 on page 88 to answer these questions.

   a) Describe what happens to the height of both sets of seedlings over the first six months and explain why the changes take place.

   b) The total wet mass of the seedlings after one month was the same whether or not they were crowded. After six months there was a big difference.

   i) Why do you think both types of seedling had the same mass after one month?

   ii) Explain why the seedlings that were more spread out each had more wet mass after six months?

   c) When scientists carry out experiments such as the one described on page 88, they try to use large sample sizes. Why?

   d) i) Name the control variable mentioned in the caption to Figure 1.

   ii) Why were these variables kept constant?

EXAM-STYLE QUESTIONS

1 Which of the following adaptations would **not** reduce the rate of transpiration in a plant?

   A Fleshy succulent leaves   **B** Thick waxy cuticle

   C Few stomata   **D** Reduced leaf area   (1)

2 Which of the following will **not** help an arctic mammal, such as a polar bear, to survive cold temperatures?

   A Thick fur coat

   B Thick layer of fat beneath the skin

   C Small extremities such as ears

   D A large surface area : volume ratio   (1)

3 To investigate competition between species, a series of 20 enclosures were set up to look at the effect of kangaroo rats on other, smaller rodents. Kangaroo rats were removed from half of the enclosures. Over three years the numbers of the smaller rodents were counted in both sets of enclosures.

   The results are shown on the graph.

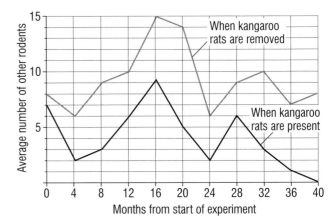

   (a) In the enclosure with kangaroo rats, how many other rodents, on average, were present after 12 months?

   A 4   **B** 6   **C** 8   **D** 10   (1)

   (b) After how many months was there the biggest difference between the average number of other rodents in the two sets of enclosures?

   A 16   **B** 20   **C** 32   **D** 40   (1)

   (c) The conclusion that can be drawn from the experiment is that:

   A kangaroo rats eat other rodents.

   B kangaroo rats compete with other rodents for food.

   C kangaroo rats compete with other rodents and limit their population size.

   D kangaroo rats make it more difficult for other rodents to breed.   (1)

(d) How might the reliability of the results of this experiment be improved?

A By increasing the number of enclosures of each type.

B By using the same number of enclosures but increasing the number of kangaroo rats and other rodents in each type.

C By using just one enclosure of each type rather than ten of each.

D By having a third set of ten enclosures with only kangaroo rats present. (1)

4 Animals that live in the arctic have a range of adaptations that allow them to survive.

(a) Explain why the coat of an arctic fox is brown in summer and white in winter. (3)

(b) If the arctic fox adapts by changing its coat colour between summer and winter, why then does the polar bear remain white throughout the year? (1)

5 The gemsbok is a large herbivore living in dry desert regions of South Africa. It feeds on grasses that are adapted to the dry conditions by obtaining moisture from the air as it cools at night.
The table below shows the water content of these grasses and the feeding activity of the gemsbok over a 24-hour period.

Time of day	% water content of grasses	% of gemsboks feeding
03.00	18	40
06.00	23	60
09.00	25	20
12.00	8	17
15.00	6	16
18.00	5	19
21.00	7	30
24.00	14	50

(a) (i) Name the independent variable investigated. (1)
  (ii) Is this a categoric, ordered, discrete or continuous variable? (1)

(b) How does the water content of the grasses change throughout the 24 hour period? (1)

(c) Between which recorded times are more than 30% of the gemsboks feeding? (1)

(d) Suggest three reasons why the gemsboks benefit from feeding at this time. (3)

HOW SCIENCE WORKS QUESTIONS

Maize is a very important crop plant. Amongst many other foods, it is made into cornflakes. It is also grown for animal feed. The most important part of the plant is the cob which fetches the most money. In an experiment to find the best growing conditions, three plots of land were used. The young maize plants were grown in different densities in the three plots.

1st plot — 10 maize plants per square metre

2nd plot — 15 maize plants per square metre

3rd plot — 20 maize plants per square metre

The results were as follows:

	Planting density (plants/m²)		
	10	15	20
Dry mass of shoots (kg/m²)	9.7	11.6	13.5
Dry mass of cobs (kg/m²)	6.1	4.4	2.8

a) What was the independent variable in this investigation? (1)

b) Draw a graph to show the effect of the planting density on the mass of the cobs grown. (3)

c) What is the pattern shown in your graph? (1)

d) This was a fieldwork investigation. What would the experimenter have taken into account when choosing the location of the three plots? (2)

e) Did the experimenter choose enough plots? Explain your answer. (1)

f) What is the relationship between the mass of cobs and the mass of shoots at different planting densities? (1)

g) The experimenter concluded that the best density for planting the maize is ten plants per m². Do you agree with this as a conclusion? Explain your answer. (2)

B1b 6.1 Inheritance

Figure 1 This picture shows a mother pig and her offspring. They aren't exactly the same as each other, but they are obviously related!

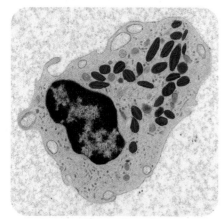

Figure 2 This micrograph shows a highly magnified human cell. In fact the nucleus of the cell would only measure about 0.005 mm! All the instructions for making you and keeping you going are inside this microscopic package. It seems amazing that they work!

Young animals and plants resemble their parents. Horses have foals and people have babies. Chestnut trees produce conkers which grow into little chestnut trees. Many of the smallest organisms that live in the world around us are actually identical to their parents. So what makes us the way we are?

Why do we resemble our parents?

Most families have characteristics which we can see clearly from generation to generation. People find it funny and interesting when one member of a family looks very much like another. Characteristics like nose shape, eye colour and dimples are *inherited* (passed on to you from your parents).

Your resemblance to your parents is the result of *genetic information* passed on to you in the sex cells (**gametes**) from which you developed. This genetic information determines what you will be like.

a) Why do you look like your parents?

Genes and chromosomes

The genetic information which is passed from generation to generation during reproduction is carried in the nucleus of your cells. Almost all of the cells of your body contain a nucleus. And it contains all the plans for making and organising a new cell. What's more, the nucleus contains the blueprint for a whole new you!

Imagine the plans for building a car. They would cover many sheets of paper! Yet in every living organism, the nucleus of the cells contains the information to build a whole new animal, plant, bacterium or fungus. A human being is far more complicated than a car. So where does all the information fit in?

b) Where is the genetic information stored?

Inside the nucleus of all your cells there are thread-like structures called **chromosomes**. The chromosomes are made up of a special chemical called **DNA** (deoxyribose nucleic acid). This is where the genetic information is actually stored.

DNA is a long molecule made up of two strands which are twisted together to make a spiral. This is known as a double helix – imagine a ladder that has been twisted round!

Each different type of organism has a different number of chromosomes in their body cells. Humans have 46 chromosomes while turkeys have 82! You inherit half your chromosomes from your mother and half from your father, so chromosomes come in pairs. You have 23 pairs of chromosomes in all your normal body cells.

Each of your chromosomes contains thousands of **genes** joined together. These are the units of inheritance.

Each gene is a small section of the long DNA molecule. Genes control what an organism is like – its size, its shape and its colour. Each gene affects a different characteristic about you.

Your chromosomes are organised so that both of the chromosomes in a pair carry genes controlling the same things in the same place. This means your genes also come in pairs, one from your father and one from your mother.

c) Where would you find your genes?

Some of your characteristics are decided by a single pair of genes. For example, there is one pair of genes which decides whether or not you will have dimples when you smile! However most of your characteristics are the result of several different genes working together. For example, your hair and eye colour are both the result of several different genes.

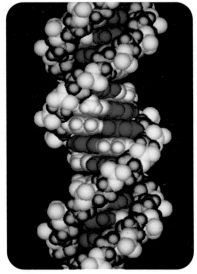

Figure 3 DNA! This huge molecule is actually made up of lots of smaller molecules joined together. Each gene is a small section of the big DNA strand.

Science pioneers: Cracking the code

For a very long time no-one knew how inheritance worked. By the 1940s most scientists thought that DNA was probably the molecule which carried inherited information from one generation to the next.

In the 1950s James Watson (a young American) and Francis Crick (from the UK) were working on the DNA problem at Cambridge. They took all the information they could find on DNA – including X-ray pictures of the molecule taken by another team, Maurice Wilkins and Rosalind Franklin in London.

Watson and Crick tried to build a model of the DNA molecule that would explain everything they knew. When they finally realised that the bases always paired up in the same way they had cracked the code. The now famous DNA double helix was seen for the first time.

GET IT RIGHT!

Make sure you are clear about the difference between a cell, the nucleus, chromosomes, genes and characteristics.

Figure 4 Francis Crick and James Watson – the two men who first showed the world how DNA works. Watson, Crick and Wilkins all received the Nobel Prize for their work. Rosalind Franklin died of cancer before the prizes were awarded.

SUMMARY QUESTIONS

1 Copy and complete using the words below:

**chromosomes   DNA   genes   genetic information
gametes   nucleus   resemble**

Offspring ...... their parents because of .... passed on to them in the ...... (sex cells) from which they developed. The information is contained in the ......, made of a chemical called ...... found in the ...... of the cell. The information is carried in small units called ....

2 a) What is the basic unit of inheritance?
   b) Offspring inherit information from their parents, but do not look exactly like them – why not?

3 a) Which molecule carries genetic information?
   b) Why do chromosomes come in pairs?
   c) Why do genes come in pairs?
   d) How many genes do scientists think human beings have?

KEY POINTS

1 Young animals and plants have similar characteristics to their parents. That's because of genetic information passed on to them in the sex cells from which they developed.

2 The nucleus of your cells contains chromosomes. Chromosomes carry the genes that control the characteristics of your body.

B1b 6.2

Types of reproduction

1 Why does asexual reproduction result in offspring that are identical to their parents?
2 How does sexual reproduction produce variety?

Figure 1 A mass of daffodils like this can contain hundreds of identical flowers. This is because they come from bulbs which reproduce asexually.

Reproduction is very important to living things. It is during reproduction that genetic information is passed on from parents to their offspring. There are two very different ways of reproducing – *asexual reproduction* and *sexual reproduction*.

Asexual reproduction

Asexual reproduction only involves one parent. The process produces more organisms completely identical to itself. There is no joining of special sex cells and there is no variety in the offspring.

Asexual reproduction gives rise to offspring known as **clones**. Their genetic material is identical both to the parent and to each other. Although there is no variety, asexual reproduction is very safe – you don't have to worry about finding a partner!

a) Why is there no variety in offspring from asexual reproduction?

Asexual reproduction is very common in the smallest animals and plants and in bacteria. However, many bigger plants like daffodils, strawberries and brambles do it too. Bulbs, corms, tubers (like potatoes), runners and suckers are all ways in which plants reproduce asexually. Asexual reproduction also takes place all the time in your own body, as cells divide to grow and to replace worn-out tissues.

Sexual reproduction

The other way of passing information from parents to their offspring is through sexual reproduction. Sexual reproduction involves the joining of a male sex cell and a female sex cell from two parents. These two special cells (gametes), one from each parent, join together to form a new individual.

If you are the result of sexual reproduction, you will inherit genetic information from both parents. You will have some characteristics from both of your parents, but won't be exactly like either of them. This introduces variety. In plants the gametes involved in sexual reproduction are found within ovules and pollen. In animals they are called ova (eggs) and sperm.

Sexual reproduction is more risky than asexual reproduction, because it relies on the sex cells from two individuals meeting. In spite of this, variety is so important to survival that sexual reproduction is seen in species of organisms ranging from bacteria to people!

b) How does sexual reproduction cause variety in the offspring?

Variation

Sexual reproduction involves the joining of different genetic information. This results in offspring which show much more variation than the offspring from asexual reproduction. This is a great advantage in making sure the species survives. That's because the more variety there is in a group of individuals, the more likely it is that at least a few of them will have the ability to survive difficult conditions.

GET IT RIGHT!

Asexual reproduction results in identical genetic information being passed on. Sexual reproduction makes sure the genetic information is mixed so there is variety in the offspring.

If we take a closer look at how sexual reproduction works, it becomes clear how variation appears in the offspring.

Each pair of genes affects a different characteristic about you. However the genes in a pair can come in different forms. These different versions of the same gene are called **alleles**. Most things about you are controlled by lots of different pairs of genes. Luckily some of your characteristics are controlled by one gene with just two possible alleles. For example, there are genes which decide whether:

- your earlobes are attached closely to the side of your head or hang freely,
- your thumb is straight or curved,
- you have dimples when you smile,
- you have hair on the second segment of your ring finger.

We can use these genes to help us understand how inheritance works.

c) Why is variety important?

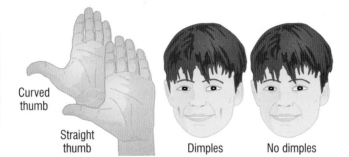

Curved thumb
Straight thumb

Unattached ear lobe   Attached ear lobe

Dimples   No dimples

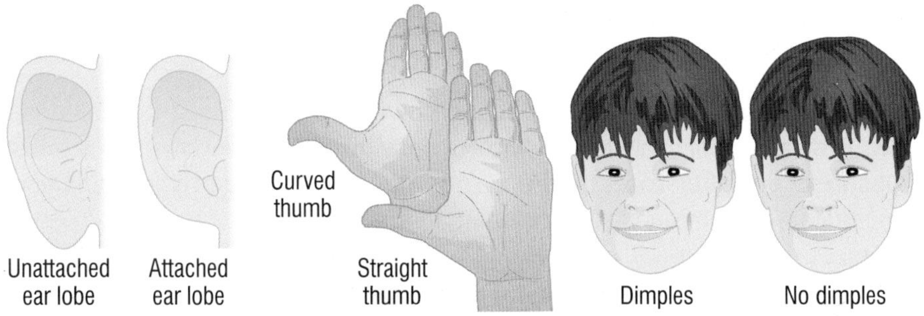

Figure 2 These are all human characteristics which are controlled by a single pair of genes, so they can be very useful in helping us to understand how sexual reproduction introduces variety and how inheritance works

The gene which controls dimples has two possible forms – an allele for dimples, and an allele for no dimples. The gene for dangly earlobes also has two possible alleles – one for dangly earlobes and one for earlobes which are attached. Some features have lots of different possible alleles.

You will get a random mixture of thousands of alleles from your parents – which is why you don't look exactly like either of them!

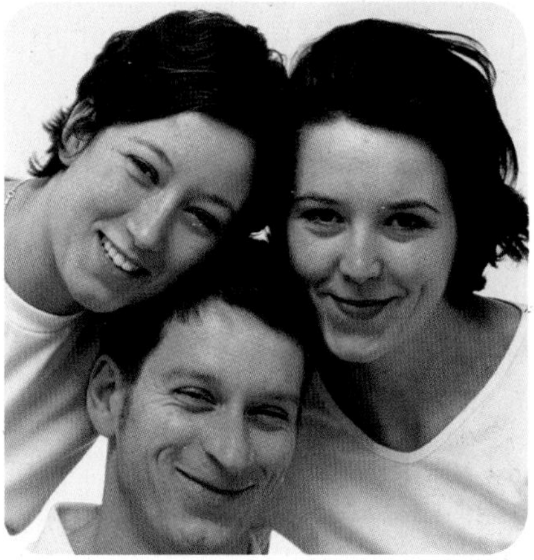

Figure 3 Although these young people have some family likenesses, the variety which results from the mixing of their parents' genetic information can clearly be seen!

SUMMARY QUESTIONS

1 Define the following words:

   a) asexual reproduction   b) sexual reproduction
   c) gamete   d) variety.

2 a) Why is sexual reproduction more risky for individuals than asexual reproduction?
   b) What is the big advantage of sexual reproduction over asexual reproduction?

3 A daffodil reproduces asexually using bulbs and sexually using flowers.

   a) How does this help to make them very successful plants?
   b) Explain the genetic differences between a daffodil's sexually and asexually produced offspring.

KEY POINTS

1 In asexual reproduction there is no joining of gametes and only one parent. There is no genetic variety in the offspring.

2 In sexual reproduction male and female gametes join. The mixture of genetic information from two parents leads to genetic variety.

B1b 6.3 Cloning

LEARNING OBJECTIVES

1 What is a clone?
2 Why do we want to create clones?

Figure1 Simple cloning by taking cuttings is a technique used by gardeners and nurserymen all around the world. It gives us plants like these.

A clone is an individual which has been produced asexually from its parent. It is therefore genetically identical to the parent. Many plants reproduce naturally by cloning, and this has been used by farmers and gardeners for many years.

Cloning plants

Gardeners can produce new plants by taking cuttings from older plants. This is a form of artificial asexual reproduction which has been carried out for hundreds of years. How do you take a cutting? First you remove a small piece of a plant – often part of the stem or sometimes just part of the leaf. If you grow it in the right conditions, new roots and shoots will form to give you a small, complete new plant.

Using this method you can produce new plants quickly and cheaply from old plants. The cuttings will be genetically identical to the parent plants.

Many growers now use hormone rooting powders to encourage the cuttings to grow. They are most likely to develop successfully if you keep them in a moist atmosphere until their roots develop. We produce plants such as orchids and many fruit trees commercially by cloning in this way.

a) Why does a cutting look the same as its parent plant?

Cloning tissue

Taking cuttings is a very old technique. In recent years scientists have come up with a more modern way of cloning plants called *tissue culture*. It is more expensive but it allows you to make thousands of new plants from a tiny piece of plant tissue. If you use the right mixture of plant hormones, you can make a small group of cells from the plant you want produce a big mass of identical plant cells.

Then, using a different mixture of hormones and conditions, you can stimulate each of these cells to form a tiny new plant. This type of cloning guarantees that the plants you grow will have the characteristics you want.

b) What is the advantage of tissue culture over taking cuttings?

Cloning animals

In recent years cloning technology has moved forward even further and now includes animals. In fact cloning animals is now quite common in farming, particularly cloning cattle embryos. Cows normally produce only one or two calves at a time. If you use embryo cloning, your very best cows can produce many more top-quality calves each year.

In embryo cloning, you give a top-quality cow fertility hormones to make her produce a lot of eggs. You then fertilise these eggs using sperm from a really good bull. Often this is done inside the cow, and the embryos which are produced are then gently washed out of her womb. Sometimes the eggs are collected and you add the sperm in a laboratory to produce the embryos.

SCIENCE @ WORK

Cloning cattle embryos and transferring them to host cattle is skilled and expensive work. Teams of scientists, technicians and vets are constantly working to improve the technique even more.

At this very early stage of development every cell of the embryo can still form all of the cells needed for a new cow. They have not become specialised.

1 Divide each embryo into several individual cells.
2 Each cell grows into an indentical embryo in the lab.
3 Transfer embryos into their host mothers, which have been given hormones to get them ready for pregnancy.
4 Identical cloned calves born. They are not biologically related to their mothers.

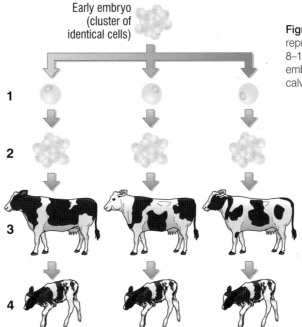

Early embryo (cluster of identical cells)

1

2

3

4

Figure 2 Using normal sexual reproduction, a top cow might produce 8–10 calves during her working life. Using embryo cloning she can produce more calves than that in a single year!

Cloning embryos in this way has made it possible for us to take high-quality embryos all around the world. We can take them to places where cattle with a high milk yield or lots of meat are badly needed for breeding with poor local stock. Embryo cloning is also used to make lots of identical copies of embryos that have been genetically modified to produce medically useful compounds. (See pages 102 and 103.)

GET IT RIGHT!

- Remember clones have identical genetic information.
- Make sure you are clear about the difference between a tissue and an embryo.

SUMMARY QUESTIONS

1 Match up the following definitions:

a) Cuttings	A Splitting cells apart from a developing embryo before they become specialised to produce several identical embryos.
b) Tissue cloning	B Taking a small piece of a stem or leaf and growing it on in the right conditions to produce a new plant.
c) Asexual reproduction	C Getting a few cells from a desirable plant to make a big mass of identical cells each of which can produce a tiny identical plant.
d) Embryo cloning	D Reproduction which involves only one parent, there is no joining of gametes and the offspring are genetically identical to the parent.

2 Tissue cloning and taking cuttings both give you plants which are identical to their parent. How do these two methods of plant cloning differ?

3 a) Why is the ability to clone cattle embryos so useful?
   b) Draw a flow diagram to show the stages of the embryo cloning of cattle.
   c) Comment on the economic and ethical issues involved in embryo cloning in cattle.

KEY POINTS

1 The genetically identical offspring produced by asexual reproduction are known as **clones**.
2 New plants can be produced quickly and cheaply by taking cuttings from older plants. The new plants are genetically identical to the older ones.
3 There are a number of more modern cloning techniques. These include tissue culture of plants and embryo cloning and transfers in animals.

B1b 6.4   New ways of cloning animals

True cloning of animals, without sexual reproduction involved at all, has been a major scientific breakthrough. The basic technique is known as *fusion cell cloning*. It is the most complicated form of asexual reproduction you can find!

Fusion cell cloning

To clone a cell from an adult animal is easy. Asexual reproduction takes place all the time in your body to produce millions of identical cells. But to take a cell from an adult animal and make an embryo or even a complete identical animal is a very different thing.

Here are the steps involved:

- The nucleus is taken from an adult cell.
- At the same time the nucleus is removed from an egg cell from another animal of the same species.
- The nucleus from the original adult cell is placed in the empty egg and the new cell is given a tiny electric shock.
- This fuses the new cell together, and starts the process of cell division.
- An embryo begins to develop which is genetically identical to the original adult animal.

Adult cell cloning

Fusion cell cloning has been used to produce whole animal clones. The first large mammal ever to be cloned from another adult animal was Dolly the sheep, born in 1997. A team of scientists in Edinburgh produced Dolly from the adult cell of another sheep.

When a new animal is produced, this is known as *adult cell* or *reproductive cloning*. It is still relatively rare. You still have to fuse the nucleus of one cell with the empty egg of another animal. Then you have to place the embryo which results into the womb of a third animal. It develops there until it is born.

a) What is the name of the technique which produced Dolly the sheep?

When Dolly was produced she was the only success from hundreds of attempts. The technique is still difficult and unreliable, but scientists hope that it will become easier in future.

Figure 1 Dolly the sheep was the first large mammal to be cloned from another adult animal. Her birth caused great excitement and many scientists have tried to clone other animals since.

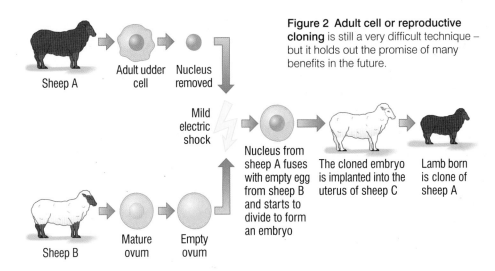

Figure 2 Adult cell or reproductive cloning is still a very difficult technique – but it holds out the promise of many benefits in the future.

Sheep A → Adult udder cell → Nucleus removed

Mild electric shock

Sheep B → Mature ovum → Empty ovum

Nucleus from sheep A fuses with empty egg from sheep B and starts to divide to form an embryo → The cloned embryo is implanted into the uterus of sheep C → Lamb born is clone of sheep A

The benefits and disadvantages of adult cell cloning

One big hope for adult cell cloning is that animals which have been genetically engineered to produce useful proteins in their milk can be cloned. This would give us a good way of producing large numbers of cloned, medically useful animals.

This technique could also be used to help save animals from extinction, or even bring back species of animals which died out years ago. The technique could be used to clone pets or prized animals so that they continue even after the original has died. However, some people are not happy about this idea. (See page 104.)

b) How might adult cell cloning be used to help people?

There are some disadvantages to this exciting science as well. Many people fear that the technique could lead to the cloning of human babies. At the moment this is not possible, but who knows what is in the future?

One big problem is that modern cloning produces lots of plants or animals with identical genes. In other words, cloning reduces variety in a population. This means the population is less able to survive any changes in their environment which might happen in the future. That's because if one of them does not contain a useful mutation, none of them will.

In a more natural population, at least one or two individuals can usually survive change. They go on to reproduce and restock. This could be a problem in the future for cloned crop plants, or for cloned farm animals.

SUMMARY QUESTIONS

1 Copy and complete:

Dolly the sheep was created from the …… cell of another sheep. She was …… to this sheep. This technique is known as …… …… …… .

2 Produce a flow chart to show how fusion cell cloning works.

3 a) List the main advantages of the development of adult cell cloning techniques and the main disadvantages.
   b) Give your own opinion about whether work on the technique should be allowed to continue. Explain your point of view.

KEY POINTS

1 Fusion cell cloning is a form of asexual reproduction

2 In adult cell cloning a whole cloned animal can result. The nucleus from a cell from an adult animal is transferred to an empty egg cell from another animal. A small electric shock fuses the cell and starts embryo development. The embryo is placed in a third animal to develop.

B1b 6.5

Genetic engineering

LEARNING OBJECTIVES

1 What is genetic engineering?
2 How are genes transferred from one organism to another?
3 What are the issues involved in using genetic engineering?

Human cell with insulin gene in its DNA

Bacterium with ring of DNA called a plasmid

Insulin gene cut out of DNA by an enzyme

Plasmid taken out of bacterium and split open by an enzyme

Insulin gene inserted into plasmid by another enzyme

Plasmid with insulin gene in it taken up by bacterium

Bacterium multiplies many times

The insulin gene is switched on and the insulin is harvested

Insulin

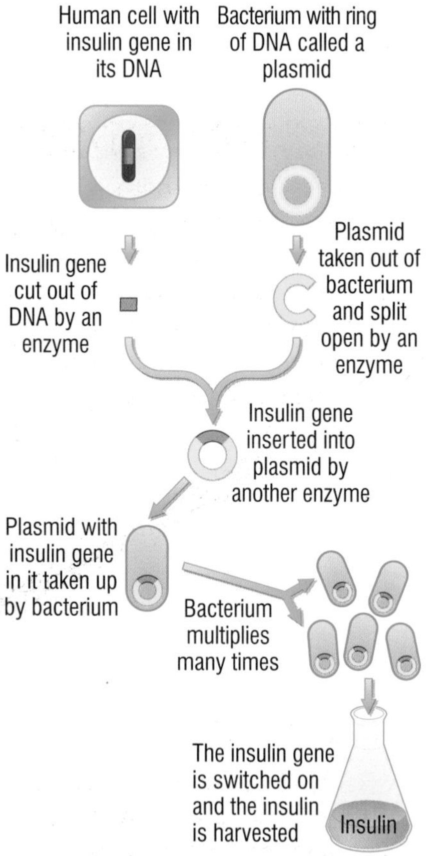

Figure 1 The principles of genetic engineering. A bacterial cell receives a gene from a human being.

DID YOU KNOW...

... glowing genes from jellyfish have been used to produce crop plants which give off a blue light when they are attacked by insects. Then the farmer knows when they need spraying!

When you clone plants and animals, you are changing the natural processes of reproduction. There is another new technology which takes the changes much further – **genetic engineering** (also known as **genetic modification**). Genetic engineering is used to change an organism and give it new characteristics which we want to see.

What is genetic engineering?

Genetic engineering involves changing the genetic material of an organism. You take a small piece of DNA – a gene – from one organism and transfer it to the genetic material of a completely different organism. So, for example, genes from the chromosomes of one of your human cells can be 'cut out' using enzymes and transferred to the cell of a bacterium. Your gene carries on making a human protein, even though it is now in a bacterium.

a) How is a gene taken out from one organism to be put into another?

If genetically engineered bacteria are cultured on a large scale they will make huge quantities of protein from other organisms. We now use them to make a number of drugs and hormones used as medicines.

Transferring genes to animal and plant cells

There is a limit to the types of proteins bacteria are capable of making. As a result, genetic engineering has moved on. Scientists have found that genes from one organism can be transferred to the cells of another type of animal or plant at an early stage of their development. As the animal or plant grows it develops with the new desired characteristics from the other organism.

b) Why are genes inserted into animals and plants as well as bacteria?

The benefits of genetic engineering

One of the biggest advantages of genetically engineered bacteria is that they can make exactly the protein needed, in exactly the amounts needed and in a very pure form. For example, people with diabetes need supplies of the hormone insulin. It used to be extracted from the pancreases of pigs and cattle but it wasn't quite the same as human insulin, and the supply was quite variable. Both of those problems have been solved by the introduction of genetically engineered human insulin. (See Figure 1.)

We can use engineered genes to improve the growth rates of plants and animals. They can be used to improve the food value of crops and to reduce the fat levels in meat. They are used to produce plants which make their own pesticide chemicals. GM food lasts longer in the supermarkets. It can also be designed to grow well in dry, hot or cold parts of the world. So it could help to solve the problems of world hunger.

A number of sheep and other mammals have also been engineered to produce life-saving human proteins in their milk. These are much more complex proteins than the ones produced by bacteria. They have the potential to save many lives.

Human engineering

If there is a mistake in your genetic material, you may have a genetic disease. These can be very serious. Many people hope that genetic engineering can solve the problem.

It might become possible to put 'healthy' DNA into the affected cells by genetic engineering, so they work properly. Perhaps the cells of an early embryo can be engineered so that the individual develops to be a healthy person. If these treatments become possible, many people would have new hope of a normal life for themselves or their children.

c) What do we mean by a 'genetic disease'?

The disadvantages of genetic engineering

Genetic engineering is still a very new science. There are many concerns about it as no-one can yet be completely sure what all of the long-term effects might be. For example, it seems possible that insects may become pesticide-resistant if they eat a constant diet of pesticide-forming plants.

Some people are concerned about the effect of eating genetically modified food on human health. Genes from genetically modified plants and animals might spread into the wildlife of the countryside. Genetically modified crops are often not fertile, which means farmers in poor countries have to buy new seed each year.

And people may want to manipulate the genes of their future children. This might be to make sure they are born healthy, but what if it is to have a child who is clever, or good-looking, or good at sport? The idea of 'designer babies' causes concern for many people. Genetic engineering raises issues for us all to think about. (See page 105.)

Figure 3 You can't tell what is genetically modified and what isn't just by looking at it! In the UK very few genetically modified foods are sold. The ones that are have to be clearly labelled. But many other countries, including the USA, are far less worried and use GM food quite widely.

Figure 2 These sheep look very normal – but they are a genetically engineered flock producing human proteins in their milk. The proteins are used in life-saving medicines.

KEY POINTS

1 In genetic engineering, genes from the chromosomes of humans and other organisms can be 'cut out' using enzymes and transferred to the cells of bacteria.
2 Genes can also be transferred to the cells of animals and plants at an early stage of their development.
3 There are many potential advantages and disadvantages to the use of genetic engineering.

SUMMARY QUESTIONS

1 Copy and complete using the words below:

cell   engineering   enzymes   gene   genetic   transfer

Genetic ...... involves changing the ...... material of an organism. You cut a ...... from one organism using .... Then ...... it to the ...... of a completely different organism.

2 a) Make a flow diagram that explains the stages of genetic engineering.
b) Make two lists, one to show the possible advantages of genetic engineering and the other to show the possible disadvantages.
c) Do you think genetic engineering is a good idea? Should it be allowed? Justify your views.

B1b 6.6   Making choices about technology

Cc – A REAL COPYCAT!

Cc, or Copycat, was the first cloned cat to be produced. Born in 2002, she was a change of direction. Most of the research into cloning had been focused on farm and research animals – but cats are thought of first and foremost as pets.

Much of the funding for cat cloning in the US comes from companies who are hoping to be able to clone people's dying or dead pets for them. It has already been shown that a succesful clone can be produced from a dead animal. Cells from beef from a slaughterhouse were used to create a live cloned calf.

But to make Cc, 188 attempts were made producing 87 cloned embryos, only one of which resulted in a kitten. Cloning your pet won't be easy or cheap. The issue is, should people be cloning their dead pets, or should they

Cc the cloned cat is unaware of the stir she has caused. Cats like this one are often well-loved pets – but should we really be cloning our old friends?

be learning to grieve, appreciate the animal they had and give a home to one of the thousands of unwanted cats already in existence? Even if a favourite pet is cloned, it may look nothing like the original because the coat colour of many cats is the result of random gene switching in the skin cells. The markings would never be the same again, even if the DNA was!

THE FOAL WHO COULD CHANGE RACING – FOR GOOD! By David Turf

This little foal with her mum looks just like any other, but in fact she's made history!

The foal in this photo with her mother is no ordinary young horse. Prometea is the first cloned horse – and her surrogate mother is also her identical twin, because the foal is a clone of the mare who gave birth to her. This new technology has in turn led to a breakthrough which could change the breeding of racehorses forever.

Pieraz is a famous Arab horse who has been world endurance racing champion several times. Pieraz 2, a foal born in 2005, is his closest relative. 'So what?', I hear you say. It is very common for successful racehorses to be used for breeding. The difference here is that Peiraz was neutered when he was still a youngster – and Pieraz 2 is not his son, but his clone!

Changing the genes – right or wrong?

Our first daughter has been affected by a dreadful genetic disease. You don't know what it's like until it happens to you. If we had another baby, I'd want them to change the genes if they could when the embryo was really tiny. Then we'd know the baby wasn't going to be ill – and any grandchildren we had in the future would be alright as well.

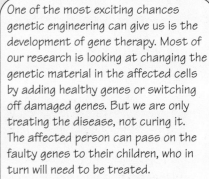

I think it is wrong to interfere with nature. It must be awful if your family is affected, but if they start fiddling about with the genes of a tiny embryo where will it all end? It'll be designer babies next, you mark my words. If they can change the genes that can make you ill, you can't tell me they won't be able to make your baby really clever or good-looking if you're prepared to pay enough.

One of the most exciting chances genetic engineering can give us is the development of gene therapy. Most of our research is looking at changing the genetic material in the affected cells by adding healthy genes or switching off damaged genes. But we are only treating the disease, not curing it. The affected person can pass on the faulty genes to their children, who in turn will need to be treated.

Gene therapy actually offers a way of curing genetic diseases. It would involve changing the genes of a fertilised egg or very early embryo so that the baby is born with healthy genes in all its cells. This is known as germ line gene therapy. We know it raises some major ethical issues and opinion across the world is strongly divided about it. In fact most countries, including the UK, have so far completely banned germ line gene therapy.

I'm more concerned about GM foods. Who knows what we're all eating nowadays. I don't want strange genes inside me, thank you very much. We've got plenty of fruit and vegetables as it is – why do we need more?

I think GM food is such a good idea. If the scientists can modify crops so they don't go off so quickly, food should get cheaper, and there will be more to go around. And what about these plants that produce pesticides – that'll stop a lot of crop spraying, so that should make our food cleaner and cheaper. It's typical of us in the UK that we moan and panic about it all.

We have some real worries about the GM crops which don't form fertile seeds. It does mean the growers in the countries where we do a lot of work are going to struggle. In the past they just kept seeds from the previous year's crops, so it was cheap and easy. On the other hand, these GM crops don't need spraying very much. They grow well in our dry conditions and they keep well too – so there are some advantages.

ACTIVITY

You are going to produce a 10-minute slot for a daytime television show entitled 'Genetic engineering – a good thing or not?' You can ask any of the people shown here to come on your show and express their views. Using this and the information about genetic engineering on pages 102 and 103 to help you, plan out the script for your time on air. Remember that you have to inform the public about genetic engineering, entertain them and make them think about the issues involved.

SUMMARY QUESTIONS

1

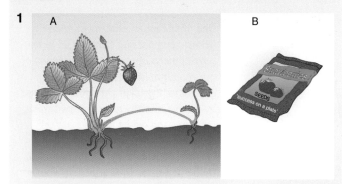

A

B

a) How has the small plant shown in diagram A been produced?

b) What sort of reproduction is this?

c) How were the seeds in B produced?

d) How are the new plants which you would grow from the packet of seeds shown in B different from the new plants shown in A?

2 Tissue culture techniques mean that 50 000 new raspberry plants can be gained from one old one instead of 2 or 3 taking cuttings. Cloning embryos from the best bred cows means that they can be genetically responsible for thirty or more calves each year instead of two or three.

a) How does tissue culture differ from taking cuttings?

b) How can one cow produce thirty or more calves in a year?

c) What are the similarities between cloning plants and cloning animals in this way?

d) What are the differences in the techniques for cloning animals and plants?

e) Why do you think there is so much interest in finding different ways to make the breeding of farm animals and plants increasingly efficient?

3 Human growth is usually controlled by the pituitary gland in your brain. If you don't make enough hormone, you don't grow properly and remain very small. This condition affects 1 in every 5000 children. Until recently the only way to get growth hormone was from the pituitary glands of dead bodies. Genetically engineered bacteria can now make plenty of pure growth hormone.

a) Draw and label a diagram to explain how a healthy human gene for making growth hormone can be taken from a human chromosome and put into a working bacterial cell.

b) What are the advantages of producing substances like growth hormone using genetic engineering?

EXAM-STYLE QUESTIONS

1 Young plants and animals resemble their parents. This is because characteristics are inherited by the young from their parents.

Match words **A**, **B**, **C** and **D** with the spaces **1** to **4** in the sentences.

A Chromosomes   **B** DNA

C Gametes   **D** Genes

Genetic information is passed from parents to offspring in sex cells called …**1**… The sex cells contain thread-like structures known as …**2**… .

The thread-like structures contain thousands of …**3**…. that are the units of inheritance.

These units of inheritance are small sections of a double helix called …**4**… .   (4)

2 The table is about the production of offspring.
Match words **A**, **B**, **C** and **D** with the processes **1** to **4** in the table.

A Sexual reproduction   **B** Asexual reproduction

C Inheritance   **D** Cloning

	Process
1	Joining of male and female sex cells (gametes) to produce young
2	Making young that are identical to both their parents and to each other, especially in agriculture
3	Producing offspring without sex cells (gametes)
4	Passing on characteristics from parents to offspring

(4)

3 Plant tissue culture is a method used to create new plants. One method is described below:

- A small piece of tissue is removed from a plant.
- Under sterile conditions, the tissue is placed in a vessel containing nutrients.
- A mass of identical plant cells develops.
- These cells are placed in a medium containing nutrients and plant growth regulators (hormones).
- Young plants develop that are separated and grown to maturity.

(a) What type of reproduction is involved in plant tissue culture?   (1)

(b) Why is a disease more likely to kill every one of a group of plants produced by tissue culture than a group grown from seeds?   (2)

(c) Suggest one advantage of growing plants from tissue culture over growing them from cuttings or seeds.   (1)

4 Some humans suffer from diabetes. One form of diabetes is caused by the inability to make the hormone insulin. Diabetics can lead normal lives providing they can inject insulin into their bloodstream at regular intervals. The diagram shows how insulin can be made using genetic engineering.

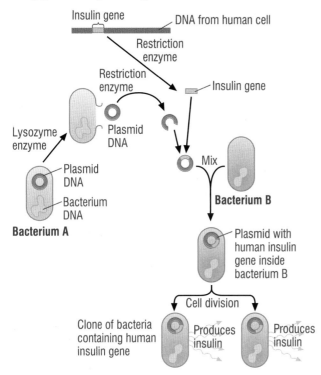

(a) **Using only the information in the diagram** suggest what the following enzymes do:

    (i) lysozyme enzyme (1)

    (ii) restriction enzyme. (1)

(b) **Using only the information in the diagram**, describe how the human insulin gene is transferred into a bacterium which then makes insulin. (6)

(c) Insulin is produced from cloned bacteria that contain the human insulin gene.

    (i) What is a clone? (1)

    (ii) Do you think that cloning is the result of sexual or asexual reproduction? Explain your answer. (1)

    (iii) A bacterium can divide every 20 minutes. Starting with 50 bacteria, how many bacteria would there be after 2 hours? Show your working. (1)

(d) Insulin to treat diabetes used to be extracted from the pancreases of pigs and cattle. Give two advantages of insulin produced by genetic engineering over extracting it from animals. (2)

HOW SCIENCE WORKS QUESTIONS

1 DNA outside the nucleus

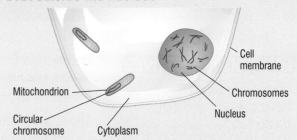

Mitochondria are sometimes referred to as the powerhouse of the cell. They carry out much of the cell's respiration. They contain their own DNA as a circular chromosome. Research has shown that this DNA is passed almost completely from the mother to the child. This is because the mitochondria are to be found in the cytoplasm. Some very rare diseases can be caused by the mitochondrial DNA mutating. These diseases can cause blindness or a disease very much like type 2 diabetes and deafness. Some children who inherit mutated mitochondrial DNA die at a very young age.

Use your scientific knowledge of cell structure and your understanding of the technology of cloning to answer the following questions.

a) Suggest how women who are at risk of passing on this mutated mitochondrial DNA might be helped. (1)

b) What are the ethical issues around this research? (2)

c) Who should be making decisions about whether or not families should be helped in this way? (3)

d) What is the role of the scientist in helping to make these decisions? (1)

2 Genetically modified (GM) crops

The first trials of GM crops were destroyed during the summer of 1999. Protesters, concerned that pollen from the crop could affect other local plants, were very pleased. The government, who started the trials, were not pleased! The government's Food Safety minister said:

'We can't operate food safety policy on a *hunch* – we have to have the science and that's why we need the trials.'

a) Explain the difference between a 'hunch' and 'science'. (2)

b) Do you think the protesters had based their ideas on a 'hunch'? Explain your answer. (1)

c) One of the purposes of the trial was to find out how far pollen from the GM crop could travel. Describe how the trial might be set up. (5)

B1b 7.1   The origins of life on Earth

We are surrounded by an amazing variety of life on planet Earth. Questions like 'Where has it all come from?' and 'When did life on Earth begin?' have puzzled people for generations.

There is no record of the origins of life on Earth – it is a puzzle which can never be completely solved. No-one was there to see it and there is no direct evidence for what happened. We don't even know when life on Earth began. However, most scientists think it was somewhere between 3 billion and 4.5 billion years ago!

There are some interesting ideas and well-respected theories which explain most of what you can see around you. The biggest problem we have is finding the evidence to support the ideas.

a) When do scientists think life on Earth began?

What can we learn from fossils?

We share the Earth with millions of different species of living organisms. But this is tiny compared to the 4 billion species that scientists believe have lived on Earth during its history.

Most of these species have disappeared again in the mists of time. Some of them have gone completely. Others have left living relatives. The fossil record gives us an insight into how much – and how little – organisms have changed since life developed on Earth.

b) How many species of living organisms are thought to have existed on Earth over the years?

Fossils are the remains of plants or animals from many thousands or millions of years ago which are found in rocks. You have probably seen a fossil in a museum or on TV or – if you are really lucky – found one yourself.

The fossil record is not complete because so much rock has been broken down, worn away, buried or melted over the years. In spite of this, it can still give us a 'snapshot' of life millions of years before we were born.

Fossils can be formed in a number of ways:

- Many fossils were formed when harder parts of the animal or plant were replaced by other minerals over long periods of time. These are the most common fossils.
- Another type of fossil was formed when an animal or plant did not decay after it died. Sometimes the temperature was too low for decay to take place and the animals and plants were preserved in ice. However, these fossils are rare.

Some of the fossils we find are not of actual animals or plants, but rather of traces they have left behind. Fossil footprints and droppings all help us to build up a picture of life on Earth long ago.

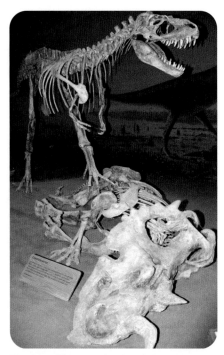

Figure 1 This amazing fossil shows two dinosaurs – prehistoric animals which died out millions of years before we appeared on Earth. Fossils can only give us a brief glimpse into the past. We will never know exactly what snuffed out the life of this spectacular reptile all those years ago.

Often the fossil record is very limited. Small bits of skeletons are found, or little bits of shells. Luckily we have a very complete fossil record for a few animals, including the horse. What's more, fossils show us that not all animals have changed over time. For example, fossil sharks from millions of years ago look very like their modern descendants.

c) Why do ice fossils give us clear evidence of animals that lived in the past?

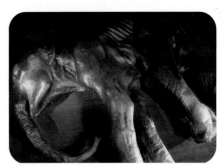

Figure 2 This baby mammoth has been preserved in ice for about 40 000 years. Ice fossils are very rare. They can give us an amazing glimpse of what a prehistoric animal looked like, the colour of its skin or fur and even what it had been eating.

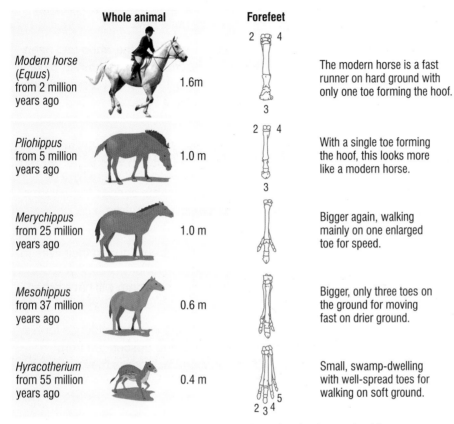

Whole animal	Forefeet	
Modern horse (*Equus*) from 2 million years ago — 1.6m	2 ⬚ 4 / 3	The modern horse is a fast runner on hard ground with only one toe forming the hoof.
Pliohippus from 5 million years ago — 1.0 m	2 ⬚ 4 / 3	With a single toe forming the hoof, this looks more like a modern horse.
Merychippus from 25 million years ago — 1.0 m		Bigger again, walking mainly on one enlarged toe for speed.
Mesohippus from 37 million years ago — 0.6 m		Bigger, only three toes on the ground for moving fast on drier ground.
Hyracotherium from 55 million years ago — 0.4 m	2 3 4 5	Small, swamp-dwelling with well-spread toes for walking on soft ground.

Figure 3 The story of the horse. The horse as we know it today has evolved from some very different animals. We know they existed from the very clear record left in the fossils.

SUMMARY QUESTIONS

1 Copy and complete using the words below:

animal   decay   evidence   fossils   ice fossils   minerals   plant

One important piece of …… for how life has developed on Earth are ……. The most common type are formed when parts of the …… or …… are replaced by …… as they decay. Some fossils were formed when an organism did not …… after it died. These …… …… are very rare.

2 a) There are several theories as to how life on Earth began. Why is it impossible to know for sure?
   b) Make a timeline to show how our ideas about the age of the Earth have changed since the 17th century as more information has become available.
   c) Why are fossils such important evidence for the way life has developed?

3 Look at the evolutionary tree of the horse in Figure 3. Explain how the fossil evidence of the legs helps us to understand what the animals were like and how they lived.

KEY POINTS

1 Fossils provide us with evidence of how much – or how little – different organisms have changed since life developed on Earth.

2 It is very difficult for scientists to know exactly how life on Earth began because there is no direct evidence.

B1b 7.2 Theories of evolution

1 What is the theory of evolution?
2 What is the evidence that evolution has taken place?

Figure 1 In Lamarck's model of evolution, giraffes have long necks because each generation stretched up to reach the highest leaves. So each new generation had a slightly longer neck!

Figure 2 Darwin was very impressed by the giant tortoises he found on the Galapagos Islands. The tortoises on each island had different shaped shells and a slightly different way of life – and Darwin made careful drawings of them all.

The theory of *evolution* tells us that all the species of living things alive today have evolved from the first simple life forms that existed on Earth. Most of us take these ideas for granted – but they are really quite new.

Up to the 18th century most people in Europe believed that the world had been created by God. They thought it was made, as described in the Christian Bible, a few thousand years ago. But by the beginning of the 19th century, scientists were beginning to come up with new ideas.

Lamarck's theory of evolution

Jean-Baptiste Lamarck, a French biologist, suggested that all organisms were linked by what he called a 'fountain of life'. His idea was that every type of animal evolved from primitive worms. He thought that the change from worms to other organisms was caused by the *inheritance of acquired characteristics*.

The theory was that useful changes, developed by parents during their lives to help them survive, are passed on to their offspring. In other words, if you do lots of swimming and develop broad shoulders, your children will have broad shoulders as well!

Lamarck's theory fell down for several reasons. There was no evidence for his 'fountain of life' and people didn't like the idea of being descended from worms. People could also see quite clearly that characteristics they acquired were not passed on to their children.

a) What is meant by the phrase 'inheritance of acquired characteristics'?

Charles Darwin and the origin of species

Our modern ideas about evolution began with the work of one of the most famous scientists of all time – *Charles Darwin*. Darwin set out in 1831 as the ship's naturalist on *HMS Beagle.* He was only 22 years old and the voyage to South America and the South Sea Islands would take five years.

Darwin planned to study geology on the trip. But as the voyage went on he became as excited by his collection of animals and plants as by his rock samples.

b) What was the name of the ship that Darwin sailed on?

In South America, Darwin discovered a new form of the common rhea, an ostrich-like bird – but not until his party had cooked and eaten one of them! Two different types of the same bird living in slightly different areas set Darwin thinking.

On the Galapagos Islands he was amazed by the variety of species and the way they differed from island to island. Darwin found strong similarities between types of finches, iguanas and tortoises on the different islands. Yet each was different and adapted to make the most of local conditions.

Darwin collected huge numbers of specimens of animals and plants during the explorations of the *HMS Beagle*. He also made detailed drawings and kept written observations. The long voyage home gave him plenty of time to think about what he had seen. Charles Darwin returned home after five years with some new and different ideas forming in his mind.

c) What is the name of the famous islands where Darwin found so many interesting species?

After Darwin returned to England he spent the next 20 years working on his ideas. He knew they would meet a lot of opposition. He realised he would need lots of evidence to support his theories. He used the amazing animals and plants he had seen on his journeys as part of that evidence.

He also built up evidence from breeding experiments on pigeons to support his theory. In 1859, he published his famous book *On the Origin of Species by means of Natural Selection* (often known as *The Origin of Species*).

Darwin's central theory is that all living organisms have evolved from simpler life forms. This evolution has come about by a process of **natural selection**. Reproduction always gives more offspring than the environment can support. Only those which are most suited to their environment – the 'fittest' – will survive. When they breed, they pass on those useful characteristics to their offspring. Darwin suggested that this was how evolution takes place.

d) What was the name of Darwin's famous book?

The evidence for evolution

The fossil record shows us how species have evolved over millions of years, and how different species are related to each other. Studying the similarities and differences between species, can also help us to understand the evolutionary relationships between them. Observing how changes in the genes can be passed from one generation to another gives us more evidence still.

GET IT RIGHT!

Don't get confused between:
- the **theory of evolution** and
- the **process of natural selection**.

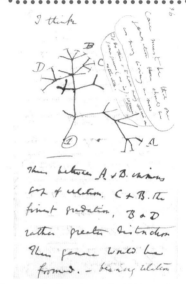

Figure 3 As Darwin studied his specimens, he began to build up a branching picture of evolution, which he tried out in different forms in his notebooks

KEY POINTS

1 The **theory of evolution** states that all the species which are alive today – and many more which are now extinct – evolved from simple life forms which first developed more than three billion years ago.
2 Darwin's theory is that evolution takes place through **natural selection**.
3 Studying the similarities and differences between species helps us to understand how they have evolved and how closely related they are to each other.

SUMMARY QUESTIONS

1 What is meant by the following terms:

  a) evolution?   b) natural selection?
  c) inheritance of acquired characteristics?

2 Why was Darwin's theory of natural selection only accepted very gradually?

3 What was the importance of the following in the development of Darwin's ideas?

  a) South American rheas.
  b) Galapagos tortoises, iguanas and finches.
  c) The long voyage of *HMS Beagle*.
  d) The twenty years from his return to the publication of his book *The Origin of Species*.

4 Suggest reasons why Lamarck and Darwin were convinced by their theories explaining life on Earth.

B1b 7.3   Natural selection

Figure 1 If all of these dandelions seeds developed to become adults and then breed themselves, there would be problems! In fact, very few of them will make it. The survivors will have a combination of genes that gives them a competitive edge over all the others.

Figure 2 The natural world is often brutal. Only the best adapted predators capture prey – and only the best adapted prey animals escape!

Scientists explain the variety of life today as the result of a process called *natural selection*. The idea was first suggested 150 years ago by Charles Darwin.

The natural world is a harsh place, and as you saw in Chapter 5 animals and plants are always in competition with each other. Sometimes an animal or plant gains an advantage in the competition. This might be against other species or against other members of its own species. That individual is more likely to survive and breed – and this is known as *natural selection*.

a) Who first suggested the idea of natural selection?

Survival of the fittest

Charles Darwin was the first person to describe natural selection as the 'survival of the fittest'. Reproduction is a very wasteful process. Animals and plants always produce more offspring than the environment can support.

Fruit flies can produce 200 offspring every two weeks. The yellow star thistle, an American weed, produces around 150 000 seeds per plant per year! If all those offspring survived we'd be overrun with fruit flies and yellow star thistles!

But the individual organisms in any species show lots of variation. This is because of differences in the genes they inherit. Only the offspring with the genes best suited to their habitat manage to stay alive and breed successfully. This is natural selection at work.

Think about rabbits. The rabbits with the best all-round eyesight, the sharpest hearing and the longest legs will be the ones which are most likely to escape being eaten by a fox. They will be the ones most likely to live long enough to breed. What's more, they will pass those useful genes on to their babies. The slower, less alert rabbits will get eaten and their genes will be digested with the rest of them!

b) Why would a rabbit with sharp hearing be more likely to survive than one with less keen hearing?

The part played by mutation

New forms of genes (new alleles) result from changes in existing genes. These changes are known as **mutations**. They are tiny changes in the long strands of DNA.

Mutations occur quite naturally through mistakes made in copying your DNA when your cells divide. Mutations introduce more variety into the genes of a species. In terms of survival, this is very important.

c) What is a mutation?

Many mutations have no effect on the characteristics of an organism, and some mutations are harmful. However, just occasionally a mutation has a good effect. It produces an adaptation which makes an organism better suited to its environment. This makes it more likely to survive and breed.

Whatever the adaptation, if it helps an organism survive and reproduce it will get passed on to the next generation. The mutant gene will gradually become more common in the population. It will cause the species to evolve.

When new forms of a gene arise from mutation, it may cause a more rapid change in a species. This is particularly true if circumstances change as well.

Natural selection in action

Have you ever eaten oysters? They are an expensive treat! They are collected from special oyster beds under the sea. Malpeque Bay in Canada has some very large oyster beds. In 1915, the oyster fishermen noticed a few oysters which were small and flabby with pus-filled blisters.

By 1922 the oyster beds were almost empty. The oysters had been wiped out by a new and devastating disease (soon known as Malpeque disease).

Fortunately a few of the shellfish carried a mutation which made them resistant to the disease. Not surprisingly, these were the only ones to survive and breed. The oyster beds filled up again and by 1940 they were producing more oysters than ever.

But the new population of oysters had evolved. As a result of natural selection, almost every oyster in Malpeque Bay now carries the allele which makes them resistant to Malpeque disease. So the disease is no longer a problem.

d) What is Malpeque disease?

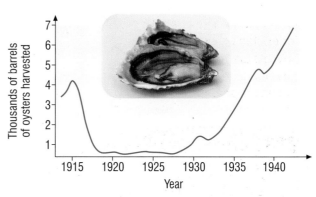

Figure 3 Oyster yields from Malpeque Bay 1915–40. As you can see, disease devastated the oyster beds. However thanks to the process of natural selection, a healthy population of oysters managed to survive and reproduce again.

GET IT RIGHT!

Make sure you learn the main points of natural selection:

mutation → variation → adaptation → survival → genes passed on to the next generation

---

SUMMARY QUESTIONS

1 Copy and complete using the words below:

adaptation   breed   environment   generation   mutation
natural selection   organism   survive

When a ……has a good effect it produces an …… which makes an ……better suited to its ……. This makes it more likely to …… and ……. The mutation then gets passed on to the next ……. This is …… …….

2 Give three examples from this spread of characteristics which are the result of natural selection, e.g. all-around eyesight in rabbits.

3 Explain how the following characteristics of animals and plants have come about in terms of natural selection.

a) Male peacocks have large and brightly coloured tails.
b) Cacti have spines instead of leaves.
c) Camels can tolerate their body temperature rising far higher than most other mammals.

4 Explain how mutation affects the evolution of a species.

KEY POINTS

1 New forms of genes result from changes (mutations) in existing genes.

2 Different organisms in a species show a wide range of variation because of differences in their genes.

3 The individuals with the characteristics most suited to their environment are most likely to survive and breed successfully.

4 The genes which have produced these successful characteristics are then passed on to the next generation.

B1b 7.4 Extinction

LEARNING OBJECTIVES

1. Why do species become extinct?
2. What killed off the dinosaurs?

Millions of years ago	Plants	Animals
NOW	Flowering plants dominant	Humans
	Conifers and flowering plants dominant	Many mammals
65	MAJOR EXTINCTION EVENT Dinosaurs extinct	
136		Bony fish spread
		Dinosaurs dominant
		Modern crustaceans
190	Conifers and ferns dominant	Dinosaur ancestors
		First mammals and birds
225	MAJOR EXTINCTION EVENT Many amphibians and invertebrates extinct	
	Conifers appeared	
280	Fern forests	First reptiles
		Many amphibians
345	Plants with veins	Many fish
		First insects
	MAJOR EXTINCTION EVENT 70% of species lost	
		Sharks and amphibians
395	Algae common	Fish with jaws
		First land arthropods
430	MAJOR EXTINCTION EVENT	
	First land plants with veins	Jawless fish and molluscs
500	FIRST MAJOR EXTINCTION EVENT	
	Algae dominant	Trilobites common
570		Worm-like animals
2500	Origins of life	

Figure 1 A summary of the main events in the evolution of life on Earth

Throughout the history of life on Earth, we think a total of about 4 billion different species have existed. Yet only a few million species of living organisms are alive today. The rest have become ***extinct***.

Extinction is the permanent loss of all the members of a species from the face of the Earth.

As conditions change, new species evolve which are better fitted to survive the new conditions. At the same time older species which cannot cope with the changes, and which do not compete so well for food and other resources, gradually die out. This is how evolution takes place and the balance of species on Earth gradually changes. Extinction is very important.

a) What is extinction?

Environmental changes

Throughout history, the climate and environment of the Earth has been changing. At times the Earth has been very hot. At other times, temperatures have fallen and the Earth has been in the grip of an Ice Age.

Organisms which do well in the heat of a tropical climate won't thrive in an icy landscape. Many of them become extinct through lack of food or being too cold to breed. New species, which cope well in cold climates, evolve and thrive by natural selection.

Changes to the climate or the environment are the main cause of extinction throughout history. For example, most scientists think it was a big climate change that caused the dinosaurs to become extinct millions of years ago. This was possibly caused by a giant meteorite crashing into the Earth, creating drastic changes to the climate.

There have been five occasions during the history of the Earth when big climate changes have led to extinction on an enormous scale. Look at the major extinction events in Figure 1. These are part of the process by which evolution takes place.

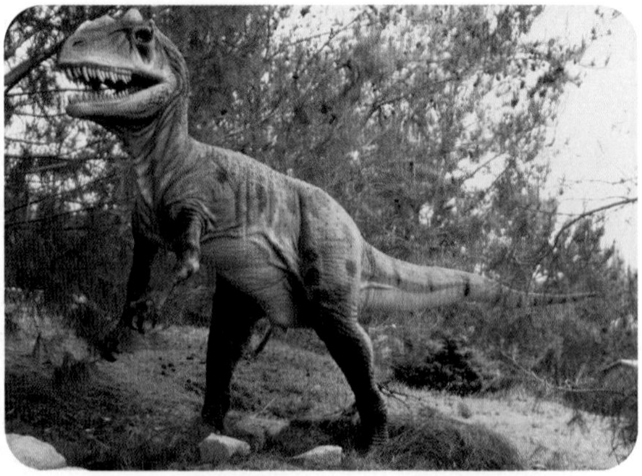

Figure 2 The dinosaurs ruled the Earth for millions of years, but when the whole environment changed, they could not adapt and died out. By the time things began to warm up again, mammals, which could control their own body temperature, were becoming dominant. The age of reptiles was over.

b) How does the climate change during an Ice Age?

Organisms which cause extinction

The other main cause of extinction is other living things. This can take place in several different ways:

- If a new **predator** turns up in an area, it can wipe out unsuspecting prey animals very quickly. That's because the prey do not have adaptations to avoid it.
  A new predator may evolve, or an existing species might simply move into new territory. Sometimes it is our fault. The brown tree snake from Australia was brought to Guam by people following World War II. By the 1960s many bird species on Guam were becoming extinct at a rapid rate – eaten by the snakes which attacked their nests at night!
  The birds had no chance to evolve a defence to this new night-time predator. Because many of the birds of Guam are now extinct, the snakes have started eating lizards instead!

- New *diseases* (caused by microorganisms) can drive a species to the point of extinction. They are most likely to cause extinctions on islands, where the whole population of an animal or plant is close together.
  The Tasmanian devil in New Zealand is one example where this may happen. These rare animals are dying from a new form of cancer, which seems to attack and kill them very quickly.

- Finally, one species can drive another to extinction through successful *competition*. You may see a new mutation which gives one type of organism a real advantage over another, or you may find people have introduced a new species by mistake.
  If the new species is really successful it may take over from the original animal or plant and make it extinct. In Australia the introduction of rabbits has been a nightmare because they eat so much and breed so fast! Other native Australian animals are dying out because they cannot compete.

SUMMARY QUESTIONS

1 Copy and complete using the words below:

| climate | competitors | diseases | Earth | environment |
| Extinction | predators | species |

...... is the permanent loss of all the members of a ...... from the .... It may be caused by changes to the ...... or ......, to new ......, new ...... or possibly new ....

2 Explain how we think the dinosaurs became extinct.

3 a) Explain how each of the following situations might cause a species of animal or plant to become extinct.
  i) Mouse Island has a rare species of black-tailed mice. They are preyed on by hawks and owls, but there are no mammals which eat them. A new family bring their pregnant pet cat to the island.
  ii) English primroses have quite small leaves. Several people bring home packets of seeds from a European primrose which has bigger leaves and flowers very early in the spring.
 b) Why is extinction an important part of evolution?

GET IT RIGHT!

Always mention a **change** when you suggest reasons for extinction.
Watch your timescales – remember humans were not responsible for the extinction of the dinosaurs!

Figure 3 The Scottish island of North Uist has a similar problem to Guam. Someone brought a few hedgehogs onto the island to tackle garden slugs. The hedgehogs bred rapidly and are eating the eggs and chicks of the many rare sea birds which breed on the island. Now people are trying to kill the hedgehogs to save the birds!

KEY POINT

1 Extinction may be caused by changes to the environment, new predators, new competitors and possibly new diseases.

B1b 7.5   Evolution – the debate goes on

In some states of America fundamentalist Christians have a powerful voice. They believe that the Earth and everything in it was created exactly as it is described in the Bible. They would like to prevent the theory of evolution from being written about in school text books or taught in schools. At the very least they would like to see the Creationist view given equal emphasis.

Science versus religion?

When Charles Darwin published his book *The Origin of Species*, he knew that it would cause trouble between the scientific community and the Church. He wanted to put forward his ideas, but he did not want to unsettle faithful Christians – his beloved wife was one! Of course the book caused an uproar, and the debate still continues in some places today.

Darwin's basic principles are not in dispute among scientists, although they often like to discuss the fine details of evolution. But not everyone agrees. For some people there is no conflict between a deep faith in God and an acceptance of evolution. Others find this a problem. Religion is a system of faith and unquestioned belief. It deals with spiritual things which cannot be explained simply by using scientific methods, based on collecting evidence and data on the natural universe.

One area where the Church and science have clashed is on the age of the Earth itself. Here are some of the stages in the debate!

The age of the Earth

Fossils provide evidence that animals and plants have changed and developed over a very long time. This process is known as **evolution**. The idea of evolution suggests that the Earth itself must also have existed for billions of years. This view of the origins of life on Earth is quite recent.

In the 17th century the story of the creation of the Earth in the Christian Bible was still largely unquestioned. One famous historian, Archbishop Usher, used it to calculate that the Earth was less than 6000 years old!

During the 18th century people began to travel more. They not only discovered new lands, but amazing plants and animals as well. Our ancestors unearthed the fossil remains of massive creatures and strange plants. They began to build up evidence that the Earth had changed dramatically during its history.

By the beginning of the 19th century, the evidence for evolution was building up. Sir Charles Lyell was a British geologist. He showed that the Earth was very ancient, and was shaped by rivers and 'subterranean fires'. He estimated that the Earth was several hundred million years old and published his ideas from 1830 onwards.

Lyell's work was important to Darwin. He came up with a theory that all living organisms have arisen from evolution by natural selection. This would have taken many millions of years. Fossils also helped to confirm Darwin's ideas. They showed some of the stages at which the different animals and plants appeared and how they have changed over time. However, some people still believe that fossils were put into rocks by God to test our faith.

By the 1890s, Arthur Holmes was using radioactivity to date rocks. He established the age of the Earth as around 4.6 billion years old, giving plenty of time for evolution to have happened!

ACTIVITY

You have been asked to give evidence at the School Board meeting in the United States. The panel is trying to decide whether to allow the scientific theory of evolution to be taught in their school.

Put together a short report on why it is important that students know about the theory of evolution. Bring together some evidence of evolution occurring in the world around them. Show that it is important for students to understand the scientific view which is held by the majority of their peers in the developed world. This is true even if they choose not to accept the evidence and to hold their own beliefs.

Making extinction extinct?

So many fascinating animals have become extinct. I mean, I'd love to see a dodo, wouldn't you? And going further back, what about a woolly mammoth walking the Earth again? These examples are a bit extreme! But there are teams around the world who are using modern technology to save species which are on the brink of extinction, or have just become extinct.

We hope cloning will be the key to the future for some species. Even if we can't clone them successfully now, by saving some of their tissue we may be able to bring the species back in the future!

For example, Banteng, the wild cattle of Java, Burma and Borneo are endangered in the wild. In 2003 a team in the US managed to produce one healthy cloned Banteng calf, and another that was abnormal. Before that, in 2000, we cloned a healthy Guar calf. Guar are very rare cattle indeed so this was quite a triumph. Sadly little Noah died from an infection just two days after he was born – but we can and will try again!

We've got some tissue from the ear of a Pyranean Ibex. The last one died in 2000. It would be fantastic to bring them back soon. And going back to mammoths ... They are very like our modern elephants in many ways and we've found very well preserved cells in some of the ice fossils. I'm sure that one day someone will clone an extinct animal and bring the species back into existence. What's more, I'm sure prehistoric animals might even be cloned one day – and I hope I'm here to see it!

Plant species are under threat all around the world – but we're fighting back. We have a plan to protect 24 000 plant species by storing their seeds! It's a massive international project masterminded from the Royal Botanic Gardens at Kew in the UK.

In the UK we have collected seeds from all the wild flowering plants for the Millennium Seed Bank Project (MSBP). Once we've collected them they are put into storage. These UK seeds are not alone in their storage jars – seeds from around the world are sent to join them.

Sixteen countries from Jordan to Madagascar, and from Botswana to Mexico are involved in the Millennium Seed Bank project.

South Western Australia in particular has many, many plants – at least 12 000 species are known! It is one of the world's top 'hot spots' for biodiversity and seeds from all of the threatened species are being collected. Once the seeds are stored, even if the plants become extinct in the wild, we have the chance of introducing them again in the future! These plants are so important – they could be a source of new medicines in the future.

Once the seeds for the MSBP have been collected, they are checked, dried and then stored at about −20°C. Under these conditions they can survive for decades – or even centuries!

New species are being cloned all the time. In future we may be able to clone species of animals and plants that are threatened with extinction (like this banteng) and keep them going. We may even be able to clone species which became extinct some time ago using ice fossils or material found on dried specimens in museums.

ACTIVITIES

a) Extinction!

Your task is to make a poster titled 'Extinction!' for display in the school science department. You can choose what to highlight in your poster. You can use information from this chapter and you might like to use other resources as well. The following might give you some ideas:

- How extinction comes about and why it is important.
- Comparing extinction rates in the past and now – why is there so much concern?
- How can extinction be prevented/undone?

b) Interfering with nature?

Write a letter to *The Times*:

Either: express your enthusiasm for using new technologies to prevent extinction, keep species going and bring back extinct species.

Or: express your disgust at using new technologies to prevent extinction, keep species going and bring back extinct species.

SUMMARY QUESTIONS

1 a) How are rock fossils formed?

b) How are ice fossils formed?

c) What evidence do fossils give us about how life on Earth has developed?

d) Why is the fossil record not complete?

2 a) Summarise the similarities and differences between Darwin's theory of evolution and Lamarck's.

b) Why do you think Lamarck's theory was so important to the way Darwin's theory was subsequently received?

3

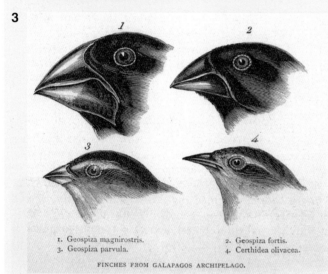

1. Geospiza magnirostris.
2. Geospiza fortis.
3. Geospiza parvula.
4. Certhidea olivacea.

FINCHES FROM GALAPAGOS ARCHIPELAGO.

Darwin's finches – more evidence for evolution

Look at the birds in the picture. They are known as Darwin's finches. They live on the Galapagos Islands. Each one has a slightly different beak and eats a different type of food.

Explain carefully how natural selection can result in so many different beak shapes from one original type of founder finch.

4 a) Why is extinction important to the success of evolution?

b) Why are scientists so worried about the rate at which extinction is occurring now?

c) Find out and write about one animal or plant which has become extinct in the last twenty years.

d) Find out and write about one animal or plant which is close to extinction now. Explain what, if anything, is being done to help prevent it from becoming extinct.

e) Many groups are keen to prevent animals and plants becoming extinct at all costs. Why is it not necessarily a good idea to prevent extinction?

EXAM-STYLE QUESTIONS

1 The list contains factors that have played a part in the development of the species we see on Earth today:

A Extinction   B Evolution

C Natural selection   D Mutation

Match words **A**, **B**, **C** and **D** with the processes **1, 2, 3** and **4** in the table.

	Process
1	Change and development of organisms over a long period of time
2	Change to the amount or arrangement of genetic material within a cell
3	Permanent loss of all members of a species
4	Passing of genes to offspring by the organisms most suited to their environment

(4)

2 Which of the following does **not** play a part in evolution by natural selection?

A Inheritance of acquired characteristics.

B Mutation of existing genes.

C Variation of individuals within a species.

D Production of offspring by individuals best suited to their environment. (1)

3 Which of the following would **not** normally cause the extinction of a species?

A Environmental change

B Fewer competitors

C New diseases

D More predators (1)

4 Not all scientists agree on the exact evolutionary relationship between different primates. The diagram shows a timeline for one version of this relationship.

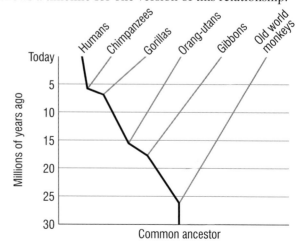

(a) Which group is the closest relative to humans?

A Old world monkeys

B Orang-utans

C Gorillas

D Chimpanzees

(b) How many million years after the old world monkeys evolved from the common ancestor did the gorillas evolve?

A 27　　**B** 20　　**C** 10　　**D** 7

(c) How many of the primate groups shown in the diagram were on Earth 20 million years ago?

A 6　　**B** 5　　**C** 3　　**D** 1

(d) Many of the ancestors of the present-day primates are now extinct. How do we know these ancestors once lived?

A By studying DNA samples.

B By studying fossil records.

C By studying blood samples.

D By studying cell structure.　　(4)

5　The Galapagos Islands are a group of islands in the Pacific Ocean. The nearest country on the mainland is Ecuador, 1 000 km away. By some means, a few seed-eating finches were the first birds to reach the islands. This single ancestral species has since evolved into many different species. Charles Darwin visited the islands and noted that each species had a beak adapted to the type of food it ate.

Using the theory of natural selection, explain how the ancestral species might have evolved into birds with different-shaped beaks.　　(6)

HOW SCIENCE WORKS QUESTIONS

It is difficult to gather data that illustrates evolution. It is possible to gather data to show natural selection, but this usually takes a long time. Simulations are useful because, while they are not factually correct, they do show how natural selection might work.

Darwin used evidence from his visit to the Galapagos Islands to show how natural selection might have worked. He used this as evidence for evolution, by natural selection. Some of the evidence he gathered was about the size and shape of the beaks of the finches on the different islands.

A class decided to simulate natural selection, by seeing if they could use different tools to pick up seeds. This is what they did.

Four students each chose a particular tool to pick up seeds. The teacher then scattered hundreds of seeds onto a patch of grass outside the lab. The four students were then given five minutes to pick up as many seeds as they could.

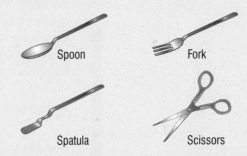

James, who was using a spoon, picked up 23 seeds, whilst Farzana, using a fork, could only pick up two. Claire managed seven seeds with the spatula, but Jenny struggled to pick up her two seeds with a pair of scissors.

a) Put the essential data into a table.　　(3)

b) How would this data be best presented?　　(1)

c) Was this a fair test? Explain your answer.　　(1)

d) What conclusion can you draw from this simulation?　　(1)

B1b 8.1

The effects of the population explosion

1 What effect is the growth in the number of people having on the Earth and its resources?

We have only been around on the surface of the Earth for a relatively short time – less than a million years. Yet our activity has changed the balance of nature on the planet enormously. Some of the changes we have made seem to be driving many other species to extinction. Some people worry that we may even threaten our own survival.

Human population growth

For many thousands of years people lived on the Earth in quite small numbers. There were only a few hundred millions of us! We were scattered all over the world, and the effects of our activity were usually small and local. Any changes could easily be absorbed by the environment where we lived.

But in the last 200 years or so, the human population has grown very quickly. At the end of the 20th century the human population was over 6 billion, and it is still growing

If the **population** of any other species of animal or plant had suddenly increased in this way, natural predators, lack of food, build up of waste products or diseases would have reduced it again. But we have discovered how to grow more food than we could ever gather from the wild. We can cure or prevent many killer diseases. We have no natural predators. This helps to explain why the human population has grown so fast.

Figure 1 The Earth from space. As the human population of the Earth grows, our impact on the planet gets bigger every day.

Not only have our numbers grown hugely, but in large parts of the world our standard of living has also improved enormously. In the UK we use vast amounts of electricity and fuel to heat and light our homes and places of work. We use fossil fuels like oil to produce this electricity. We also use it to move about in cars, planes, trains and boats at high speed and to make materials like plastics. We have more than enough to eat and if we are ill we can often be made better.

a) Approximately how many people are living on the Earth today?

The effect on land and resources

The increase in the numbers of people has had a big effect on our environment. All these billions of people need land to live on. More and more land is used for the building of houses, shops, industrial sites and roads. Some of these building projects destroy the habitats of rare species of other living organisms.

We use billions of acres of land around the world for farming, to grow food and other crops for human use. Wherever people farm, the natural animal and plant population is destroyed.

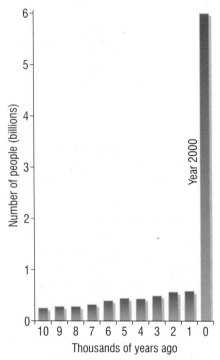

Figure 2 This record of human population growth shows the massive increase during the last few hundred years

In quarrying we dig up great areas of land for the resources it holds such as gravel, metal ores and diamonds. This also reduces the land available for other organisms.

b) How do people reduce the amount of land available for other animals and plants?

Figure 3 In the UK alone hundreds of thousands of new houses are being built, and miles of new road systems. Every time we clear land like this, the homes of countless animals and plants are destroyed.

The huge human population is an enormous drain on the resources of the Earth. Raw materials are rapidly being used up. This includes non-renewable energy resources such as oil and natural gas and metal ores which cannot be replaced.

Pollution

The growing human population also means vastly increased amounts of waste. This is both human bodily waste and the rubbish from packaging, uneaten food and disposable goods. The dumping of this waste makes large areas of land unavailable for any other life except scavengers.

There has also been an enormous increase in manufacturing and industry to produce the goods we want. This in turn has led to industrial waste.

The waste we produce presents us with some very difficult problems. If it is not handled properly it can cause serious **pollution**. Our water may be polluted by sewage, by fertilisers from farms and by toxic chemicals from industry. The air we breathe may be polluted with smoke and poisonous gases such as sulfur dioxide. (See page 122.)

The land itself can be polluted with toxic chemicals from farming such as pesticides and herbicides, and with industrial waste such as heavy metals.

These chemicals in turn can be washed from the land into the water. If the ecology of the Earth is affected by our population explosion, our use of resources and our waste, everyone will pay the price.

c) What substances commonly pollute
   i) water, ii) air, and iii) land?

GET IT RIGHT!

Make sure you know exactly which pollutants affect air, land and water!

SCIENCE @ WORK

Scientists working for groups as diverse as the United Nations and Greenpeace are involved both in monitoring the world population growth and in measuring and controlling levels of pollution.

Figure 4 In countries like ours, we have a very high standard of living. But a kitchen like this uses lots of resources – wood, metals, plastics and energy. This all results in pollution and the removal of resources which can never be replaced.

KEY POINTS

1. The human population is growing rapidly and the standard of living is increasing.
2. More waste is being produced. If it is not handled properly it can pollute the water, the air and the land.
3. Humans reduce the amount of land available for other animals and plants.
4. Raw materials, including non-renewable resources, are being used up rapidly.

SUMMARY QUESTIONS

1 Copy and complete using the words below:

     diseases   farming   food   increase   population
           predators   treat   two hundred

    The human ...... has increased dramatically in the last ...... years. Better ...... methods mean we have more .... We can ...... and prevent many .... We have no natural .... All this has allowed the numbers to ....

2 a) How has the standard of living increased over the last hundred years?
   b) Give three ways in which people have used up different resources.

3 Write a clear paragraph explaining how the ever-increasing human population causes pollution in a number of different ways.

B1b 8.2   Acid rain

Figure 1 Air pollution is usually invisible. Just occasionally the level is so high it can actually be seen. The brown haze you can see over this city is caused by high levels of nitrogen oxides produced from car exhausts.

Figure 2 These trees should be covered in leaves and full of insects, birds and other animal life. Instead they are dead and bare, killed by the action of acid rain.

Human activities can have far-reaching effects on the environment and all the other living things which share the Earth. One of the biggest problems is the way we produce pollution.

Everybody needs air – so when the air we breathe is polluted, no-one escapes the effects. One of the major sources of air pollution is the burning of fossil fuels. We are using more and more oil, coal and natural gas. We also burn huge amounts of the fuels made from them, such as petrol, diesel and aviation fuel for planes. Fossil fuel is a non-renewable resource – there is a limited amount of it on Earth and eventually it will all be used up.

a) Name three fossil fuels.

The formation of acid rain

When fossil fuels are burned, carbon dioxide is released into the atmosphere as a waste product. However, carbon dioxide is not the only waste gas produced. Fossil fuels often contain sulfur impurities. When these burn they react with oxygen to form sulfur dioxide gas. At high temperatures, for example in car engines, nitrogen oxides are also released into the atmosphere.

These gases pollute the air and can cause serious breathing problems for people if the concentration gets too high. They are also involved in the formation of acid rain. This pollutes land and water over a wide area.

The sulfur dioxide and nitrogen oxides dissolve in the rain and react with oxygen in the air to form dilute sulfuric acid and nitric acid. This makes the rain more acidic – it is known as **acid rain**.

b) What are the main gases involved in the formation of acid rain?

The effects of acid rain

Not surprisingly, acid rain has a damaging effect on the environment. If it falls onto trees, the acid rain can cause direct damage. It may kill the leaves and, as it soaks into the soil, even the roots of the trees may be destroyed. In some parts of Europe and America, huge areas of woodland are dying as a result of acid rain.

Acid rain has an indirect effect on our environment, as well as its very direct effect on plants. As acid rain falls into lakes, rivers and streams the water in them becomes acidic. If the concentration of acid gets too high, plants and animals can no longer survive. Many lakes and streams have become dead, no longer able to support life.

c) How does acid rain kill trees?

Acid rain is a difficult form of air pollution to pin down and control. It is formed by pollution from factories. It also comes from the cars and other vehicles we use every day. The source of the gases is pretty widespread. Many Western countries have worked hard to stop their factories and power stations from producing these acidic gases. Unfortunately there are still many places in the world where these gases are not controlled.

The worst effects of acid rain are often not felt by the country which produced the pollution in the first place. The sulfur and nitrogen oxides are carried high in the air by the prevailing winds. As a result, it is often relatively 'clean' countries which get the pollution and the acid rain from their dirtier neighbours. Their own clean air goes on to benefit someone else!

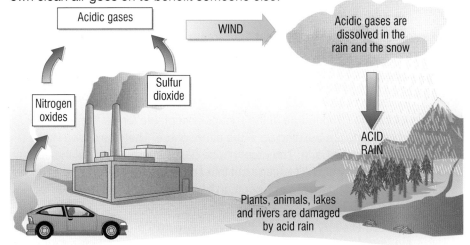

Figure 3 Air pollution in one place can cause acid rain – and serious pollution problems – somewhere else entirely. Depending on the prevailing winds, it can even be in another country!

People have become more aware of the problems caused by acid rain. The UK and other countries have introduced measures to reduce the levels of sulfur dioxide and nitrogen oxides in the air. More and more cars are fitted with catalytic converters. Once hot, these remove the acidic gases before they are released into the air. There are strict rules about the levels of sulfur dioxide and nitrogen oxides in the exhaust fumes of new cars.

Power stations are one of the main sources of acidic gases. In the UK we have introduced cleaner, low-sulfur fuels and started generating more electricity from gas and nuclear power. We have also put systems in power station chimneys to clean the flue gases before they are released into the atmosphere. As a result, the levels of sulfur dioxide in the air, and of acid rain, have fallen steadily over the last 40 years. (See Figure 4.)

SUMMARY QUESTIONS

1 Copy and complete using the words below:

**acid rain   carbon dioxide   fossil   nitric   nitrogen oxides
sulfur   sulfuric**

When ...... fuels are burned the pollutant gases ...., ... dioxide and .... are released into the atmosphere. The sulfur dioxide and nitrogen oxides dissolve in the rain to form dilute ...... acid and ...... acid. This is known as ....

2 Explain how pollution from cars and factories burning fossil fuels pollute:

a) the air   b) the water   c) the land.

3 To get rid of acid rain it is important that all the countries in an area control their production of sulfur and nitrogen oxides. If only one or two clean up their factories and cars it will not be effective. Explain why this is.

4 Use Figure 4 to help you answer this question.

a) Produce a bar chart to show the approximate levels of sulfur dioxide in the air in the UK at five-year intervals from 1980 to 2000.

b) Explain the trend you can see on your chart.

GET IT RIGHT!

Be clear about the effects of the different combustion gases produced – sulphur dioxide does **not** affect global warming but it **does** cause acid rain!

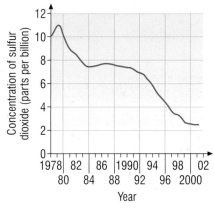

Figure 4 Graph to show levels of sulfur dioxide concentrations in the air in the UK, 1978–2002

KEY POINTS

1 When we burn fossil fuels, carbon dioxide is released into the atmosphere.

2 Sulfur dioxide and nitrogen oxides can be released when fossil fuels are burnt. These gases dissolve in the rain and make it more acidic.

3 Acid rain may damage trees directly. It can make lakes and rivers too acidic so plants and animals cannot live in them.

B1b 8.3   Global warming

Many scientists are very worried that the climate of the Earth is getting warmer. This is often called **global warming**.

The greenhouse effect

Normally the Earth radiates back much of the heat energy it absorbs from the Sun. This keeps the temperature at the surface acceptable for life. Now carbon dioxide and methane are building up in the atmosphere. They are acting like a greenhouse around the Earth. The greenhouse gases absorb much of the energy which is radiated away. It can't escape out into space. As a result, the Earth and its surrounding atmosphere are warmer than they should otherwise be.

The effect is to raise the temperature of the Earth's surface. The change is very small, about 0.06°C every ten years at the moment. Not much – but an increase of only a few degrees Celsius could cause quite large changes in the Earth's climate.

Many scientists think that an increase in severe and unpredictable weather will be one of the changes we see due to global warming. Some people think the very high winds and extensive flooding seen around the world in the 21st century are early examples of the effects of global warming.

If the Earth warms up, the ice caps at the north and south poles will melt. This will cause sea levels to rise. There is evidence that this is already happening. It will mean more flooding for low-lying shores of all countries all over the world. Eventually parts of countries, or even whole countries, will disappear beneath the seas.

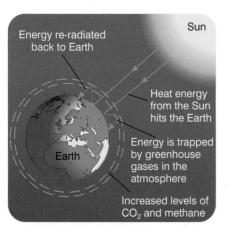

Figure 1 Many scientists believe that this simple warming effect could, if it is not controlled, change life on Earth, as we know it

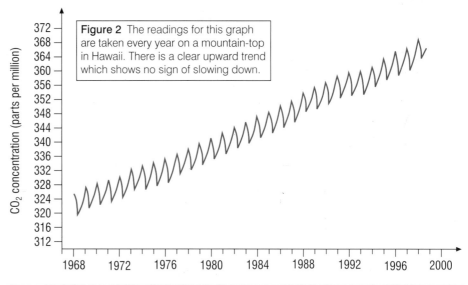

Figure 2 The readings for this graph are taken every year on a mountain-top in Hawaii. There is a clear upward trend which shows no sign of slowing down.

a) Name two greenhouse gases.

The effects of combustion

Carbon dioxide is made when we burn fossil fuels in cars, in our homes and in power stations. The number of cars and power stations around the world is steadily increasing. And respiration by all the living organisms on Earth produces carbon dioxide as well! So carbon dioxide levels are rising.

GET IT RIGHT!

Respiration and combustion **produce** carbon dioxide, photosynthesis **removes** it from the atmosphere. Greenhouse gases in the atmosphere **re-radiate** heat energy back to the surface of the Earth.

The effects of deforestation

All around the world large-scale deforestation is taking place. We are cutting down trees over vast areas of land for timber and to clear the land for farming. The trees are felled and burned in what is known as 'slash-and-burn' farming. The land produced is only fertile for a short time, after which more forest is destroyed. No trees are planted to replace those cut down.

Deforestation increases the amount of carbon dioxide released into the atmosphere. Burning the trees leads to an increase in carbon dioxide levels from combustion. The dead vegetation left behind decays. It is attacked by decomposing microorganisms, which releases more carbon dioxide.

Normally trees and other plants use carbon dioxide in photosynthesis. They take it from the air and it gets locked up in plant material like wood for years. So when we destroy trees we lose a vital carbon dioxide 'sink'. Dead trees don't take carbon dioxide out of the atmosphere.

For millions of years the levels of carbon dioxide released by living things into the atmosphere have been matched by the plants taking it out and the gas dissolving in the seas. As a result the levels in the air stayed about the same from year to year. But now the amount of carbon dioxide produced is increasing fast as a result of human activities. This speed means that the natural sinks cannot cope, and so the levels of carbon dioxide are building up.

b) What is deforestation?

Cows, rice and methane

Methane levels are rising too. It has two major sources. As rice grows in swampy conditions, known as paddy fields, methane is released. As the population of the world has grown so has the farming of rice, the staple diet of many countries.

The other source of methane is cattle. Cows produce methane during their digestive processes and release it at regular intervals.

In recent years the number of cattle raised to produce cheap meat for fast food, such as burgers, has grown enormously. So the levels of methane are rising. Many of these cattle are raised on farms produced by deforestation.

c) Where does the methane that is building up in the atmosphere come from?

Figure 3 Tropical rainforests are being destroyed at an alarming rate to supply the developed world with goods like mahogany toilet seats and cheap burgers

FOUL FACTS

When we lose forests, we lose biodiversity. Lots of plants and animals die out. We may well be destroying a source of new medicines or food for the future!

KEY POINTS

1 There is large-scale deforestation in tropical areas.
2 Large-scale deforestation has led to an increase of carbon dioxide into the atmosphere (from burning and the actions of microorganisms). It has also reduced the rate at which carbon dioxide is removed by plants.
3 More rice fields and cattle have led to increased levels of methane in the atmosphere.
4 Increased levels of the greenhouse gases carbon dioxide and methane may be causing global warming as a result of the greenhouse effect.

SUMMARY QUESTIONS

1 Define the following terms:
   global warming; greenhouse gases; deforestation; a carbon sink.

2 a) Why are the numbers of
      i) rice fields, and   ii) cattle
      in the world increasing?
   b) Why is this a cause for concern?

3 Give three reasons why deforestation increases the amount of greenhouse gases in the atmosphere.

4 a) Use the data in Figure 2 to produce a bar chart showing the maximum recorded level of carbon dioxide in the atmosphere every tenth year from 1970 to the year 2000.
   b) Explain the trend you can see on your chart.
   c) Explain the greenhouse effect. How might it affect the conditions on Earth?

B1b 8.4   Sustainable development

LEARNING OBJECTIVES

1 What do we mean by sustainable development?
2 How can families help to conserve resources?

GET IT RIGHT!

Be clear about the meaning of the term 'sustainable'. Be able to give examples of how families can help conserve natural resources.

As our world gets more and more crowded, we are becoming increasingly aware of the need for **sustainable development**. This combines human progress and environmental stability. It improves the quality of our lives without risking the future of generations to come.

Sustainable development

Sustainable development means looking after the environment. We need to conserve natural resources. So, for example, farmers need to look after the land. They can plough the remains of crop plants into the soil, and use animal waste instead of chemical fertilisers. They can also replant hedgerows to prevent soil erosion and avoid deforestation. These will help to make sure that growing crops will be possible for years to come.

We can see another example of sustainable development in our woodlands. We use an enormous amount of wood and paper, but we have fewer trees than almost all of our European neighbours.

However, over the last 80 years or so the Forestry Commission has developed sustainable commercial woodlands. Felling can only take place as long as replanting replaces the felled trees. Now farmed woodlands not only provide a sustainable resource but also a rich environment for a wide variety of species.

Sustainable development has to be a global idea – it is no good if it only happens in the UK. Everywhere that deforestation occurs, replanting needs to take place.

Figure 1 Sustainable woodlands have become an important and attractive part of sustainable development in the UK

Another example of the importance of sustainable development is in the management of our fishing stocks. In many areas we have taken so many fish from the sea that the populations can no longer replace themselves. The numbers of fish like cod are dropping fast.

To save the fish stocks, the numbers of fish caught **must** be reduced. But it needs agreement by fishermen everywhere for this sustainable development of the sea to become a reality.

a) What is sustainable development?

Conserving resources

An important part of sustainable development is using natural resources wisely. We must use only what we need, and conserve natural resources as much as possible. You and your family can help to do this in lots of different ways.

We live in a throw-away society. We use something – and then put it in the bin. This uses up resources and means land is wasted under rubbish tips. But if we recycle our waste, we save resources, use land wisely and use less energy. You can recycle your old newspapers (saving trees), glass bottles (saving energy) and aluminium cans (saving aluminium ore and energy!).

Figure 2 Our throw-away society causes problems in lots of ways. Not only do we waste resources, but a tip like this uses lots of land, and pollutes the area all around it.

Another way we can help is to make our homes more energy-efficient. Energy is one of our most important resources. Unfortunately, using electricity, gas or oil in our homes uses up some of these valuable resources. It adds to the carbon dioxide in the atmosphere as well, so the less we use the better.

We use huge amounts of energy heating our homes – and then lose it through the windows, roof and gaps around the doors. Making our homes energy-efficient helps save resources and prevent global warming. By insulating your roof spaces and the walls of your homes, and having double glazing, you will save a lot of energy. Energy-efficient boilers make the best use of your fuel. Switching things off when you have finished using them helps as well!

Finally we can look at our transport. Use your car less, and walk or cycle to places nearby. This will help you save petrol (a non-renewable resource) and also avoid adding more carbon dioxide to the atmosphere. Public transport can help as well, carrying lots of people at the same time. It may not be as convenient as using your own car, but it is certainly better for the environment.

Figure 3 Just changing to energy-efficient light bulbs like these will help to conserve resources and reduce carbon dioxide levels

SUMMARY QUESTIONS

1 Copy and complete using the words below:

   cars   conserving   energy efficient   recycle   resources   sustainable

   An important part of …… development is using natural …… wisely. This means using only what we need, and …… natural resources as much as possible. We can …… waste, make our homes more …… …… and make less use of our …… .

2 List as many ways as possible in which you and your family could help to conserve natural resources.

3 Choose one aspect of sustainable development and produce a report on how it works and why it is so important.

KEY POINTS

1 Improving your quality of life without compromising future generations is known as sustainable development.
2 Sustainable development involves using natural resources wisely.
3 We can help by recycling, making our homes energy-efficient and avoiding using our cars when possible.

B1b 8.5   Planning for the future

PRACTICAL

Pollution indicators

Have a look at your local area and see how many different types of lichens you can find. How clean do you think your air is?

- Is the information we gain from lichens reliable and valid evidence of pollution?
- Compare them to using data-logging equipment to monitor pollution.

Figure 1 Lichens grow well where the air is clean. In a polluted area there would be far fewer species of lichen growing. This is why they are useful bioindicators.

Figure 2 This water looks clean and inviting – a look at some pollution indicator species would tell us if it is as clean as it looks!

Most of the population of the world lives in the developing countries. Their way of life has not changed much over the centuries. There are many people, but each one uses few resources.

There are relatively few people in the developed world. However, those people are surrounded by technology, almost all of which uses a great deal of energy. Planning is needed at local, regional and global levels to make sure that the resources of the world are used in a fair and sustainable way.

a) Why does the developed world use so many more resources than the developing world?

Making local planning decisions

All around the UK people want to build homes, shopping centres, roads and factories. Before any building can take place, you have to apply for planning permission. For sustainable development to work, we need to consider the environment every time a planning decision is made.

To help us make the right decision we need information from field studies, carried out by scientists. They can give us information about the environment at the site of a proposed development. They can also show us the impact of a similar development elsewhere. One important tool we have in these field studies is to use living organisms as indicators of pollution.

Lichens grow on places like rocks, roofs and the bark of trees. They are very sensitive to air pollution. When the air is clean, scientists will find many different types of lichen growing. The more polluted the air, the fewer lichen species there will be. So a field check on lichen levels will give us the information we need about possible air pollution from a factory or road.

In the same way we can use invertebrate animals as water pollution indicators. The cleaner the water, the more species you will find. There are some species which are only found in the cleanest water. Others can be found even in very polluted waters.

By using living organisms as pollution indicators in this way we can reach sensible decisions about the impact a new development might have on the environment. This helps us make planning decisions which allow for the sustainable development of an area.

b) How can we use living organisms as indicators of pollution?

Brown field or green field?

As the UK population grows we need more houses. We can build them on 'green field sites' or 'brown field sites'.

Green field sites are countryside which has not been built on before. Brown field sites are usually within towns and cities, and have already been used. They are often the sites of old industrial buildings, factories, petrol stations or even rubbish tips.

Using brown field sites has many advantages for building. Water, sewage, electricity, gas and transport systems are usually already in place or easily available. No farmland or countryside is spoilt or lost.

However, brown field sites are often contaminated with chemicals and can suffer from subsidence. It is often expensive to use brown field sites because they have to be cleaned up (decontaminated) before they can be used. What's more, they can be a valuable environment in their own right.

Although it is a good idea to build on brown field sites when it is possible, the decision is not always an easy one!

Protecting SSSIs

Around the country there are many areas which are **SSSIs** (**Sites of Special Scientific Interest**). These sites may have a particularly interesting or unique landscape. They may be home to rare species of plants or animals. They may be important in bird migration or as a breeding ground.

Environmentalists believe it is very important that these SSSIs are not disturbed and – most importantly – not built on. When planning decisions are made, the environmental importance of an area needs to be considered carefully.

Figure 3 People will soon be living in these homes – yet only a short time ago this brown field site was an old bus station

The idea of protected areas is common now in Europe, and is spreading around the world. In terms of global planning, it is important that governments everywhere take their environmental responsibilities seriously. That is the only way we can continue to develop while sustaining the variety of life around us – and protecting the Earth from the worst effects of pollution and waste.

Figure 4 Rare plants such as these Spring Squill flowers often grow in SSSIs

SUMMARY QUESTIONS

1 Define the following terms: pollution indicator; brown field site; green field site; SSSI.

2 Think about your local area. Choose a brown field site which you think could be used for building. Write a proposal which puts forward your plans, explaining why you have chosen the site rather than a local farmer's field.

3 a) Here is some information about four different species of living things. Explain how you would use each one as an indicator to help make planning decisions about sustainable development.
   i) Salmon are fish which are only found in clean water with lots of oxygen in it.
   ii) Lichens are very sensitive to air pollution.
   iii) Lizards are badly affected by pesticides in their insect prey.
   iv) Blood worms can survive in very polluted water.
   b) Evaluate different methods you could use to collect environmental data. Comment on their reliability and usefulness as sources of evidence.

KEY POINTS

1 We can use information gained from field studies to help us make planning decisions about sustainable development. One type of information is to use living organisms as indicators of pollution levels.

2 Brown field sites are thought to be more suitable than green field sites for new buildings.

3 Environmentalists believe it is important not to build on Sites of Special Scientific Interest (SSSIs).

B1b 8.6   Environmental issues

How can we be sure?

The build-up of greenhouse gases cannot be denied, because there is hard evidence for it. However, there is still debate among scientists about whether these really cause the problems which are blamed on the 'greenhouse effect'. The majority of scientists now believe that global warming is at least partly linked to human activities such as burning of fossil fuels and deforestation. But not everyone agrees.

Some extreme weather patterns have certainly been recorded in recent years. But go back in history and it is possible to find other equally violent periods of weather long before fossil fuels were used so heavily and deforestation was happening.

The temperature of the Earth has varied greatly over millions of years. We have had both Ice Ages and times when almost the entire Earth was covered in tropical vegetation, all long before humans evolved. So some scientists argue that what we are seeing is the result of natural changes rather than a direct result of human activities.

ACTIVITY

You are going to present a 3–4 minute slot on global warming for a news programme for young people. You need to explain the different scientific views and why it is so hard to be certain what is happening. You also need to present some of the scientific evidence and plan some informative and attention-grabbing graphics. Make sure you present a balanced picture, with plenty of facts – but make it interesting too or they'll all change channels!

Scientists suggest that flooding is due to global warming

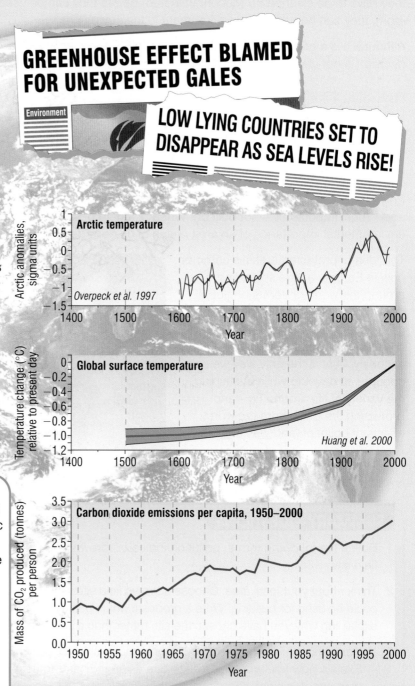

GREENHOUSE EFFECT BLAMED FOR UNEXPECTED GALES

Environment

LOW LYING COUNTRIES SET TO DISAPPEAR AS SEA LEVELS RISE!

Arctic temperature

Arctic anomalies, sigma units

Overpeck et al. 1997

Global surface temperature

Temperature change (°C) relative to present day

Huang et al. 2000

Carbon dioxide emissions per capita, 1950–2000

Mass of CO_2 produced (tonnes) per person

What can I do?

What people do in their lives really does affect our planet. More and more evidence shows that we cannot go on using the resources of the Earth in the way we do. But there are lots of different ways of looking at the problems. Here are just a few of them.

Many people are now trying to control and reduce the amount of greenhouse gases we produce. The problem is that the whole world has to agree because the Earth only has one atmosphere and this is affected by things which happen everywhere in the world. There are enormous problems with this.

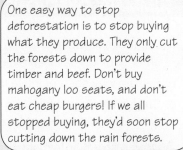

One easy way to stop deforestation is to stop buying what they produce. They only cut the forests down to provide timber and beef. Don't buy mahogany loo seats, and don't eat cheap burgers! If we all stopped buying, they'd soon stop cutting down the rain forests.

It's the wealthy developed nations who have to take responsibility for most of the carbon dioxide emissions. We need to persuade our citizens to cut back a lot on using their cars. They won't do it because they're afraid if would affect their jobs and industry. I mean, America wouldn't even sign up to the Kyoto agreement, and they produce more pollution than anyone else!

I just want to earn a living to support my family. We have no proper schools, hospitals and roads. I must work at the logging camp – it's the only work for miles around and I want my children to have an education.

I've got so little money, I need to buy things as cheaply as I can. I don't really care where the beef comes from – I can feed my kids really cheaply and give them meat nearly every day if we eat burgers and things. We're never going to see a rainforest, are we?

It looks a bit different from where I live. If we can't expand our industries, what chance do we have of getting richer? We need to make more money to improve the life and health of our people. Why should we have to sacrifice our development to help solve a problem caused by other, much richer nations?

There are a lot of us who are trying to monitor and support a more responsible use of the forests. The trouble is, it's very difficult to enforce guidelines, and there are always people ready to make money fast and illegally.

ACTIVITY

Problems like greenhouse emissions, global warming and the use of non-renewable resources can seem overwhelming. But each one of us can make a difference in the choices we make in our everyday lives.

You are going to develop some web pages to be used by your school to help everyone recognise some of the problems there are. More importantly, you are going to suggest ways in which they can help to conserve resources and change attitudes.

Design a flyer to be handed out at registration time that will give everyone the web address of your site and make them want to visit it.

Think of some targets you can set for your school, or ways in which students can set targets for their own families. Make a web site – and make a difference!

SUMMARY QUESTIONS

1 a) List the main ways in which humans reduce the amount of land available for other living things.

 b) Explain why each of these land uses is necessary.

 c) Suggest ways in which two of these different types of land use might be reduced.

2 a) Draw a flow diagram showing acid rain formation.

 b) Look at Figure 4, on page 123.
   i) What was the level of sulfur dioxide in the air in the UK in 1980?
   ii) What was the approximate level of sulfur dioxide in the air in the UK in the year that you were born? (Make sure you give your birth year.)
   iii) What was the level of sulfur dioxide in the air in 2001?

 c) Explain how the levels of sulfur dioxide have been reduced in the UK since 1978.

3 In Figure 2, on page 124 you can see clearly annual variations in the levels of carbon dioxide recorded each year. These fluctuations are thought to be due to seasonal changes in the way plants are growing and photosynthesising through the year.

 a) Explain how changes in plant growth and rate of photosynthesis might affect carbon dioxide levels.

 b) Explain how you could use this as evidence to try and prevent the loss of plant life by deforestation.

 c) How is the ever-increasing human population affecting the build-up of greenhouse gases – and what effect are they having on world climate?

4 Explain carefully how the following measures can help to support sustainable development.

 a) Encouraging farmers to plough the remains of crop plants into the soil, to use animal waste as well as chemical fertilisers and to replant hedgerows.

 b) Recycling all the aluminium drinks cans you use.

 c) Monitoring the types of invertebrates found in a river both above and below the sewage outfall.

 d) Setting up a 'walking bus' for a primary school.

 e) Enforcing standards for the building trade. These include the thickness of the insulation, the materials which can be used and the fittings (e.g. boilers, light bulbs) to be put in new houses.

EXAM-STYLE QUESTIONS

1 The diagram shows a town and some of its surroundings.

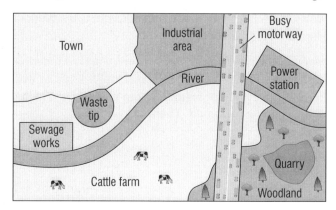

(a) A site that reduces the level of carbon dioxide in the atmosphere is:

  A Quarry   B Waste tip

  C Woodland   D Town

(b) Methane is another greenhouse gas. The site that produces the most methane is:

  A Industrial area

  B Cattle farm

  C Power station

  D Woodland

(c) Three sites produce gases that contribute to acid rain. These are:

  A Town, industrial area, quarry

  B Industrial area, busy motorway, power station

  C Busy motorway, power station, waste tip

  D Town, quarry, waste tip

(d) The two sites most likely to be damaged by acid rain falling on them are:

  A Cattle farm and quarry

  B Quarry and woodland

  C Woodland and river

  D River and waste tip

(e) Air pollution in the town is monitored continuously. This is most likely to be done by:

  A A large team of council technicians working shifts.

  B A small team of elite scientists on call 24 hours a day.

  C Local people using biological indicators such as lichen.

  D Electronic sensors attached to data logging equipment.   (5)

2 In 1997 the World Climate Summit took place in Kyoto. Agreement was reached to control global warming by cutting emissions of greenhouse gases.

(a) List two possible consequences of global warming. (2)

(b) One of the most important greenhouse gases is methane.
  (i) What are the two main sources of methane? (2)
  (ii) Why has the amount of methane produced increased steadily over the past 200 years? (2)

(c) Another group of greenhouse gases are the oxides of nitrogen. State the main source of these gases. (1)

(d) State two ways in which deforestation may also contribute to global warming. (2)

3 Lichens are plants that do not grow well where there is air pollution. The number of lichen species growing along a 15 km line from the centre of a UK city was recorded. The results were plotted on a graph.

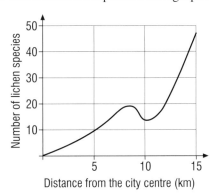

Distance from the city centre (km)

(a) (i) Name the independent variable shown above. (1)
  (ii) Is this variable categoric, ordered, discrete or continuous? (1)
  (iii) Is the dependent variable categoric, ordered, discrete or continuous? (1)

(b) Draw the headings for the table used to record the data for the graph. (1)

(c) How many species of lichen are found at a distance of 5 km from the city centre? (1)

(d) At what distance from the city centre is the least polluted air found? (1)

(e) What is the relationship between the number of lichen species and the distance from the city centre to a point 8 km from the centre? (1)

(f) Give a possible reason for the fall in the number of lichen species at a distance of 10 km from the city. (2)

HOW SCIENCE WORKS QUESTIONS

Kim needs some help to experiment on the effect of acid rain on some radish plants. She is not too certain of exactly what to do. She has the idea that if she set up some dishes with some soil and planted some radish seeds, she could then put different amounts of vinegar onto them to see what happens.

a) What advice would you give Kim as to how many different dishes she should use? (2)

b) How would you decide how many seeds to plant in each dish? (2)

c) Kim said that she wanted to keep the type of soil the same in each dish. Is this a good idea? Explain your answer. (1)

d) Kim was interested in the effects of acid rain. Do you think that vinegar was a good choice to simulate the effects of acid rain? Explain your answer. (2)

e) You could use universal indicator papers or a pH meter to measure the pH of the soil and the acid. Which would you choose . . . and why? (2)

f) Name the independent variable in Kim's investigation. (1)

g) Suggest to Kim what might be a suitable dependent variable. (1)

h) Where would you suggest Kim keeps the dishes? (2)

i) It's always a good idea to prepare a table before you start the investigation. Prepare a table for Kim to collect her results. (3)

EXAMINATION-STYLE QUESTIONS

Biology A

Questions **1** and **2**

In these questions match the letters with the numbers.
Use each answer only once.

1   To survive, animals and plants must be adapted to live in their environments.
    Four different adaptations **A**, **B**, **C** and **D** are listed below.
    Match words **A**, **B**, **C** and **D** with the spaces **1** to **4** in the sentences.

    A   Water storage tissue

    B   Reduced surface area

    C   Thick coat

    D   Camouflage

    Polar bears are insulated by a ......**1**......
    The white coat of an arctic fox acts as ......**2**......
    The stem of a cactus is used as ......**3**......
    To reduce water loss some desert plants have a ......**4**......   *(4 marks)*

See pages 82–5

2   Animals and plants can now be produced with features that we choose.
    Match **A**, **B**, **C** and **D** with the spaces **1** to **4** in the sentences.

    A   Tissue culture

    B   Genetic engineering

    C   Embryo transplantation

    D   Cloning

    Producing genetically identical individuals from a single parent is called
    ......**1**...... Transferring genes from one organism to another is known as
    ......**2**......
    Using small groups of cells to grow new organisms is called ......**3**......
    Splitting apart cells from an organism before they specialise and placing them in
    a host mother is known as ......**4**......   *(4 marks)*

See pages 98–103

3   A scientist wished to find out which plants are best to grow on soil near the sea.
    He performs an experiment to measure the mass of crop produced for different
    plants at a range of salt concentrations.

    What would be the dependent variable in this experiment?

    A   Varieties of plant used.

    B   Number of plants used.

    C   Salt concentration.

    D   Mass of crop produced.   *(1 mark)*

See page 7

Biology B

1 The rock shown in the diagram contains a
fossil of a plant which lived millions of years ago.

See pages 108–11

(a) Describe **one** way in which this fossil might
have been formed. *(2 marks)*

(b) (i) Evidence from fossils supports the theory
of evolution. What is the theory of
evolution? *(2 marks)*

(ii) How does fossil evidence support this
theory? *(2 marks)*

2 The concentration of gases in glacier ice gives a measure of the composition of
the air at the time the ice was formed. Because air moves rapidly around the
Earth, this measure also applied to areas far away from where the measurements
were taken. The graph shows the concentration, in parts per million, of carbon
dioxide which has dissolved in glacier ice over the past 250 years.

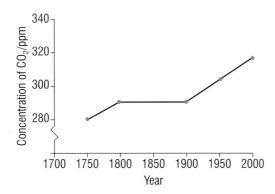

(a) Using the graph, describe the changes in concentration of carbon dioxide
in the glacier ice from 1750 to 2000. *(3 marks)*

(b) Give one piece of evidence that suggests the instruments used in this
investigation had to be very sensitive. *(1 mark)*

(c) List two sources of the carbon dioxide found in the atmosphere. *(2 marks)*

(d) Explain why there has been an increase in carbon dioxide in the
atmosphere since 1900. *(5 marks)*

(e) Thirty-eight nations signed the Kyoto Treaty and agreed to cut their carbon
dioxide emissions by just over 5% by 2012. Suggest three different actions
that governments could take to help achieve their targets. *(3 marks)*

(f) If the concentration of carbon dioxide in the atmosphere increases further,
how might this affect the temperature of the air at the Earth's surface?
(1 mark)

(g) Explain **how** carbon dioxide in the atmosphere might cause this
temperature change. *(2 marks)*

GET IT RIGHT!

'Using the graph' is how
question 2 part (a) begins.
Another expression used by
examiners is 'using only
information in the diagram'.
In all these cases you should
only use the information
provided. Do not panic if you
have not met the information
before. Simply work logically
through the facts provided.
In this case start at year 1750
and describe (say what is
happening) as you work
along the plotted line to the
year 2000.
For example, 'from 1750 to
1800 there is an increase in
CO_2 . . . etc.' No explanations
are required – not in this
stage at least.

B2 | Additional biology

What you already know

Here is a quick reminder of previous work that you will find useful in this unit:

- Both plant and animal cells have a cell membrane, cytoplasm and a nucleus. Plants cells also have cell walls and chloroplasts.

- Some cells, such as sperm, ova and root hair cells, are specially adapted to carry out particular functions in an organism.

- Enzymes play an important part in breaking down large molecules into smaller ones during digestion.

- Food is used as a fuel during respiration to keep your body activity levels up. You also need it as the raw material for growth and repair of your body cells.

- Plants and animals all carry out aerobic respiration.

- Aerobic respiration involves a reaction in our cells between oxygen and food. Glucose is broken down into carbon dioxide (CO_2) and water (H_2O).

- Plants need carbon dioxide, water and light for photosynthesis. They produce biomass, in the form of new plant material, and oxygen.

- Plants also need nitrogen and other elements to grow.

RECAP QUESTIONS

1 a) Make a list of the things all living things need or do.

b) Write down three differences between animals and plants.

c) What are the jobs of:
  i) the nucleus,
  ii) the cell membrane,
  iii) the cytoplasm,
  in a cell?

2 a) Why do we need food?

b) What has to happen to the food you eat before it can be useful to your body?

c) What is an enzyme?

d) Why are enzymes so important in digestion?

3 a) Respiration takes place in all living cells. Why is it so important?

b) Write a word equation for what happens during respiration in your cells.

4 a) What would happen if you put a plant in a dark cupboard and left it for several weeks?

b) There is more carbon dioxide in the air people breathe out than in the air they breathe in. Some people claim that talking to house plants makes them grow better. What might be a scientific explanation for this claim?

c) Sunlight and water are not enough for plants to grow well.
What else do they need – and why?

Making connections

The plant production line!

Plants produce food for all the animals that live on Earth, including us. They do this through the process of photosynthesis. They use carbon dioxide, water, and energy from light, to make sugars and oxygen.

Feeding the world

Plants could provide enough material to feed everyone in the world. If everyone understood how pyramids of biomass work, perhaps we would all eat differently and no-one would starve!

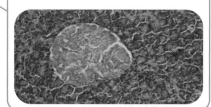

Enzymes

The food you eat is made up of big molecules. They can't get out of your gut and into your bloodstream. So they can't reach the cells where they are needed. Fortunately your body makes digestive enzymes. They work in your gut to break your food down into much smaller molecules, which your body can use.

Food – vital for life!

Specialised cells

The cells in your pancreas are very specialised. Some of them (stained pink in this photo) produce enzymes needed to break down your food. Others (stained purple) make the hormone insulin which controls your blood sugar levels.

Balancing blood sugar

After you have eaten and digested a meal, the levels of sugar in your blood shoot up. You need to be able to take this sugar into your cells so they can use it. You also need to store some of the sugar to use later. The hormone insulin is vital for you to balance your blood sugar.

Inheriting problems

Most babies are born with guts that work perfectly. But some inherit genes which mean they can't feed properly. With pyloric stenosis, the baby vomits all its food back. It needs surgery to correct the fault in its gut. In cystic fibrosis the glands that make many of the digestive enzymes get clogged up with thick sticky mucus. Then they don't work at all.

ACTIVITY

Lots of what you will learn in this unit is linked in some way to food. Every living thing needs food to survive. List, draw or find images of as many different types of food as you can.

Think about the food eaten by different types of animals and by different people around the world. There are some amazing sources of energy out there – see how many you can think of!

Chapters in this unit

Cells • How plants produce food • Energy flows • Enzymes • Homeostasis • Inheritance

B2 1.1   Animal and plant cells

The Earth is covered with a great variety of living things. The one thing all these living organisms have in common is that they are all made up of cells. Most cells are very small. You can only see them using a microscope.

The **light microscopes** you will use in school may magnify things several hundred times. Scientists have found out even more about cells using **electron microscopes** which can magnify more than a hundred thousand times!

Animal cells – structure and function

All cells have some features in common. We can see these clearly in animal cells. The cells of your body have these features, just like the cells of every other living thing!

- A **nucleus,** which controls all the activities of the cell. It also contains the instructions for making new cells or new organisms.
- The **cytoplasm**, a liquid gel in which most of the chemical reactions needed for life take place. One of the most important of these is respiration, where oxygen and sugar react to release the energy the cell needs.
- The **cell membrane**, which controls the passage of substances in and out of the cell.
- The **mitochondria**, structures in the cytoplasm where most of the energy is released during respiration.
- **Ribosomes**, where protein synthesis takes place. All the proteins needed in the cell are made here.

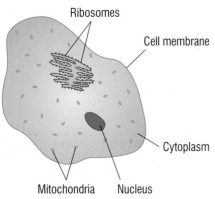

Figure 1 A simple animal cell like this shows the features which are common to all living cells

a) What are the main features found in all living cells?

Plant cells – structure and function

Plants are very different from animals, as you may have noticed! They make their own food by photosynthesis and they do not move their whole bodies about. So while plant cells have all the features of a typical animal cell, they also contain structures which are needed for their very different way of life.

All plant cells have:

- a cell wall made of cellulose which strengthens the cell and gives it support.

Many (but not all) plant cells also have these other features:

- chloroplasts, found in all the green parts of the plant. They are green because they contain the green substance chlorophyll which gives the plant its colour. They absorb light energy to make food by photosynthesis.
- a permanent vacuole (a space in the cytoplasm filled with cell sap), which is important for keeping the cells rigid to support the plant.

b) How do plant cells differ from animal cells?

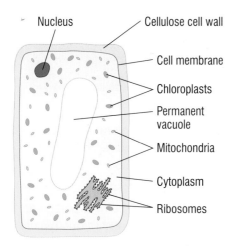

Nucleus
Cellulose cell wall
Cell membrane
Chloroplasts
Permanent vacuole
Mitochondria
Cytoplasm
Ribosomes

Figure 2 A plant cell has many features in common with an animal cell, but others which are unique to plants

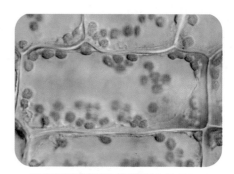

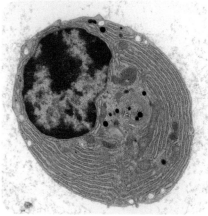

Figure 3 Diagrams of cells are much easier to understand than the real thing seen under a microscope. These pictures show a magnified plant cell and animal cell.

PRACTICAL

Looking at cells

Set up a microscope to look at plant cells, e.g. from onions and rhubarb. You should see the cell wall, the cytoplasm and sometimes a vacuole but you won't see chloroplasts.

● Why won't you see any chloroplasts?

Chemical reactions in cells

Imagine 100 different reactions going on in a laboratory test tube. Chemical chaos and probably a few explosions would be the result! But this is the level of chemical activity going on all the time in your cells.

Cell chemistry works because each reaction is controlled by an enzyme. Each enzyme is a protein which controls the rate of a very specific reaction. It makes sure that the reaction takes place without becoming mixed up with any other reaction.

We find enzymes throughout the structure of a cell, but particularly in the mitochondria (and the chloroplasts in plants).

c) What are enzymes made of?

The enzymes involved in different chemical processes are usually found in different parts of the cell. So, for example, most of the enzymes controlling the reactions of:

● respiration are found in the mitochondria,
● photosynthesis are found in the chloroplasts,
● protein synthesis are found on the surface of the ribosomes.

These cell compartments help to keep your cell chemistry well under control.

DID YOU KNOW?

Although most cells are so small we can only see them under the microscope, the largest cells in the world weigh 1.35 kg and are easily visible with the naked eye. The largest single cell is . . . an ostrich egg!

SUMMARY QUESTIONS

1 a) List the main structures you would expect to find in an animal cell.
  b) You would find all of these things in a plant cell. There are three extra features which are found in plant cells. What are they?
  c) What are the main functions of these three extra structures?

2 Root cells in a plant do not have chloroplasts. Why?

3 A nucleus and mitochondria are important structures in almost all cells. Why are they so important?

4 Explain how enzymes control the chemistry of your cells.

KEY POINTS

1 Most animal cells contain a nucleus, cytoplasm, cell membrane, mitochondria and ribosomes.
2 Plant cells contain all the structures seen in animal cells as well as a cell wall and, in many cases, chloroplasts and a permanent vacuole filled with sap.
3 Enzymes control the chemical reactions inside cells.

B2 1.2   Specialised cells

LEARNING OBJECTIVES

1 What different types of cells are there?
2 How is the structure of a specialised cell related to its function?

The smallest living organisms are single cells. They can carry out all of the functions of life, from feeding and respiration to excretion and reproduction. Most organisms are bigger and are made up of lots of cells. Some of those cells become **specialised** in order to carry out particular jobs.

When a cell becomes specialised its structure is adapted to suit the particular job it does. As a result, specialised cells often look very different to our 'typical' plant or animal cell. Sometimes cells become so specialised that they only have one function within the body. Good examples of this include sperm, eggs, red blood cells and nerve cells.

PRACTICAL

Observing specialised cells

Try looking at different specialised cells under a microscope.

When you look at a specialised cell there are two useful questions you can ask yourself:

● How is this cell different in structure from a generalised cell?
● How does the difference in structure help it to carry out its function?

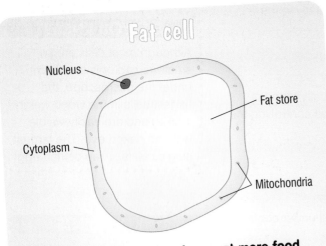

Fat cell

Nucleus
Fat store
Cytoplasm
Mitochondria

Fat cells are storage cells. If you eat more food than you need, your body makes fat and fills up the fat cells. They are important for helping animals, including us, to survive when food is in short supply. They have three main adaptations:

★ They have very little normal cytoplasm – this leaves plenty of room for large amounts of fat.
★ They have very few mitochondria as they use very little energy.
★ They can expand – a fat cell can end up 1000 times its original size as it fills up with fat.

Cone cell from human eye

Cone cells are in the light-sensitive layer of your eye (the retina). They make it possible for you to see in colour. They have three main adaptations:

Outer segment – containing visual pigment
Middle section – many mitochondria
Nucleus
Connections to nerve cells in optic nerve

★ The outer segment is filled with a special chemical known as a *visual pigment*. This changes chemically in coloured light. It then has to be changed back to its original form. This uses up energy.
★ The middle segment of the cell is packed full of mitochondria. They produce lots of energy. This means the visual pigment can reform and so the eye can see continually in colour.
★ The final part of the cell is a specialised nerve ending or synapse. This connects to the optic nerve which carries impulses to your brain. When coloured light makes your visual pigment change, an impulse is triggered which crosses the synapse. This is how the response of the cone cell to coloured light passes to your brain.

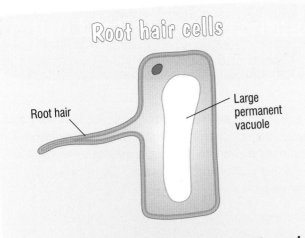

Root hair cells

Root hair

Large permanent vacuole

We find root hair cells close to the tips of growing roots. Their function is to enable plants to take in the water which they need. Root hair cells have three main adaptations:

★ The root hairs themselves, which increase the surface area for water to move into the cell.
★ A large permanent vacuole, which affects the movement of water from the soil across the root hair cell.
★ Root hair cells are always positioned close to the xylem tissue that carries water up into the rest of the plant.

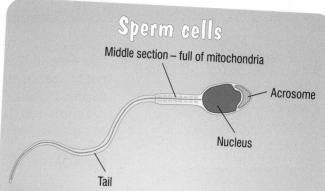

Sperm cells

Middle section – full of mitochondria

Acrosome

Nucleus

Tail

Sperm cells are usually released a long way from the egg they are going to fertilise. They contain the genetic information from the male parent to pass on to the offspring. They need to move through the female reproductive system to reach an egg. Then they have to break into the egg. They have several adaptations to make all this possible:

★ Long tails with muscle-like proteins so they can swim towards the egg.
★ The middle section is full of mitochondria, which provide the energy for the tail to work.
★ The acrosome, which stores digestive enzymes for breaking down the outer layers of the egg.
★ A large nucleus, which contains the genetic information to be passed on.

Organised cells

Specialised cells are often grouped together to form a **tissue**. Connective tissue joins bits of your body together. Nervous tissue carries information around your body and muscles move your body about.

Similarly in plants photosynthetic tissues make food by photosynthesis while storage tissues store any extra food made as starch.

In many bigger living organisms there is another level of organisation. Several different tissues work together to do particular jobs. They form an **organ** such as the heart, the kidneys or the leaf. In turn, different organs are combined in **organ systems** to carry out major functions in the body, such as transporting the blood or reproduction.

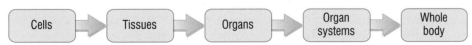

Cells → Tissues → Organs → Organ systems → Whole body

SUMMARY QUESTIONS

1 Explain how the structure of each cell on this spread is adapted to its functions.

2 Think back to two other types of specialised cells you have met in biology, e.g. motor neurones, photosynthetic cells in plants or white blood cells.
Draw the cells you have chosen. Label them fully to show how the structures you can see are related to the function of the cells.

KEY POINTS

1 Cells may be specialised to carry out a particular function.
2 Examples of specialised cells are fat cells, cone cells, root hair cells, sperm cells.

B2 1.3 How do substances get in and out of cells?

LEARNING OBJECTIVES

1 What is diffusion?
2 What affects the rate of diffusion?

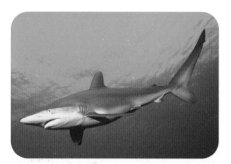

Figure 1 Everyone knows that bleeding in the sea when there are sharks around is a bad idea. Sharks are sensitive to just a few particles of blood in the water. Blood from an injury spreads quickly through the sea by diffusion – and brings the sharks to investigate!

Your cells need to take in substances such as oxygen and glucose. They also need to get rid of waste products and chemicals that are needed elsewhere in your body. Dissolved substances move into and out of your cells across the cell membrane. They can do this in three different ways – by **diffusion**, by **osmosis** and by **active transport**.

Diffusion

Sharks can smell their prey from a long way away – the smell reaches them by **diffusion**. Diffusion happens when the particles of a gas, or any substance in solution, spread out.

It is the net movement of particles from an area of high concentration to an area of lower concentration. It takes place because of the random movement of the particles of a gas or of a substance in solution in water. All the particles are moving and bumping into each other and this moves them all around.

a) Why do sharks find an injured fish – or person – so easily?

Imagine a room containing a group of boys and a group of girls. If everyone closes their eyes and moves around briskly but randomly, people will bump into each other. They will scatter until the room contains a mixture of boys and girls. This gives you a good working model of diffusion.

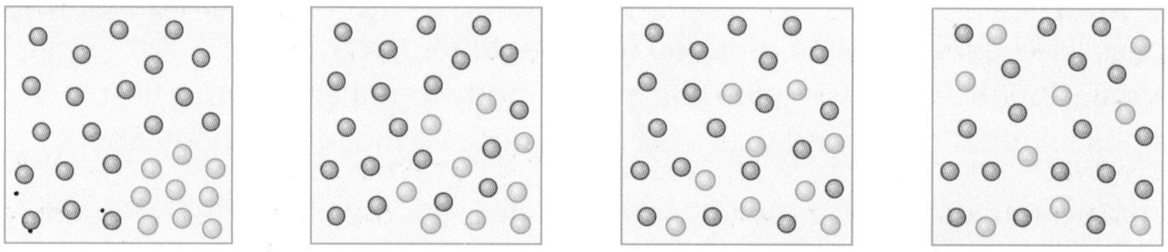

At the moment, when the blue particles are added to the red particles they are not mixed at all

As the particles move randomly, the blue ones begin to mix with the red ones

As the particles move and spread out, they bump into each other. This helps them to keep spreading randomly

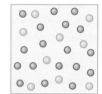

Eventually, the particles are completely mixed and diffusion is complete

Figure 2 The random movement of particles results in substances spreading out or diffusing from an area of higher concentration to an area of lower concentration

Rates of diffusion

If there is a big difference in concentration between two areas, diffusion will take place quickly. However when a substance is moving from a higher concentration to one which is just a bit lower, the movement toward the less concentrated area will appear to be quite slow. This is because although some particles move into the area of lower concentration by random movement, at the same time other identical particles are leaving that area by random movement.

The overall or **net** movement = particles moving in − particles moving out

In general the bigger the difference in concentration, the faster the rate of diffusion will be. This difference between two areas of concentration is called the **concentration gradient**. The bigger the difference, the steeper the gradient will be.

b) What is meant by the net movement of particles?

SCIENCE @ WORK

Cell biologists and biochemists work hard to discover the secrets of transport systems in your cells. They can use their findings both to understand human diseases at a cellular level and to change the nature of some organisms using genetic engineering. A salt-tolerant tomato plant which can actively move salt out of its cytoplasm into its vacuoles is just one example. This enables the plant to grow on salty ground.

Both types of particles can pass through this membrane – it is freely permeable

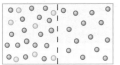

Beginning of experiment

Steep concentration gradient

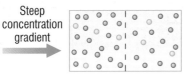

Random movement means three blue particles have moved from left to right by diffusion

Beginning of experiment

Shallow concentration gradient

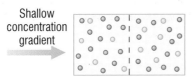

Four blue particles have moved as a result of random movement from left to right – but two have moved from right to left. There is a **net** movement of **two** particles to the right by diffusion

Figure 3 This diagram shows us how the overall movement of particles in a particular direction is more effective if there is a big difference (a steep concentration gradient) between the two areas. This is why so many body systems are adapted to maintain steep concentration gradients.

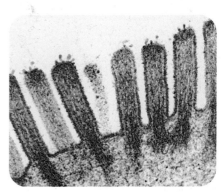

GET IT RIGHT!

Diffusion is **passive** – it takes place along a concentration gradient from high to low concentration and uses up no energy.

Concentration isn't the only thing that affects the rate of diffusion. An increase in temperature means the particles in a gas or a solution move more quickly. This in turn means diffusion will take place more rapidly as the random movement of the particles speeds up.

Diffusion in living organisms

Many important substances can move across your cell membranes by diffusion. Water is one. Simple sugars, such as glucose and amino acids from the breakdown of proteins in your gut, can also pass through cell membranes by diffusion. The oxygen you need for respiration passes from the air into your lungs and into your cells by diffusion.

Individual cells may be adapted to make diffusion easier and more rapid. The most common adaptation is to increase the surface area of the cell membrane over which diffusion occurs. Increasing the surface area means there is more room for diffusion to take place. By folding up the membrane of a cell, or the tissue lining an organ, the area over which diffusion can take place is greatly increased. So the amount of substance moved by diffusion is also greatly increased.

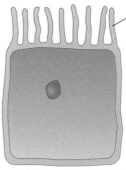

Infoldings of the cell membrane form microvilli, which increase the surface area of the cell

Figure 4 An increase in the surface area of a cell membrane means more diffusion can take place

---

SUMMARY QUESTIONS

Copy and complete using the words below:

Diffusion   gas   high   low   random   solute

1 …… is the net movement of particles of a …… or a …… from an area of …… concentration to an area of …… concentration as a result of the …… movement of the particles.

2 Explain why a cut in water looks much worse than a cut on land in terms of diffusion and the movement of particles.

3 a) Explain why diffusion takes place faster as the temperature increases.
  b) Explain in terms of diffusion why so many cells have folded membranes along at least one surface.

KEY POINTS

1 Dissolved substances move in and out of cells by diffusion, osmosis and active transport.
2 Diffusion is the net movement of particles from an area where they are at a high concentration to an area where they are at a lower concentration.

B2 1.4 Osmosis

Diffusion takes place where particles can spread freely from one place to another. However the solutions inside cells are separated from those outside by the cell membrane which does not let all types of particles through. Because it only lets some types of particles through, it is known as **partially permeable**.

Osmosis

Partially permeable cell membranes will allow water to move across them. It is important to remember that a dilute solution of, for example sugar, contains a *high* concentration of water (the **solvent**) and a *low* concentration of sugar (the **solute**). A concentrated sugar solution contains a relatively *low* concentration of water and a *high* concentration of sugar.

A cell is basically some chemicals dissolved in water inside a partially permeable bag of cell membrane. The cell contains a fairly concentrated solution of salts and sugars. Water will move from a high concentration of water particles (in a dilute solution) to a less concentrated solution of water particles (in a concentrated solution) across the membrane of the cell.

This special type of diffusion, where only water moves across a partially permeable membrane, is known as **osmosis**.

a) What is the difference between diffusion and osmosis?

PRACTICAL

Investigating osmosis

You can make model cells using bags made of partially permeable membrane. Figure 1 shows you some of these model cells. You can see what happens to them if the concentrations of the solutions inside or outside of the cell change.

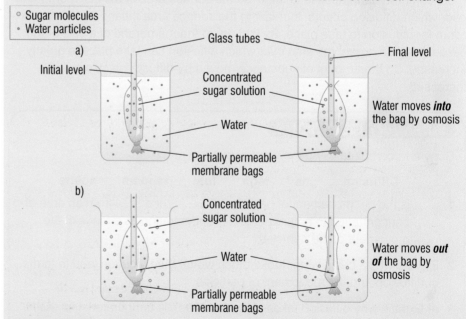

Figure 1 Using bags of partially permeable membrane to make model cells, we can clearly see the effect of osmosis as water moves across the membrane from a dilute to a concentrated solution

The internal concentration of your cells needs to stay the same all the time for the reactions of life to take place. Yet animal and plant cells are bathed in liquid which can be at very different concentrations to the inside of the cells. This can make water move into or out of the cells by osmosis.

Osmosis in animals

If a cell uses up water in its chemical reactions, the cytoplasm becomes more concentrated and more water will immediately move in by osmosis. Similarly if the cytoplasm becomes too dilute because water is produced during chemical reactions, water will leave the cell by osmosis, restoring the balance.

However osmosis can also cause some very serious problems in animal cells. (See Figure 2.) If the solution outside the cell is more dilute than the cell contents, then water will move into the cell by osmosis. The cell will swell and may burst.

On the other hand, if the solution outside the cell is more concentrated than the cell contents, then water will move out of the cell by osmosis. The cytoplasm will become too concentrated and the cell will shrivel up. Once you understand the effect osmosis can have on cells, the importance of homeostasis and maintaining constant internal conditions becomes very clear!

b) How does osmosis help maintain the body cells at the same concentration?

Osmosis in plants

Plants rely on well-regulated osmosis to support their stems and leaves. Water moves into plant cells by osmosis, making the cytoplasm swell and press against the plant cell walls. The pressure builds up until no more water can physically enter the cell. This makes the cell hard and rigid.

This swollen state keeps the leaves and stems of the plant rigid and firm. So for plants it is important that the fluid surrounding the cells always has a higher concentration of water (it is a more dilute solution of chemicals) than the cytoplasm of the cells. This keeps osmosis working in the right direction.

But sometimes plant and animal cells need to move substances such as glucose against a concentration gradient. For this there is another method of transport known as **active transport** which uses energy from respiration.

When the concentration of your body fluids is the same as in your red blood cell contents, equal amounts of water enter and leave the cell by random movement and the cell keeps its shape

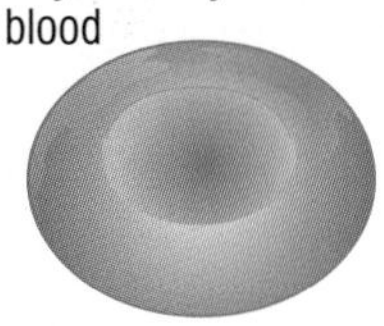

If the concentration of the solution around the red blood cells is higher than the concentration of substances inside the cell, water will leave the cell by osmosis. This makes it shrivel and shrink so it can no longer carry oxygen around your body.

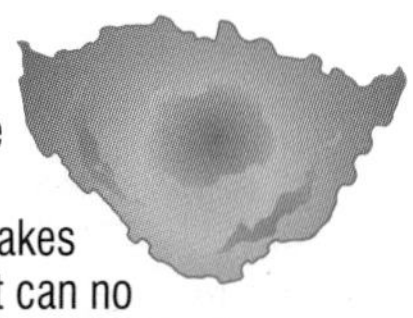

If the concentration of your body fluids is lower than in your red blood cell contents, water enters the cells by osmosis so your red blood cells swell up, lose their shape and eventually burst!

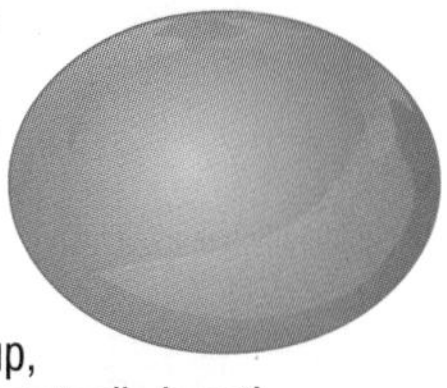

Figure 2 The impact of osmosis on your red blood cells can be devastating – so keeping your body fluids at the right concentration is vital

GET IT RIGHT!

Take care with your definition of osmosis. Make it clear that it is only water which is moving across the membrane, and get your concentrations right!

SUMMARY QUESTIONS

1 Define the following words: **diffusion**; **osmosis**; **partially permeable membrane**

2 Explain using a diagram what would happen:

   a) if you set up an experiment with a partially permeable bag containing strong sugar solution in a beaker full of pure water.

   b) if you set up an experiment using a partially permeable bag containing pure water in a beaker containing strong sugar solution.

3 Animals that live in fresh water have a constant problem with their water balance. The single-celled organism called *Amoeba* has a special vacuole in its cell. It fills with water and then moves to the outside of the cell and bursts. A new vacuole starts forming straight away. Explain in terms of osmosis why the *Amoeba* needs one of these vacuoles.

KEY POINTS

1 Osmosis is a special case of diffusion.

2 Osmosis is the diffusion/movement of water from a high water concentration (dilute solution) to a low water concentration (concentrated solution) through a partially permeable membrane.

Discovering cells

Over the past three centuries our ideas about cells have developed as our ability to see them has improved. In 1665, the English scientist Robert Hooke designed the first working microscope and saw cells in cork.

At around the same time a Dutchman, Anton van Leeuwenhoek, also produced a microscope. It enabled him to see bacteria, microscopic animals and blood cells for the first time ever.

Almost two centuries later, by the 1840s, scientists had accepted that cells are the basic units of all living things. From then on, as optical microscopes improved, more details of the secret life inside a cell were revealed as cells were magnified up to 1000 times.

With the invention of the electron microscope in the 1930s it became possible to magnify things much more. We can now look at cells magnified up 500 000 times!

Anton van Leeuwenhoek (1632–1723)

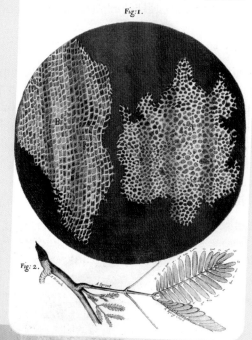

Cork cells drawn by Robert Hooke

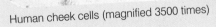

Human cheek cells (magnified 3500 times)

ACTIVITY

Produce a timeline to show how microscopes have developed since they were first invented. Annotate your timeline to show how important our discoveries about cells have been.

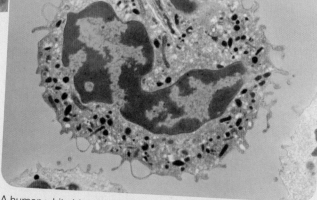

A human white blood cell at high magnification

The ability to see cells and the secret worlds inside them has developed in an amazing way since the days of the early microscopes

Beating osmosis

The cells of all living organisms contain sodium chloride and other chemicals in solution. This means they can always be prone to water moving into them by osmosis. If they are immersed in a solution with a lower concentration of salts than the body cells they will tend to gain water. If in a more concentrated solution, water is lost. Either way can spell disaster. Here are just a few of the different ways in which living things attempt – largely successfully – to beat osmosis!

No contest!

For many marine invertebrates like this jellyfish, osmosis causes no problems because the concentration of solutes in the cells of their bodies is exactly the same as the sea water. So there is no net movement of water in or out of the cells.

Copy cats!

Living on land causes all sorts of problems for the cells, particularly if water is lost and the body fluids get concentrated. Then water will leave the body cells by osmosis fast. Many insects have taken a leaf out of the plant's book – they have a tough, waterproof outer layer which prevents water loss from the body surface. They even have breathing holes known as *spiracles*. These can be closed up when they aren't needed – very like the stomata on the leaves of plants.

Flooding in

Fish that live in fresh water have a real problem. They need a constant flow of water over their gills to get the oxygen they need for respiration. But water moves into their gill cells and blood by osmosis at the same time. Like all vertebrates, fish have kidneys which play a big part in using osmosis to regulate their internal environment. So freshwater fish produce huge amounts of very dilute urine, which gets rid of the excess water that gets into their bodies. They also have special salt-absorbing glands. These use active transport to move salt against the concentration gradient from the water into the fish – rather like the situation in plant root cells.

The big ones

Marine vertebrates like this whale are constantly drinking salty water. The salt loading would cause water to move out of their body cells and kill them if they couldn't deal with it. Fortunately whales have extremely efficient kidneys. When a whale drinks $1000\,cm^3$ of sea water, it produces $670\,cm^3$ of very concentrated urine – and gains $330\,cm^3$ of pure water.

SUMMARY QUESTIONS

1

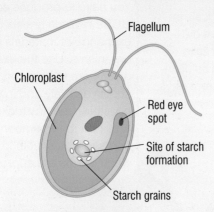

Flagellum

Chloroplast

Red eye spot

Site of starch formation

Starch grains

Chlamydomonas is a single-celled organism which lives under water. It can move itself to the light to photosynthesise, and stores excess food as starch.

a) What features does it have in common with most plant cells?

b) What features are not like plant cells and what are they used for?

c) Would you class *Chlamydomonas* as a plant cell or an animal cell? Explain why.

2

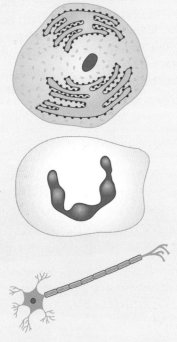

Each of these cells is specialised for a particular function in your body.

a) Copy each of these diagrams and label the cells carefully. Carry out some research if necessary.

b) Describe what you think is the function of each of these cells.

c) Explain how the structure of the cell is related to its function.

EXAM-STYLE QUESTIONS

1 The diagram is of a cell from the leaf of a plant.

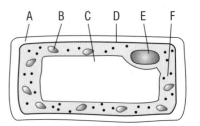

A   B   C   D   E   F

(a) Name the structures **D**, **E** and **F**. (3)

(b) (i) What is the name of structure **A**? (1)

(ii) What material is structure **A** made of? (1)

(c) (i) What is the name of structure **C**? (1)

(ii) What is the liquid it contains called? (1)

(d) Structure **B** is a chloroplast. What is its function? (2)

(e) Name two different structures that are found within the material labelled **F**. (2)

(f) (i) A different type of plant cell is a root hair cell. What is the function of this type of cell? (1)

(ii) State one way in which a root hair cell differs from the leaf cell shown in the diagram. (1)

2 Copy the table below. Look at the structures listed in the first column. Fill in the empty columns by putting a tick (✓) if you think it is present and a cross (✗) if you think it is absent. (6)

Structure	Animal cell	Plant cell
Nucleus		
Cytoplasm		
Cell wall		
Cell membrane		
Chloroplast		
Permanent vacuole		

3 A student noticed that different trees give different amounts of shade on a sunny day. She decided to investigate three species of tree – oak, sycamore and ash. She thought that the more shading, the better the tree was at gathering light for photosynthesis. She would use a light meter to record the light levels. The student had many things to consider when deciding on a method.

(a) Should she take readings in direct sunlight as well as under the trees? Explain your answer. (2)

(b) Describe the weather that would be most appropriate when collecting the data. (1)

(c) Should the student collect data from one or more than one position? Explain your answer. (1)

(d) Explain why it would be necessary for the student to take as many readings as she could under the trees. (1)

(e) What type of independent variable has the student decided to use? (1)

(f) What type of dependent variable has she decided to use? (1)

(g) How should the student calculate the mean for each set of results? (1)

(h) Suggest how she should present her data. (1)

4 List **A** gives the names of different types of cells found in plants and animals. List **B** gives one special feature of each of these cells. Match each cell type with its feature by writing the relevant letter and number next to one another. (6)

List A	List B
A Fat cell	**1** Has a long tail with muscle-like proteins
B Root hair cell	**2** Can divide and change into many different types of cell
C Sperm cell	**3** Contains chloroplasts
D Leaf cell	**4** Can expand up to 1 000 times its original size
E Stem cell	**5** Contains a chemical called visual pigment
F Cone cell (in eye)	**6** Has extension to increase its surface area

HOW SCIENCE WORKS QUESTIONS

Spinning cells!

It is possible to separate the different parts of a cell using a centrifuge. Your teacher might be able to show you one of these. They really are very simple. They spin around rather like a very fast spin dryer.

They are used to separate structures that might be mixed together in a liquid. One of their uses is to separate the different parts of a cell.

The cells are first broken open so that the contents spill out into the liquid. The mixture is then put into the centrifuge. The centrifuge starts to spin slowly and a pellet forms at the bottom of the tube. This is removed. The rest is put back into the centrifuge at a higher speed and the next pellet removed and so on.

Here are some results:

Centrifuge speed (rpm)	Part of cell in pellet
3 000	Nuclei
10 000	Mitochondria
12 000	Ribosomes

(rpm = revolutions per minute)

a) From these observations can you suggest a link between the speed of the centrifuge and the size of the part of the cell found in the pellet? (1)

b) What apparatus would you need to test your suggestion? (1)

c) i) What was your independent variable? (1)
   ii) Is your independent variable best described as categoric, discrete or continuous? (1)

d) What was your dependent variable? (1)

e) If your suggestion is correct, what results would you expect? (1)

f) What would be the easiest measurement to make to show the size of the mitochondria? (1)

g) Suggest how many mitochondria you might measure. (1)

h) How would you calculate the mean for the measurements you have taken? (2)

B2 2.1

Photosynthesis

1 What is photosynthesis?
2 What are the raw materials for photosynthesis?
3 Where does the energy for photosynthesis come from and how do plants absorb it?

Like all living organisms, plants need food. It provides them with the energy for respiration, growth and reproduction. But plants aren't like us – they don't need to eat.

Plants can make their own food! They do it by **photosynthesis**. This takes place in the green parts of plants (especially the leaves) when it is light.

The process of photosynthesis

Photosynthesis can be summed up in the following equation:

$$\textbf{carbon dioxide} + \textbf{water} \; (+ \, \textbf{light energy}) \rightarrow \textbf{glucose} + \textbf{oxygen}$$

The cells in the leaves of a plant are full of small green parts called **chloroplasts**. They contain a green substance called **chlorophyll**.

During photosynthesis, light energy is absorbed by the chlorophyll in the chloroplasts. This energy is then used to convert carbon dioxide from the air plus water from the soil into a simple sugar called **glucose**. The chemical reaction also produces oxygen gas. This is released into the air.

a) What is the word equation for photosynthesis?

Some of the glucose produced during photosynthesis is used immediately by the cells of the plant. However, a lot of the glucose made is converted into starch for storage.

Iodine solution is a yellowy-brown liquid which turns dark blue when it reacts with starch. You can use this *iodine test for starch* to show that photosynthesis has taken place in a plant.

PRACTICAL

Producing oxygen

You can show a plant is photosynthesising by collecting the oxygen given off as a by-product. It is very difficult to see oxygen, a colourless gas, being given off by land plants. But if you use water plants you can collect the gas which they give off when they are photosynthesising. It will relight a glowing splint, showing that it is oxygen gas.

PRACTICAL

Testing for starch

Chlorophyll is vital for photosynthesis to take place. It absorbs the light which provides the energy for the plant to make glucose and convert it into starch.

Take a leaf from a variegated plant (partly green and partly white). After treating the leaves, you use iodine solution to show how important chlorophyll is. (See Figure 1.)

● What happens in the test? Explain your observations

Figure 1 These leaves came from a plant which had been kept in the light for several hours. Leaves have to be specially prepared so the iodine solution can reach the cells. The one on the right has been tested for starch, using iodine solution. Only the green parts of the leaf made their own starch which turns the iodine solution blue-black.

b) What is chlorophyll?

The leaves of plants are perfectly adapted because:

● most leaves are broad, they have a big surface area for light to fall on,
● they contain chlorophyll in the chloroplasts to absorb the light energy,
● they have air spaces which allow carbon dioxide to get to the cells, and oxygen to leave them,
● they have veins, which bring plenty of water to the cells of the leaves.

All of these adaptations mean the plant can carry out as much photosynthesis as possible whenever there is light available.

c) How does the broad shape of leaves help photosynthesis to take place?

PRACTICAL

Observing leaves

Look at a whole plant leaf and then a section of a leaf under a microscope. You can see how well adapted it is.

- Compare what you can see with Figure 2 below.
- What magnification did you use?

Learn the equation for photosynthesis.
Be able to explain the results of experiments on photosynthesis.

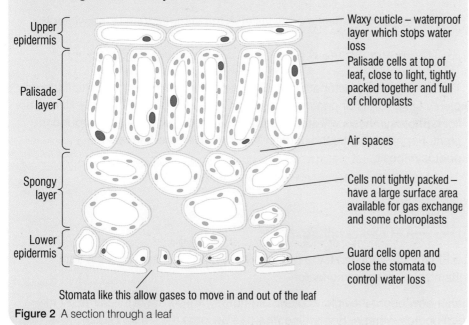

Upper epidermis

Palisade layer

Spongy layer

Lower epidermis

Waxy cuticle – waterproof layer which stops water loss

Palisade cells at top of leaf, close to light, tightly packed together and full of chloroplasts

Air spaces

Cells not tightly packed – have a large surface area available for gas exchange and some chloroplasts

Guard cells open and close the stomata to control water loss

Stomata like this allow gases to move in and out of the leaf

Figure 2 A section through a leaf

NEXT TIME YOU...

... breathe in, remember that the oxygen in the air you are breathing was produced as a by-product of photosynthesis by plants. Luckily for us, the world's plants produce about 368 000 000 000 tonnes of oxygen every year!

SUMMARY QUESTIONS

1 Copy and complete using the words below:

**carbon dioxide   chlorophyll   energy   gas   glucose
light   Oxygen   water**

During photosynthesis …… energy is absorbed by ……, a substance found in the chloroplasts. This …… is then used to convert …… …… from the air and …… from the soil into a simple sugar called …… . …… is also produced and released as a ……

2 a) Where does a plant get the carbon dioxide, water and light that it needs for photosynthesis?
   b) Work out the path taken by a carbon atom as it moves from being part of the carbon dioxide in the air to being part of a starch molecule in a plant.

3 Design experiments to show that plants need a) carbon dioxide and b) light for photosynthesis to take place. For each experiment explain what your control would be and how you would show that photosynthesis has taken place.

KEY POINTS

1 Photosynthesis can be summed up by the equation:

carbon dioxide + water
[+ light energy] →
glucose + oxygen

2 During photosynthesis light energy is absorbed by the chlorophyll in the chloroplasts. It is used to convert carbon dioxide and water into sugar (glucose). Oxygen is released as a by-product.

3 Leaves are well adapted to allow the maximum photosynthesis to take place.

B2 2.2   Limiting factors

You may have noticed that plants grow quickly in the summer, and hardly at all in the winter. Plants need certain things like light, warmth and carbon dioxide if they are going to photosynthesise as fast as they can.

If any of these things are in short supply they may limit the amount of photosynthesis a plant can manage. This is why they are known as **limiting factors**.

a) Why do you think plants grow faster in the summer than in the winter?

Light

The most obvious factor affecting the rate of photosynthesis is light. If there is plenty of light, lots of photosynthesis can take place. If there is very little or no light, photosynthesis will stop regardless of the other conditions around the plant. For most plants, the brighter the light, the faster the rate of photosynthesis.

PRACTICAL

How does the intensity of light affect the rate of photosynthesis?

We can look at this experimentally. (See Figure 1.) At the start, the rate of photosynthesis goes up as the light intensity increases. This tells us that light intensity is a limiting factor.

However, we reach a point when no matter how bright the light, the rate of photosynthesis stays the same. At this point, light is no longer limiting the rate of photosynthesis. Something else has become the limiting factor.

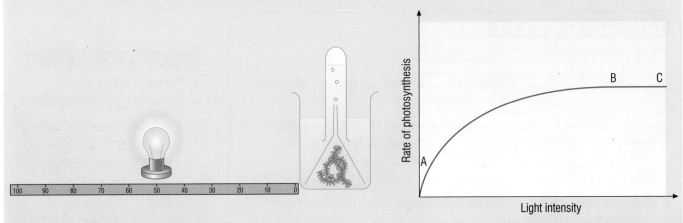

Figure 1 When the light is moved away from this water plant, the rate of photosynthesis falls – shown by a slowing in the stream of oxygen bubbles being produced. If the light is moved closer (keeping the water temperature constant) the stream of bubbles becomes faster, showing an increased rate of photosynthesis. The results can be plotted on a graph like this which shows the effect of light intensity on the rate of photosynthesis.

- Why is light a limiting factor for photosynthesis?
- Name the independent and the dependent variables in this investigation. (See page 7.)

Temperature

Temperature affects all chemical reactions, including photosynthesis. As the temperature rises, the rate of photosynthesis will increase as the reaction speeds up. However, because photosynthesis takes place in living organisms it is controlled by enzymes. Enzymes are destroyed once the temperature rises to around 40 to 50°C. This means that if the temperature gets too high, the rate of photosynthesis will fall as the enzymes controlling it are denatured.

b) Why does temperature affect photosynthesis?

Carbon dioxide levels

Plants need carbon dioxide to make glucose. The atmosphere only contains about 0.04% carbon dioxide, so carbon dioxide levels often limit the amount of photosynthesis which can take place. Increasing the carbon dioxide levels will increase the rate of photosynthesis.

For the plants you see around you on a sunny day, carbon dioxide levels are the most common limiting factor. The carbon dioxide levels around a plant tend to rise in the night as it respires but doesn't photosynthesise. Then as the light and temperature levels increase in the morning, the carbon dioxide all gets used up.

However, in a laboratory or in a greenhouse the levels of carbon dioxide can be increased artificially. This means they are no longer limiting, and the rate of photosynthesis increases with the rise in carbon dioxide.

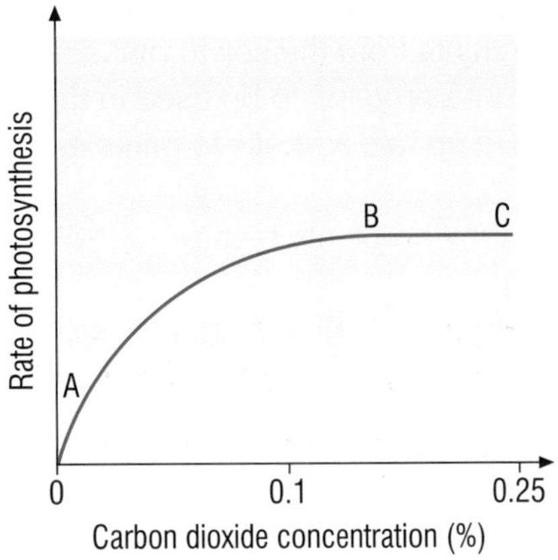

Figure 3 This graph shows the effect of increasing carbon dioxide levels on the rate of photosynthesis at a particular light level and temperature. Eventually one of the other factors becomes limiting.

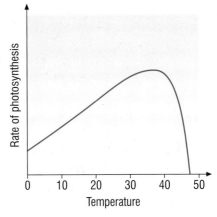

Figure 2 The rate of photosynthesis increases steadily with a rise in temperature up to a certain point. After this the enzymes are destroyed and the reaction stops completely.

DID YOU KNOW?

There are a few plants that live in very shady areas which have evolved to photosynthesise at their maximum at relatively low levels of light. For them, too much light causes the rate of photosynthesis to drop!

GET IT RIGHT!

Make sure you can explain limiting factors.
Learn to interpret graphs which show the effect of limiting factors on photosynthesis.

KEY POINTS

1 There are three main factors that limit the rate of photosynthesis – light, temperature and carbon dioxide levels.
2 We can artificially change the environment in which we grow plants. We can use this to observe the effect of different factors on the rate of photosynthesis. We can also use it to control their rate of photosynthesis.

SUMMARY QUESTIONS

1 a) What is photosynthesis?
  b) What are the three main limiting factors that affect the rate of photosynthesis in a plant?

2 Which factors do you think would be limiting photosynthesis in the following situations? In each case, explain why the rate of photosynthesis is limited.

  a) Plants growing on a woodland floor in winter.
  b) Plants growing on a woodland floor in summer.
  c) A field of barley first thing in the morning.
  d) The same field later on in the day.

3 Look at the graph in Figure 1.

  a) Explain what is happening between points A and B on the graph.
  b) Explain what is happening between points B and C on the graph.
  c) Look at Figure 2. Explain why it is a different shape to the other two graphs on this spread.

B2 2.3

How plants use glucose

LEARNING OBJECTIVES

1 What do plants do with the glucose they make?
2 How do plants store food?

NEXT TIME YOU...

...tuck into a plate of chips or a pile of mashed potato, remember that you are eating the winter food store of a potato plant! The starch you are enjoying was formed from glucose made in the leaves of the potato plant by photosynthesis. It was transported down from the leaves to the roots to form a tasty tuber!

GET IT RIGHT!

Remember:
- Plants respire 24 hours a day to release energy.
- Glucose is soluble in water, but starch is insoluble.

Plants make glucose when they photosynthesise. This glucose is vital for their survival. Some of the glucose produced during photosynthesis is used immediately by the cells of the plant. They use it for respiration and to provide energy for cell functions, growth and reproduction.

Respiration

Plants cells, like any other living cells, respire all the time. They break down glucose using oxygen to provide energy for their cells. Carbon dioxide and water are the waste products of the reaction.

The energy released in respiration is then used to build up smaller molecules into bigger molecules. Some of the glucose is converted into starch for storage (see below). Plants also build up sugars into more complex carbohydrates like cellulose. They use this to make new plant cell walls.

Plants use some of the energy from respiration to combine sugars with other nutrients from the soil to make amino acids. These amino acids are then built up into proteins to be used in the cells. Energy from respiration is also used to build up fats and oils to make a food store in the seeds.

a) Why do plants respire?

Transport and storage

Plants make food by photosynthesis in their leaves and other green parts. However, the food is needed all over the plant. It is moved around the plant in a special transport system.

There are two separate transport systems in plants. The **phloem** is made up of living tissue. It transports sugars made by photosynthesis from the leaves to the rest of the plant. They are carried to all the areas of the plant. These include the growing regions where the sugars are needed for making new plant material, and the storage organs where they are needed to provide a store of food for the winter.

The **xylem** is the other transport tissue. It carries water and mineral ions from the soil around the plant.

A vascular bundle. It contains **xylem** and **phloem** with **cambium cells** between them.

Phloem tubes – they have thin walls and living cells

Phloem

Xylem

Cambium cells grow into new xylem and phloem

Xylem vessels – they have thick, strong walls and are not living

Figure 1 A look at a section of a plant stem shows you how the transport system of a plant is arranged

Plants convert some of the glucose produced in photosynthesis into starch to be stored. Glucose is **soluble** (it dissolves in water). If it was stored in plant cells it could affect the way water moves into and out of the cells. Large amounts of glucose stored in the plant cells could affect the water balance of the whole plant. Starch is **insoluble** (it doesn't dissolve in water). This means that plants can store large amounts of starch in their cells without it having any effect on the water balance of the plant.

So the main energy store in plants is starch and it is found all over a plant. It is stored in the leaves to provide an energy store for when it is dark or when light levels are low.

PRACTICAL

Making starch

You can use the presence of starch in a leaf as evidence that photosynthesis has been taking place. It is no good just adding iodine to a leaf – the waterproof cuticle and the green chlorophyll will prevent it reacting clearly with the starch. But once you have treated the leaf, adding iodine will show you clearly if the leaf has been photosynthesising or not. Look at Figure 2.

Figure 2 We use the iodine test for the presence of starch to show us that photosynthesis has taken place. The leaf on the right has been kept in the dark. It has made no glucose to turn into starch, and has used up any starch stores it had for respiration. The leaf on the left has been in the light and been able to photosynthesise. The glucose has been converted to starch which is clearly visible when it reacts with iodine and turns blue-black. The colour is removed from the leaves before testing by boiling them in ethanol.

Starch is also stored in special storage areas of a plant. Many plants produce tubers and bulbs to help them survive through the winter. These are full of stored starch. We often take advantage of these starch stores and eat them ourselves. Potatoes, carrots and onions are all full of starch to keep a plant going until spring comes again!

b) What is the main storage substance in plants?

Figure 3 Trees like this giant redwood can be up to 30 metres tall – and then the roots spread out in all directions underground. Plants need a very effective transport system to move the food they make in their leaves distances like these.

SUMMARY QUESTIONS

1 Copy and complete using the words below:

**energy   glucose   growth   photosynthesise   respiration
reproduction   starch   storage   twenty-four**

Plants make ...... when they .... Some of the glucose produced is used by the cells of the plant for ...... which goes on .... hours a day. It provides ...... for cell functions, ... and .... Some is converted to ...... for ....

2 List as many ways as possible in which a plant uses the glucose produced by photosynthesis.

3 a) Why is the glucose made by photosynthesis converted to starch to be stored in the plant?
  b) Where might you find starch in a plant?
  c) How could you show that a potato is a store of starch?

KEY POINTS

1 Plant cells use some of the glucose they make during photosynthesis for respiration.
2 Some of the soluble glucose produced during photosynthesis is converted into insoluble starch for storage.

B2 2.4

Why do plants need minerals?

LEARNING OBJECTIVES

1   What happens if plants don't get enough nitrates?
2   Why do fertilisers make your vegetables grow so well?

Figure 1 The plants on the left of this picture have been grown in a mixture containing all the minerals they need. The experimental plants on the right have been grown without nitrates. The difference in their rate of growth is clear to see.

If you put a plant in a pot of water, and give it plenty of light and carbon dioxide, it won't survive for very long! Although plants can make their own food by photosynthesis, they cannot survive long on photosynthesis alone.

Just as you need minerals and vitamins for healthy growth, so plants need more than simply carbon dioxide, water and light to thrive. They need mineral salts from the soil to make the chemicals needed in their cells.

Why do plants need nitrates?

The problem with the products of photosynthesis is that they are all carbohydrates. Carbohydrates are very important. Plants use them for energy, for storage and even for structural features like cell walls. However, a plant can't function without proteins as well. It needs proteins to act as enzymes and to make up a large part of the cytoplasm and the membranes.

a) What are the products of photosynthesis?

Glucose and starch are made up of carbon, hydrogen and oxygen. Proteins are made up of amino acids which contain carbon, hydrogen, oxygen and **nitrogen**. Plants need **nitrates** from the soil to make proteins.

These nitrates, dissolved in water, are taken up from the soil by the plant roots. If a plant is deficient in nitrates (doesn't have enough) it doesn't grow properly. It is small and stunted. So nitrates are necessary for healthy growth.

When plants die and decay the nitrates and other minerals are returned to the soil to be used by other plants.

b) Why do plants need nitrates?

Why do plants need magnesium?

It isn't only nitrates that plants need to grow well. There is a whole range of *mineral ions* they need. For example, plants need **magnesium** to make chlorophyll.

Chlorophyll is vital to plants. It is chlorophyll which absorbs the energy from light which makes it possible for plants to photosynthesise. So if the plant can't make chlorophyll, it can't make food and it will die. This is why magnesium ions are so important for plants – they make up part of the chlorophyll molecule.

Plants only need a tiny amount of magnesium. However, if they don't get enough, they have pale, yellowish areas on their leaves where they cannot make chlorophyll.

c) Why do plants need magnesium ions?

If any of the mineral salts that a plant needs are missing it will begin to look very sickly. This is true in the garden and for houseplants just as much as for crops in a farmer's field.

PRACTICAL

Investigating the effect of minerals

You can grow young plants in water containing different combinations of minerals and see the effect on their growth.

- Why are some plants grown in water with no minerals added and some with all the minerals they need?

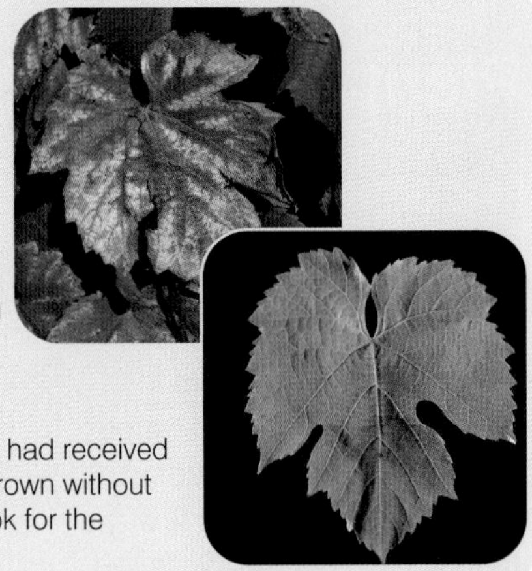

Figure 2 The leaf on the right came from a plant that had received all the minerals it needed. The plant on the left was grown without magnesium. It is easy to see which is which – just look for the yellow patches!

GET IT RIGHT!

Make sure you know the roles of nitrate and magnesium ions – and the deficiency symptoms of each of them.

If there are not enough mineral ions in the soil, your plants cannot grow properly. They will show the symptoms of mineral deficiencies. If you can pick up the symptoms soon enough and give them the mineral ions that they need, all will be well. If not, your plants will die!

Mineral ion	Why needed?	Deficiency symptoms
nitrate	making protein	stunted growth
magnesium	making chlorophyll	pale, yellow leaves

The most recent development in growing crops is **hydroponics**. You don't plant your crops in soil. Instead you plant them in water to which you add the minerals your plants need to grow as well as possible.

Hydroponic crops are usually grown in massive greenhouses where all the other factors can be controlled as well. Everything is monitored and controlled by computers 24 hours a day. The crops are very clean – no mud on the roots! And you can grow crops very quickly, and even out of their usual season.

All this means you get a good price for them. The downside is that it is an expensive way to farm, and it uses a lot of resources.

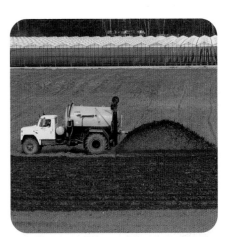

Figure 3 If you are a farmer you harvest the crops that you grow and sell them. They are not left to die and decay naturally, returning minerals to the soil. So farmers add fertiliser to the soil to replace the minerals lost, ready for the next crop. The fertiliser may be a natural one like manure or an artificial mixture of the minerals that plants need to grow.

SUMMARY QUESTIONS

1 a) Why do plants need mineral ions?
b) Where do they get mineral ions from?
c) Which mineral ion is needed by plants to form proteins?

2 a) Look at the plants in Figure 1. Describe how the plants grown without nitrates differ from the plants grown with all the mineral ions they need. Why are they so different?
b) Look at the plants in Figure 2. Describe how the plants grown without magnesium differ from the plants grown with all the mineral ions they need. Why are they so different?

3 Explain the following in terms of the mineral ions needed by plants and how they are used in the cells:
a) Farmers spread animal manure on their fields.
b) Gardeners recommend giving houseplants a regular mineral feed containing nitrates and magnesium.
c) If the same type of crop is grown in the same place every year it will gradually grow less well and becomes stunted, with pale, patchy leaves.

KEY POINTS

1. Plant roots absorb mineral salts including nitrate needed for healthy growth.
2. Nitrates and magnesium are two important mineral ions needed for healthy plant growth.
3. If mineral ions are deficient, a plant develops symptoms because it cannot grow properly.

B2 2.5 — Plant problems?

Smallholder

'In days gone by most farms were small. Farmers fed their own families and hoped to make enough profit to survive. Different crops had to be grown each year (crop rotation) and the land was rested between crops. Fields lay fallow (no crops were grown) every few years to let the land recover. Manure from their own animals was the main fertiliser.

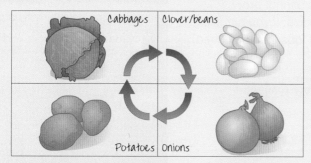

'We're trying to stick to the old ways on our little small-holding. We rotate our crops – you can see from my field plan how I do it. It helps to make sure that the minerals in the soil don't all get used up. It helps keep diseases at bay too.

'We feed our family well, and sell our extra produce to the village shop. Of course, I earn most of my money through my computer business.'

Arable farmer

'We farm a pretty big area. I grow wheat and oil seed rape. After we have harvested, I plough the stubble back into the soil. We used to burn it off but that's not allowed now. I think it's better to plough the stuff in anyway – puts something back!

'My farm is a big business – I can't afford to have land doing nothing. So we add fertiliser to keep the mineral levels right. We need to get the best crop we can every time! Modern fertilisers mean I can plant one crop straight after the other, and I avoid fallow years altogether.

'I have to get the balance right – if I spend too much on fertiliser, I don't make enough profit. But if I don't put enough fertiliser on the fields, I don't grow enough crops! I manage to support the family pretty well with the farm, and we employ one local man as well.'

Hydroponics grower

'In the laboratory you can isolate different factors and see how they limit the rate of photosynthesis. However, for most plants a mixture of these factors affects them. Early in the morning, light levels and temperature probably limit the rate of photosynthesis. Then as the level of light and the temperature rise, the carbon dioxide becomes limiting. On a bright, cold winter day, temperature probably limits the rate of the process. There is a constant interaction between the different factors.

'In commercial greenhouses we can take advantage of this knowledge of limiting factors and leave nothing to chance. We can control the temperature and the levels of light and carbon dioxide to get the fastest possible rates of photosynthesis. This makes sure our plants grow as quickly as possible. We even grow our plants in a nutritionally balanced solution rather than soil to make sure nothing limits their rate of photosynthesis and growth.'

ACTIVITY

The National Farmers Union (NFU) wants to produce a resource for schools to show how arable (crop) farming has changed over the years. Your job is to design **either** one large poster **or** a series of smaller posters that they can send out free to science departments in schools around the country. You need to explain how plants grow, and how farmers give them what they need to grow as well as possible. Use the information on this spread and in the rest of the chapter to help you.

By controlling the temperature, light and carbon dioxide levels in a greenhouse like this we can produce the biggest possible crops – fast!

'We invested in all the computer software and control systems about two years ago. It cost us a lot of money – but we are really reaping the benefits. We can change the carbon dioxide levels in the greenhouses during the day. We control the temperature and the light levels very carefully. What's more we can change the mineral content of the water as the plants grow and get bigger.

'We sell all our stuff to one of the big supermarket chains. Our lettuces are always clean, big and crisp – and we have a really fast turnover. No more ploughing fields for us!

'Of course we don't need as many staff now. We just have lots of alarm systems in our house. Then if anything goes wrong in one of the greenhouses, day or night, we know about it straight away. The monitoring systems and computers are vital to our way of growing. As far as our plants are concerned, limiting factors are a thing of the past!'

SUMMARY QUESTIONS

1 a) Match each word related to photosynthesis to its description:

A Carbon dioxide gas	1 ...... is produced and released into the air
B Water	2 ......provides energy
C Sunlight	3 ...... from the roots moves up to the leaf through the stem
D Glucose	4 ...... is absorbed from the air
E Oxygen	5 ...... is made in the leaf and provides the plant with food

b) Write a word equation for photosynthesis.

c) Much of the glucose made in photosynthesis is turned into an insoluble storage compound. What is this compound?

2

Year	Mean height of seedlings grown in 85% full sunlight (cm)	Mean height of seedlings grown in 35% full sunlight (cm)
2000	12	10
2001	16	12.5
2002	18	14
2003	21	17
2004	28	20
2005	35	21
2006	36	23

The figures in the table show the mean growth of two sets of oak seedlings. One set was grown in 85% full sunlight, the other set in only 35% full sunlight.

a) Plot a graph to show the growth of both sets of oak seedlings.

b) Using what you know about photosynthesis and limiting factors, explain the difference in the growth of the two sets of seedlings.

3 Plants make food in one organ and take up water from the soil in another organ. But both the food and the water are needed all over the plant.

a) Where do plants make their food?

b) Where do plants take in water?

c) There are two transport tissues in a plant. One is the phloem. What is the other one?

d) Which transport tissue carries food around the plant?

e) Which transport tissue carries water around the plant?

EXAM-STYLE QUESTIONS

1 Jenny carried out an investigation to show the rate of photosynthesis in two species of plant at different light intensities.

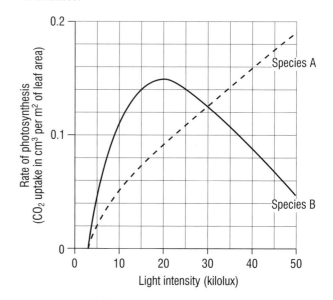

This investigation had two independent variables.

(a) Name the categoric independent variable. (1)

(b) Name the continuous independent variable. (1)

(c) Describe the pattern shown by species B. (3)
The results for species B were as follows:

Light intensity (kilolux)	CO_2 uptake (cm^3/m^2)
5	0.04
10	0.11
20	0.15
30	0.125
40	0.09
50	0.04

(d) Jenny was not sure where the peak of the graph should be drawn. Which extra measurements should she take to be sure of this? (1)

(e) At what light intensity do both species photosynthesise at the same rate? (1)

(f) If species **A** has a total leaf area of 100 m², how many cm³ of carbon dioxide will it take up at a light intensity of 10 kilolux? Show your working. (2)

(g) Which species shows the best adaptation to shade conditions? Using the information in the graph give reasons for your answer. (2)

(h) What is the name of the sugar produced during photosynthesis? (1)

(i) What is the name of the process by which this sugar is broken down to provide energy for the plant? (1)

2 The diagram below represents a section through a plant leaf showing the arrangement of cells as seen under a microscope.

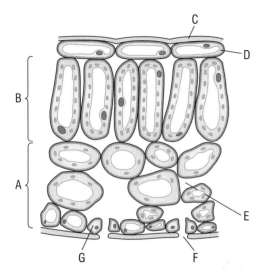

(a) Name the parts labelled **E**, **F** and **G**. (3)

(b) Give one function of the parts labelled
   (i) **C**
   (ii) **G** (2)

(c) List the four letters that indicate structures that contain chloroplasts. (4)

(d) The diagram shows only a small section through a leaf. State **FOUR** ways in which the **whole leaf** is adapted to carry out photosynthesis. In each case show how this feature helps the plant to carry out photosynthesis. (8)

3 Plants need to obtain mineral salts in order to survive.

(a) Name two mineral salts that are essential to plants and in each case give a reason why they are needed. (4)

(b) How do plants obtain the minerals they need? (2)

(c) If crops are grown for long periods on the same piece of land, they may use up some of the minerals in the soil. State two ways in which farmers can avoid these crops dying due to lack of minerals. (2)

HOW SCIENCE WORKS QUESTIONS

Water gardens – or rather hydroponics!

Ed had seen some entrepreneurs make a fortune by growing lettuce in the middle of winter. He wanted some of the action! He knew that he would have to provide heat and light as well as the nutrients and the correct pH. He knew that the plants required water and oxygen to their roots.

None of the books told him how often he should water the lettuce. Water them too often and they would not get enough oxygen. Leave them too long without watering and they would dry out. He decided on an investigation.

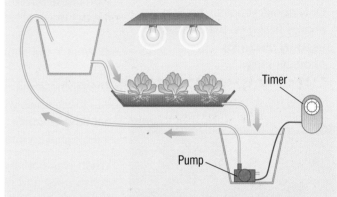

Ed set up five different trays and buckets. He set the timer differently for each tray. The lettuce would now be watered for a different number of times each day. He could therefore work out which was the best for his lettuce.

a) Suggest some time intervals for Ed to water his lettuce. (1)

b) Suggest a dependent variable he could measure. (1)

c) Explain why you have chosen this variable. (1)

d) Describe how Ed might measure this dependent variable. (1)

e) Suggest three control variables he should use. (3)

f) Explain why it would be sensible for Ed to repeat his results. (2)

Ed's first set of results showed very little difference. It did not seem to matter how often he watered them.

g) Suggest a problem he had with the design of his investigation. (1)

h) Why was it important that Ed did his own research and not ask advice from those already growing the lettuce? (2)

B2 3.1   Pyramids of biomass

Figure 1 Plants produce a huge mass of biological material in just one growing season

As you saw in the previous chapter, radiation from the Sun is the source of energy for all the groups of living things on Earth.

Light energy pours out continually onto the surface of the Earth. Green plants capture a small part of this light energy using chlorophyll. It is used in photosynthesis. So some of the energy from the Sun is stored in the substances which make up the cells of the plant. This new plant material adds to the **biomass**.

Biomass is the mass of living material in an animal or plant. Ultimately all biomass is built up using energy from the Sun. Biomass is often measured as the dry mass of biological material in grams.

a) What is the source of all the energy in the living things on Earth?

The biomass made by plants is passed on through food chains or food webs into the animals which eat the plants. It then passes on into the animals which eat other animals. No matter how long the food chain or complex the food web, the original source of all the biomass involved is the Sun.

When you look at a food chain, there are usually more producers than primary consumers, and more primary consumers than secondary consumers. If you count the number of organisms at each level you can compare them. You can show this using a **pyramid of numbers**. However, in many cases a pyramid of numbers does not accurately reflect what is happening.

b) What is a pyramid of numbers?

Pyramids of biomass

To show what is happening in food chains more accurately we can use biomass. We can draw the total amount of biomass in the living organisms at each stage of the food chain to scale and show it as a pyramid of biomass.

Figure 2 This food chain cannot be accurately represented using a pyramid of numbers. Using biomass shows us the amount of biological material involved at each level in a way that simple numbers cannot do.

Organism	Number	Biomass – dry mass in g
Oak tree	1	500 000
Aphids	10 000	1000
Ladybirds	200	50

Pyramid of numbers   Pyramid of biomass

c) What is a pyramid of biomass?

Interpreting pyramids of biomass

The biomass at each stage of a food chain is less than it was at the previous stage.

This is because:

- Not all organisms at one stage are eaten by the stage above.
- Some material taken in is passed out as waste.
- When a herbivore eats a plant, it turns some of the plant material into new herbivore. But much of the biomass from the plant is used by the herbivore in respiration to release energy for living. It does not get passed on to the carnivore when the herbivore is eaten.

So at each stage of a food chain the amount of biomass which is passed on gets less. A large amount of plant biomass supports a smaller amount of herbivore biomass. This in turn supports an even smaller amount of carnivore biomass.

In general, pyramids of biomass are drawn in proportion. Sometimes, when the biomass of one type of organism is much, much bigger than the others, this doesn't work and so the diagram can only give a rough idea.

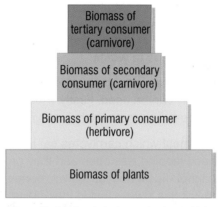

Figure 3 Any food chain can be turned into a pyramid of biomass like this

---

SUMMARY QUESTIONS

1 a) What is biomass?
   b) Why is a pyramid of biomass more useful for showing what is happening in a food chain than a pyramid of numbers?

2

Organism	Biomass, dry mass (g)
Grass	100 000
Sheep	5000
Sheep ticks	30

   a) Draw a pyramid of biomass for this grassland ecosystem.
   b) What would you expect the pyramid of numbers for this food chain to look like?
   c) Draw the pyramids of numbers and the pyramids of biomass you would expect from the following two food chains:
      i) stinging nettles → caterpillars → robin
      ii) marine plants → small fish → large fish → seals → polar bear

3 a) Explain simply why the biomass from one stage of a pyramid of biomass does not all become biomass in the next stage of the pyramid.
   b) Using the data in Figure 2, calculate the percentage biomass passed on from:
      i) the producers to the primary consumers,
      ii) the primary consumers to the secondary consumers.

KEY POINTS

1 Radiation from the Sun is the main source of energy for all living things. The Sun's energy is captured and used by plants during photosynthesis.
2 The mass of living material at each stage of a food chain is less than at the previous stage. The biomass at each stage can be drawn to scale and shown as a pyramid of biomass.

B2 3.2 Energy losses

LEARNING OBJECTIVES

1 How do we lose energy to the environment?
2 What is the effect of maintaining a constant body temperature?

Figure 1 The amount of biomass in a lion is a lot less than the amount of biomass in the grass which feeds the zebra it preys on. But where does all the biomass go?

An animal like a zebra eats grass and other small plants. It takes in a large amount of plant biomass, and converts it into a much smaller amount of zebra biomass. This is typical of a food chain.

The amounts of biomass and energy contained in living things always gets less at each stage of a food chain from plants onwards. Only a small amount of the biomass taken in gets turned into new animal material. The question is – what happens to the rest?

Energy loss in waste

The biomass which an animal eats is a source of energy, but not all of the energy can be used. Firstly, herbivores cannot digest all of the plant material they eat. The material they can't digest is passed out of the body in the faeces.

The meat which carnivores eat is easier to digest than plants, so they tend to need feeding less often and they produce less waste. But even carnivores often cannot digest hooves, claws, bones and teeth, so some of the biomass that they eat is always lost in their faeces.

When an animal eats more protein than it needs, the excess is broken down and passed out as urea in the urine. So biomass – and energy – are lost from the body.

a) Why is biomass lost in faeces?

Figure 2 Animals like horses eat very large amounts of biomass every day. However they also produce very large quantities of dung made up of all the biomass they couldn't actually digest!

GET IT RIGHT!

Make sure you can explain the different ways in which energy is lost between the stages of a food chain.
Check that you know how to use energy flow (Sankey) diagrams to tell if an animal is a herbivore or a carnivore, warm-blooded or cold-blooded.

Energy loss due to movement

Part of the biomass eaten by an animal is used for respiration in its cells. This supplies all the energy needs for the living processes taking place within the body.

Movement uses a great deal of energy. The muscles use energy to contract. So the more an animal moves about the more energy (and biomass) it uses from its food. The muscles produce heat as they contract.

b) Why do animals that move around a lot use up more of the biomass they eat than animals which don't move much?

Keeping a constant body temperature

Much of the energy animals produce from their food in cellular respiration is eventually lost as heat to the surroundings. Some of this heat is produced by the muscles as the animals move.

Heat losses are particularly large in mammals and birds because they are 'warm-blooded'. This means they keep their bodies at a constant temperature regardless of the temperature of the surroundings. They use up energy all the time, to keep warm when it's cold or to cool down when it's hot. Because of this, warm-blooded animals need to eat far more food than cold-blooded animals, such as fish and reptiles, to get the same increase in biomass.

c) What do we mean by a 'warm-blooded animal'?

PRACTICAL

Investigating the heat released by respiration

Even plants produce heat by cellular respiration. You can investigate this using germinating peas in a vacuum flask.

- What would be the best way to monitor the temperature continuously?
- Plan the investigation.

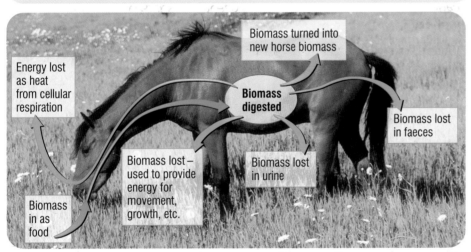

Figure 3 Only between 2% and 10% of the biomass eaten by an animal such as this horse will get turned into new horse – the rest of the stored energy will be used or lost in other ways

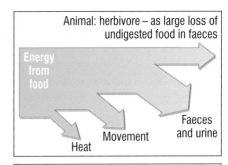

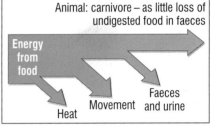

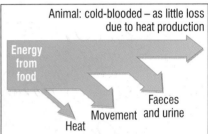

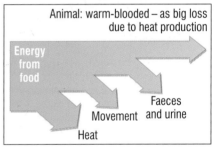

Figure 4 **Sankey diagrams** show how energy is transferred in a system. We can use them to look at the energy which goes in to and out of an animal and predict whether it eats plants or is a carnivore. You can even tell if it is warm-blooded or cold-blooded!

SUMMARY QUESTIONS

1 Copy and complete using the words below:

biomass   body temperature   energy   food chain   growth
movement   producers   respiration   waste

The amounts of …… and …… contained in living things always gets less at each stage of a …… …… from …… onwards. Biomass is lost as …… products and used to produce energy in …… . This is used for …… and to control …… …… . Only a small amount is used for …… .

2 Explain why so much of the energy from the Sun which lands on the surface of the Earth is not turned into biomass in animals.

3 Why do warm-blooded animals need to eat more food than cold-blooded ones of the same size if they are to put on weight?

KEY POINTS

1 The amount of biomass and energy gets less at each successive stage in a food chain.
2 This is because some material is always lost in waste, and some is used for respiration to supply energy for movement and for maintaining the body temperature.

B2 3.3 Energy in food production

LEARNING OBJECTIVES

1 Why do short food chains make food production more efficient?
2 How can we manage food production to reduce energy losses?

Person
| Person | Cow |
| Potato | Grass |

Figure 1 Reducing the number of stages in food chains could dramatically increase the efficiency of our food production. Eating less meat would mean more food for everyone.

Pyramids of biomass clearly show us that the organisms at each stage of a food chain contain less material and therefore less energy. This has some major implications for the way we human beings feed ourselves.

Food chains in food production

In the developed world much of our diet consists of meat or other animal products such as eggs, cheese and milk. The cows, goats, pigs and sheep that we use to produce our food eat plants. By the time it reaches us, much of the energy from the plant has been used up.

In some cases we even feed animals to animals. Ground up fish, for example, is often part of commercial pig and chicken feed. This means we have put another extra layer into the food chain – plant to fish, fish to pig, pig to people. What could have been biomass for us has been used as energy by other animals in the chain.

a) Name three animals which we use for food.

There is only a limited amount of the Earth's surface that we can use to grow food. The most efficient way to use this food is to grow plants and eat them directly. If we only ate plants, then in theory at least, there would be more than enough food for everyone on the Earth. As much of the biomass produced by plants as possible would be used to feed people.

But every extra stage we introduce – feeding plants to animals before we eat the food ourselves – means less biomass and energy getting to us at the end of the chain. In turn this means less food to go round the human population.

b) Why would there be more food for everyone if we all ate only plants?

Artificially managed food production

As you saw on the previous page, animals don't turn all of the food they eat into new animal. Apart from the food which can't be digested and is lost as waste, energy is used in moving around and maintaining a constant body temperature.

Farmers apply these ideas to food production. People want meat, eggs and milk – but they want them as cheaply as possible. So farmers want to get the maximum possible increase in biomass from animals without feeding them any more. There are two ways of doing this:

- Limiting the movement of food animals. Then they lose a lot less energy in moving their muscles and so will have more biomass available from their food for growth.

- Controlling the temperature of their surroundings. Then the animals will not have to use too much energy keeping warm. Again this leaves more biomass spare for growth.

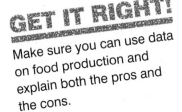

GET IT RIGHT!
Make sure you can use data on food production and explain both the pros and the cons.

This means keeping the animals inside with restricted room to move, and a constant temperature. This is exactly what happens in the massive poultry rearing sheds where the majority of the chickens that we eat are produced.

Keeping chickens in these conditions means relatively large birds can be reared to eat in a matter of weeks. When animals are reared in this way they can appear more like factory products than farm animals. That's why these intensive methods are sometimes referred to as factory farming.

Intensive farming methods are used because there has been a steady increase in demand for cheap meat and animal products. This is the only way farmers can meet those demands from consumers.

On the other hand, these animals live very unnatural and restricted lives. More people are now aware of how our cheap meat and eggs are produced. So there has been a backlash against the conditions in which intensively reared animals live.

Many people now say they would be willing to eat meat less often and pay more if the animals they eat are raised more naturally.

Figure 2 These chickens are provided with an ideal temperature, plenty of food and very little opportunity to move. They will produce meat and lay more eggs far faster than if they were moving about and keeping themselves warm.

Figure 3 Intensively reared pigs live in small stalls in a warm building with food delivered regularly for maximum growth. It makes life relatively easy for the farmer but costs money to run. Animals reared outside grow more slowly, but seem to have a much better quality of life. The farmer needs land, and has to cope with horrible weather – but it's cheaper as there is no artificial heating or lighting to pay for.

SUMMARY QUESTIONS

1 The world population is increasing and there are food shortages in many parts of the world. Explain, using pyramids of biomass to help you, why it would make better use of resources if people everywhere ate much less meat and more plant material.

2 Why are animals prevented from moving much and kept indoors in intensive farming?

3 a) What are the costs for a farmer of rearing animals intensively?
   b) What are the advantages of intensive rearing for a farmer?
   c) What are the advantages of less intensive rearing methods?
   d) What are the disadvantages of these more natural methods?

KEY POINTS

1 Biomass and energy are lost at each stage of a food chain. The efficiency of food production can be improved by reducing the number of stages in our food chains. It would be most efficient if we all just ate plants.

2 If you stop animals moving about and keep them warm, they lose a lot less energy. This makes food production much more efficient.

B2 3.4   Decay

You need to know the type of organisms that cause decay, the conditions needed for decay and the importance of decay in recycling nutrients.

Figure 1 These tomatoes are slowly being broken down by the action of decomposers. You can see the fungi clearly, but the bacteria are too small to be seen.

FOUL FACTS

There is a forensic research site in the USA known as the Body Farm where scientists have buried or hidden human bodies in many different conditions. They are studying every stage of human decay. The information is used by police forces all over the world when a body is found. It can help to pinpoint when a person died, and show if they were the victim of a crime.

Plants take minerals from the soil all the time. These minerals are then passed on into animals through the food chains and food webs which link all living organisms. If this was a one-way process the resources of the Earth would have been exhausted long ago!

Many trees shed their leaves each year, and most animals produce droppings at least once a day. Animals and plants eventually die as well. Fortunately all these materials are recycled and returned to the environment. We can thank a group of organisms known as the **decomposers** for this.

a) Which group of organisms take materials out of the soil?

The decay process

The decomposers are a group of microorganisms which include bacteria and fungi. They feed on waste droppings and dead organisms.

Detritus feeders, such as maggots and worms, often start the process, eating dead animals and producing waste material. The bacteria and fungi then digest everything – dead animals, plants and detritus feeders plus their waste. They use some of the nutrients to grow and reproduce. They also release waste products.

The waste products of the decomposers are carbon dioxide, water, and minerals which plants can use. When we say that things decay, they are actually being broken down and digested by microorganisms.

The recycling of materials through the process of decay makes sure that the soil remains fertile and plants can grow. It is also thanks to the decomposers that you aren't wading through the dead bodies of all the animals and plants that have ever lived!

b) Which type of organisms are the decomposers?

Conditions for decay

The speed at which things decay depends partly on the temperature. The chemical reactions in microorganisms are like those in most other living things. They work faster in warm conditions. (See Figure 3.) They slow down and even stop if conditions are too cold. Because the reactions are controlled by enzymes, they will stop altogether if the temperature gets too hot as the enzymes are denatured. You can investigate this in a simple experiment.

PRACTICAL

Investigating decay

Plan an investigation into the effect of temperature on how quickly things decay.

● Name the independent variable in this investigation. (See page 7.)

Most microorganisms also grow better in moist conditions. The moisture makes it easier to dissolve their food and also prevents them from drying out. So the decay of dead plants and animals – as well as leaves and dung – takes place far more rapidly in warm, moist conditions than it does in cold, dry ones.

Although some microbes work without oxygen, most decomposers respire like any other organism. This means they need oxygen to release energy, grow and reproduce. This is why decay takes place more rapidly when there is plenty of oxygen available.

c) Why are water, warmth and oxygen needed for the process of decay?

The importance of decay in recycling

Decomposers are vital for recycling resources in the natural world. What's more, we can take advantage of the process of decay to help us recycle our waste.

In **sewage treatment plants** we use microorganisms to break down the bodily waste we produce. This makes it safe to be released into rivers or the sea. These sewage works have been designed to provide the bacteria and other microorganisms with the conditions they need. That includes a good supply of oxygen.

Another place where the decomposers are useful is in the garden. Many gardeners have a **compost heap**. You put your grass cuttings, vegetable peelings and weeds on the compost heap. Then you leave it to let decomposing microorganisms break all the plant material down. It forms a fine, rich powdery substance known as compost. This can take up to a year.

The compost produced is full of mineral nutrients released by the decomposers. Once it is made you can dig your compost into the soil to act as a fertiliser.

Figure 2 The decomposers cannot function at low temperatures so if an organism – like this 4000 year old man – is frozen as it dies, it will be preserved with very little decay

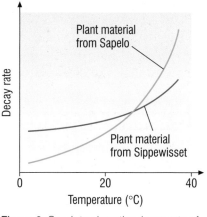

Figure 3 Graph to show the decay rate of plant material (leaves) from two different areas of the USA. The effect of temperature can be seen clearly.

Figure 4 The decomposers are all microorganisms and so they are vulnerable to drying out. Moisture is vital for decay, along with warm temperatures and plenty of oxygen.

KEY POINTS

1 Living organisms remove materials from the environment as they grow. They return them when they die through the action of the decomposers.
2 Dead materials decay because they are broken down (digested) by microorganisms.
3 Decomposers work more quickly in warm, moist conditions. Many of them also need a good supply of oxygen.
4 The decay process releases substances which plants need to grow.
5 In a stable community the processes that remove materials (particularly plant growth) are balanced by the processes which return materials.

SUMMARY QUESTIONS

1 Copy and complete using the words below:

**bacteria   carbon dioxide   dead   decomposers   digest
fungi   microorganisms   minerals   nutrients
waste droppings   water**

The ...... are a group of ...... which includes ...... and .... They feed on .... and ...... organisms. They ...... them and use some of the .... They also release waste products which include ...., ... and ...... which plants can use.

2 The following methods are all ways of preserving foods to prevent them from decaying. Use your knowledge of the decomposing microorganisms to explain how each method works:

a) Food may be frozen.
b) Food may be cooked – cooked food keeps longer than fresh food.
c) Food may be stored in a vacuum pack – with all the air sucked out.
d) Food may be tinned – it is heated and sealed in an airtight container.

B2 3.5 The carbon cycle

Imagine a stable community of plants and animals. The processes which remove materials from the environment are balanced by processes which return materials. Materials are constantly cycled through the environment. One of the most important of these is carbon.

All of the main molecules that make up our bodies (carbohydrates, proteins, fats and DNA) are based on carbon atoms combined with other elements.

The amount of carbon on the Earth is fixed. Some of the carbon is 'locked up' in fossil fuels like coal, oil and gas. It is only released when we burn them.

Huge amounts of carbon are combined with other elements in carbonate rocks like limestone and chalk. There is a pool of carbon in the form of carbon dioxide in the air. It is also found dissolved in the water of rivers, lakes and oceans. All the time a relatively small amount of available carbon is cycled between living things and the environment. We call this the **carbon cycle**.

a) What are the main sources of carbon on Earth?

Photosynthesis

Green plants use carbon dioxide from the atmosphere in photosynthesis. They use it to make carbohydrates which in turn make biomass. This is passed on to animals which eat the plants. The carbon goes on to become part of the carbohydrates, proteins and fats in their bodies.

This is how carbon is taken out of the environment. But how is it returned?

b) What effect does photosynthesis have on the distribution of carbon levels in the environment?

Figure 1 Within the natural cycle of life and death in the living world, mineral nutrients are cycled between living organisms and the physical environment

Respiration

Animals and plants respire all the time. They use oxygen to break down glucose, providing energy for their cells. Carbon dioxide is produced as a waste product and is returned to the atmosphere.

Also when plants and animals die their bodies are broken down by the decomposers. These decomposers release carbon dioxide into the atmosphere as they respire. All of the carbon dioxide released by the various types of living organisms is then available again. It is ready to be taken up by plants in photosynthesis.

Combustion

Fossil fuels contain carbon, which was locked away by photosynthesising plants millions of years ago. When we burn fossil fuels, we release some of that carbon back into our atmosphere:

Photosynthesis: carbon dioxide + water (+ light energy) → glucose + oxygen
Respiration: glucose + oxygen → carbon dioxide + water (+ energy)
Combustion: fossil fuel or wood + oxygen → carbon dioxide + water (+ energy)

The constant cycling of carbon is summarised in Figure 2.

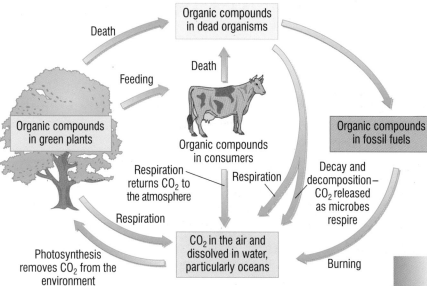

Figure 2 The carbon cycle in nature

GET IT RIGHT!

Make sure you can label the processes in a diagram of the carbon cycle.

For thousands of years the carbon cycle has regulated itself. However, as we burn more fossil fuels we are pouring increasing amounts of carbon dioxide into the atmosphere. Scientists fear that the carbon cycle may not cope. If the levels of carbon dioxide in our atmosphere increase it may lead to global warming.

Energy transfers

It isn't just carbon that passes through all the living organisms. The energy from the Sun also passes through all the different types of organisms. It starts with photosynthesis in plants, and is then transferred into animals. It is then transferred into the detritus feeders and decomposing microorganisms. They recycle the materials as plant nutrients.

All of the energy originally captured by green plants is eventually either:

- transferred back into the plants (in the minerals they take in),
- transferred into the decomposers, or
- transferred as heat into the environment by respiration.

Figure 3 Energy is transferred from one type of organism to another. Along the way large amounts are transferred as heat to the environment through the process of respiration.

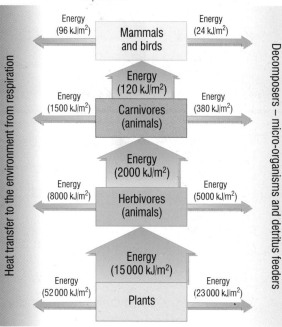

This represents the energy flow through $1\,m^2$ of an ecosystem – the figures in brackets are those recorded for one particular area

KEY POINTS

1 The constant cycling of carbon in nature is known as the carbon cycle.
2 Carbon dioxide is removed from the atmosphere by photosynthesis. It is returned to the atmosphere through respiration and combustion.
3 The energy originally captured by green plants is eventually transferred back into plants, into decomposers or as heat into the environment.

SUMMARY QUESTIONS

1 a) What is the carbon cycle?
   b) What are the main processes involved in the carbon cycle?
   c) Why is the carbon cycle so important for life on Earth?

2 Explain carefully how a) carbon, and b) energy are transferred through an ecosystem.

B2 3.6   Farming – intensive or free range?

Intensive farming – costs and benefits	'Free-range' farming – costs and benefits

Chickens for meat and eggs

Benefits:
- Lots of chickens in small space
- Little or no food wastage
- Energy wasted in movement/heat loss kept to a minimum
- Maximum weight gain/number of eggs laid
- Cheap eggs/chicken meat

Costs:
- Chickens unable to behave naturally – may be debeaked and cannot perch
- Large barns need heating and lighting
- Chickens legs may break as bones unable to carry weight of rapidly growing bodies
- Risk of disease with many birds closely packed together

Benefits:
- Chickens live a more natural life
- No heating/lighting costs
- Less food needs supplying as they find some for themselves
- Can charge more money for free-range eggs/chickens

Costs:
- Chickens more vulnerable to weather and predators
- More land needed for each bird
- Eggs cannot be collected automatically
- Fewer eggs laid, especially in the winter when it is cold and dark for longer periods of time

ACTIVITIES

1 Choose either cattle or chickens. Produce a leaflet to be handed out in your local shopping centre either supporting intensive farming methods or supporting free-range farming methods. In each case back up your arguments with scientific reasoning.

2 You are going take part in a debate on animal rights and farming methods. You have been chosen to speak *either* FOR intensive farming *or* AGAINST 'free-range' farming.
You have to think carefully about the benefits to the animals of intensive methods, and the disadvantages of free-range farming.

Intensive farming – costs and benefits

Cattle for beef

Benefits:

- Uses the male calves produced by dairy cows
- Weaning takes place by about 8 weeks and then farmers know exactly how much food each calf eats
- Balance of nutrients in food changed as calf grows to maximise growth
- Kept largely indoors, energy loss through movement and heat loss is kept to a minimum – can get weight gains of 1.5 kg a day!
- Cheap meat

Costs:

- Feedstuff must be bought and can be expensive
- Cowsheds need care and cleaning
- Cowsheds have to be heated and lit

'Free-range' farming – costs and benefits

Benefits:

- Calves are weaned naturally and stay with their mothers for up to 6 months
- Feeding on grass or food grown by farmer means no contamination, such as that which led to BSE, is possible
- Cattle behave and live relatively naturally

Costs:

- Animals may take slightly longer to gain weight as they are moving more actively – but they are less stressed
- More land is needed to provide grazing, hay and silage

ACTIVITY

3 Design a poster for the school gardening club explaining how to make compost and why it is important for the soil. Use the information in this chapter to help you get your facts right!

SUMMARY QUESTIONS

1

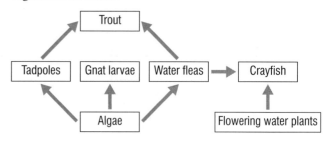

a) From this table calculate the percentage biomass passed on
   i) from producers to primary consumers,
   ii) from primary to secondary consumers,
   iii) from secondary consumers to top carnivores.

b) In any food chain or food web, the biomass of the producers is much larger than that of any other level of the pyramid. Why is this?

c) In any food chain or food web there are only a small number of top carnivores. Use your calculations to help you explain why.

d) All of the animals in the pyramid of biomass shown here are cold-blooded. What difference would it have made to the average percentage of biomass passed on between the levels if mammals and birds had been involved? Explain the difference.

2 Chicks grown for food arrive in the broiler house as one-day-old chicks. They are slaughtered at 42 days of age when they weigh about 2 kg. The temperature, amount of food and water and light levels are carefully controlled. About 20 000 chickens are reared together in one house. The table below shows their weight gain.

Age (days)	1	7	14	21	28	35	42
Mass (g)	36	141	404	795	1180	1657	1998

a) Plot a graph to show the growth rate of one of these chickens.

b) Explain why the temperature is so carefully controlled in the broiler house.

c) Explain why so many birds are reared together in a relatively small area.

d) Why are birds for eating reared like this?

e) Draw a second line to show how you would expect a chicken reared outside in a free-range system to gain weight, and explain the difference.

EXAM-STYLE QUESTIONS

1 The diagram below shows a part of a food web for organisms in a lake.

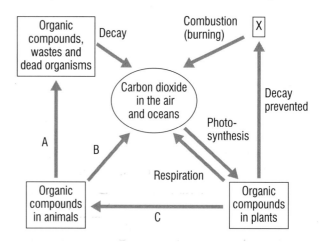

(a) Which organisms feed on algae? (1)

(b) Which organisms are producers? (1)

(c) Which organism is both a primary consumer and a secondary consumer? (1)

(d) Draw and label a pyramid of biomass for the food chain below:

Algae → Tadpole → Trout (1)

(e) If a disease suddenly killed all the water fleas explain how the population of the algae might be affected. (2)

2 The diagram below is a version of the carbon cycle.

(a) Name the three processes indicated by the three arrows labelled with the letters **A**, **B** and **C**. (3)

(b) In what form is the carbon in the box labelled **X**? (1)

(c) The organic compounds of plants and animals are mostly in the form of three groups of substances that make up the **majority** of the bodies of these organisms. What are the three groups of organic compounds? (3)

(d) The table shows the percentage of carbon cycled by some of the processes involved in the carbon cycle.

Process	Percentage of total carbon cycled
Photosynthesis	50
Respiration by animals	20
Respiration by plants	20
Respiration by microorganisms	5
Combustion/absorbed by oceans	5

    (i) Draw a pie chart of these proportions. (3)

    (ii) If the total amount of carbon that is cycled in one year across the Earth is 165 gigatonnes, calculate how much carbon is cycled by the respiration of plants. Show your working. (2)

(e) Respiration is an important process in recycling carbon. The word equation for respiration is shown below, with most words replaced by the letters **A**, **B**, **C** and **D**. Give the names of **A**, **B**, **C** and **D**. (4)

$$\mathbf{A} + \mathbf{B} \rightarrow \mathbf{C} + \mathbf{D} + \text{energy}$$

(f) The concentration of carbon dioxide in the atmosphere has increased over the past 200 years. Suggest one human activity that might have contributed to this increase. (1)

3 A factory which packaged shrimps produced tonnes of waste shrimp heads. It cost money to dump these in the local tip. The managers decided to investigate the decay of shrimp heads to see if they might be used as fertiliser. They used 80 shrimp heads in 4 sealed jars. Each jar had a different amount of water. They measured the length of the shrimp heads, left them for 60 days and then measured them again:

Amount of water (cm³)	% loss in length
40	68
50	61
60	59
70	56

(a) Explain why they decided to measure the length of shrimp heads. (1)

(b) How many shrimp heads would they have put into each jar? (1)

(c) They predicted that the more water they added the greater the breakdown of the shrimp heads. Is their prediction supported? Explain your answer. (1)

HOW SCIENCE WORKS QUESTIONS

Can this be true?

A scientist claims to have bred a featherless chicken.

The scientist says that it was the result of natural selective breeding and not genetic engineering. He claims that it will be ideal for warmer climates where the intensive breeding of chickens requires expensive air conditioning to keep the chickens at around 25°C. It will therefore be cheaper for farmers to rear these featherless chickens.

They will also be cheaper to feed as they will not need to use energy to grow feathers. Also they will cut down on the pollution caused by having to dump feathers before they are prepared for market.

'I looked to see if the date was April 1st when I read about this story,' said a geneticist.

A biologist said, 'The birds would probably find it difficult to breed without feathers.' Others claimed that it was 'ugly science' and should not be allowed.

a) Why do you think the scientist is so keen to promote his research? (1)

b) Do you think that the scientist was wrong to do the research? (1)

c) Which groups are likely to oppose such research? (1)

d) Who should make the final decision whether farmers should breed these featherless chickens or not? (1)

e) Do we know if the chickens are suffering? How could we find out? (1)

B2 4.1

Enzyme structure

LEARNING OBJECTIVES

1 What is an enzyme?
2 How do enzymes speed up
   reactions?

DID YOU KNOW?

The lack of just one enzyme in your body can have disastrous results. If you don't make the enzyme phenylalanine hydroxylase you can't break down the amino acid phenylalanine. It builds up in your blood and causes serious brain damage. All UK babies are tested for this condition soon after birth. If they are given a special phenylalanine-free diet right from the start, the risk of brain damage can be avoided.

The cells of your body are like tiny chemical factories. Hundreds of different chemical reactions are taking place all the time. These reactions have to happen fast – you need energy for your heart to beat and to hold your body upright *now*! They also need to be very controlled. The last thing you need is for your cells to start exploding!

Chemical reactions can only take place when different particles collide. The reacting particles don't just have to bump into each other. They need to collide with enough energy to react.

The minimum amount of energy particles must have to be able to react is known as the **activation energy**. So you will make the reaction more likely to happen if you can make it:

- more likely that reacting particles bump into each other,
- increase the energy of these collisions, or
- reduce the activation energy needed.

a) What is the activation energy of a reaction?

Controlling the rate of reactions

In everyday life we control the rates of chemical reactions all the time. When you cook food, you increase the temperature to speed up the chemical reactions. You lower the temperature to slow reactions down in your fridge or freezer. And sometimes we use special chemicals known as **catalysts** to speed up reactions for us.

A catalyst changes the rate of a chemical reaction, usually speeding it up. Catalysts are not used up in the reaction so you can use them over and over again. Different types of reactions need different catalysts. Catalysts work by bringing reacting particles together and lowering the activation energy needed for them to react.

b) What is a catalyst?

Enzymes – the biological catalysts

In your body chemical reaction rates are controlled by **enzymes**. These are special *biological catalysts* which speed up reactions.

Enzymes do not change the overall reaction in any way except to make it happen faster. Each one catalyses a specific type of reaction.

Enzymes are involved in:

- building large molecules from lots of smaller ones,
- changing one molecule into another, and
- breaking down large molecules into smaller ones.

Inorganic catalysts and enzymes both lower the activation energy needed for a reaction to take place.

GET IT RIGHT!

Remember that the way an enzyme works depends on the shape of the active site which allows it to bind with the substrate.

PRACTICAL

Breaking down hydrogen peroxide

Investigate the effect of i) manganese(IV) oxide, and ii) raw liver on the breakdown of hydrogen peroxide solution.

● Describe your observations and interpret the graph below.

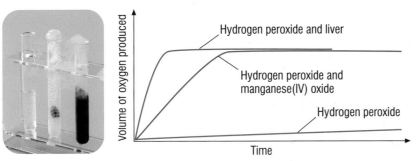

Figure 1 Hydrogen peroxide is a colourless liquid which slowly breaks down to form oxygen and water. The decomposition reaction goes much faster using manganese(IV) oxide as a catalyst. Raw liver contains an enzyme (catalase) which also speeds up the breakdown of hydrogen peroxide.

Your enzymes are large protein molecules. They are made up of long chains of amino acids, folded and coiled to give a molecule with a very special shape. The enzyme molecule usually has a hole or indentation in it. This special shape allows other molecules to fit into the enzyme. We call this the **active site**. The shape of an enzyme is vital for the way it works.

How do enzymes work?

The substrate (reactant) of the reaction fits into the shape of the enzyme. You can think of it like a lock and key. Once it is in place the enzyme and the substrate bind together. This is called the **enzyme–substrate complex**.

Then the reaction takes place rapidly and the products are released from the surface of the enzyme. (See Figure 3.) Remember that enzymes can join together small molecules as well as breaking up large ones.

Enzymes usually work best under very specific conditions of temperature and pH. This is because anything which affects the shape of the active site also affects the ability of the enzyme to speed up a reaction.

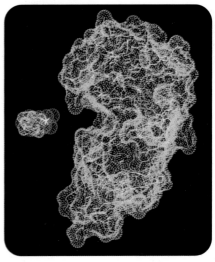

Figure 2 Enzymes have a very complex structure made up of chains of amino acids folded and coiled together. This computer-generated image shows just how complicated the structure really is!

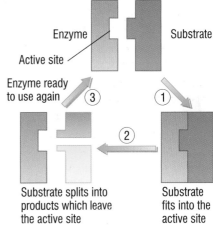

Figure 3 Enzymes have their effect as catalysts using the 'lock-and-key' mechanism shown here. You can see that anything which changes the shape of the protein molecule might change the shape of the active site and stop the enzyme from working.

SUMMARY QUESTIONS

1 Match the words and the definitions:

a) catalyst	A The special site in the structure of an enzyme where the substrate binds.
b) enzyme	B The energy needed for a chemical reaction to take place.
c) activation energy	C A substance which changes the rate of a chemical reaction without being changed itself.
d) active site	D A biological catalyst.

2 a) What is an enzyme? c) Why is their structure so important?
  b) What are enzymes made of?

3 a) How do enzymes act to speed up reactions in your body?
  b) Why are enzymes so important in your body?

KEY POINTS

1 Catalysts increase the rate of chemical reactions. Enzymes are biological catalysts
2 Enzymes are protein molecules made up of long chains of amino acids. The chains are folded to form the active site. This is where the substrate of the reaction binds with the enzyme.

B2 4.2   Factors affecting enzyme action

LEARNING OBJECTIVES

1 How does increasing the temperature affect your enzymes?
2 What effect does a change in pH have on your enzymes?

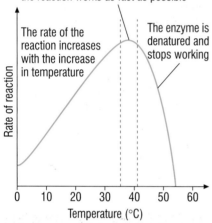

Optimum temperature – this is when the reaction works as fast as possible

The rate of the reaction increases with the increase in temperature

The enzyme is denatured and stops working

Figure 1 Like most chemical reactions, the rate of an enzyme-controlled reaction increases as the temperature rises – but only until the point where the complex protein structure of the enzyme breaks down

Leave a bottle of milk at the back of your fridge for a week or two and you'll find it is pretty disgusting. The milk will have gone off as enzymes in bacteria break down the protein structure.

Leave your milk in the Sun for a day and the same thing will happen – but much faster. Temperature affects the rate at which chemical reactions take place even when they are controlled by biological catalysts.

Biological reactions are affected by the same factors as any other chemical reactions – concentration, temperature and particle size all affect them. But in living organisms an increase in temperature only works up to a certain point.

a) Why does milk left in the Sun go off quickly?

The effect of temperature on enzyme action

The chemical reactions which take place in living cells happen at relatively low temperatures. Like most other chemical reactions, the rate of enzyme-controlled reactions increases with an increase in temperature. The enzyme and substrate particles move faster as the temperature increases, so this makes them more likely to collide with enough energy to react.

However this is only true up to temperatures of about 40°C. After this the protein structure of the enzyme is affected by the temperature. The long amino acid chains unravel. As a result the shape of the active site changes. We say the enzyme has been **denatured**. It can no longer act as a catalyst, so the rate of the reaction drops dramatically. Most enzymes work best from about 20°C to 40°C.

b) What does it mean if an enzyme is denatured?

PRACTICAL

Investigating the effect of temperature

You can show the effect of temperature on the rate of enzyme action using simple practicals like the one shown opposite.

The enzyme amylase (found in your saliva) breaks down starch into simple sugars. You mix starch solution and amylase together and keep them at different temperatures. Then you test samples from each temperature with iodine solution at regular intervals.

In the presence of starch, iodine solution turns blue-black. But when there is no starch present, the iodine stays yellowy-brown. When the iodine solution no longer changes colour you know all the starch has been broken down.

This gives you some clear evidence of the effect of temperature on the rate of enzyme controlled reactions.

- How does iodine solution show you if starch is present?
- Why do we test starch solution without amylase added?
- What conclusion can you draw from the results?

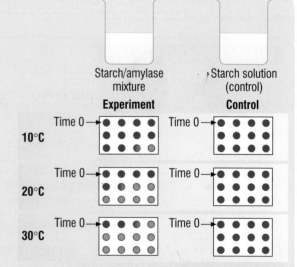

Figure 2 In each case the starch amylase mixture and the control are kept in a water bath at a given temperature. Samples are taken every five minutes and tested with iodine solution on a spotting tile.

Effect of pH on enzyme action

Enzymes have their effect by binding the reactants to a specially shaped **active site** in the protein molecule. Anything which changes the shape of this active site stops the enzyme from working. Temperature is obviously one thing which changes the shape of the protein molecule. The surrounding pH is another.

The shape of enzymes is the result of forces between the different parts of the protein molecule which hold the folded chains in place. A change in the pH affects these forces and changes the shape of the molecule. As a result, the active site is lost, so the enzyme can no longer act as a catalyst.

Different enzymes have different pH levels at which they work at their best – and a change in the pH can stop them working completely.

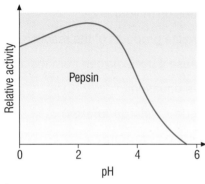
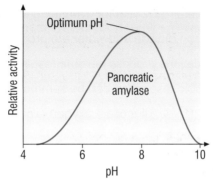

Figure 3 These two enzymes are found in quite different parts of the human gut, and they need very different conditions of pH to work at their maximum rate. Pepsin is found in the stomach, along with hydrochloric acid, while pancreatic amylase is in the small intestine along with alkaline bile.

The role of enzymes

Enzymes are vital to all living cells. They catalyse a huge range of reactions. Without them respiration, photosynthesis and protein synthesis would be impossible. This also applies to all the other reactions which take place in your cells. For the enzymes to work properly the temperature and pH must be just right. This is why it is so dangerous if your temperature goes very high when you are ill and run a fever. Once your body temperature reaches about 41°C, your enzymes start to be denatured and you will soon die.

SUMMARY QUESTIONS

1 Copy and complete using the words below:

> **active site   cells   denatured   enzyme   increases**
> **protein   reactions   shape   temperatures   40°C**

The chemical ...... which take place in living ...... happen at relatively low .... The rate of these ...... controlled reactions ...... with an increase in temperature. However this is only true up to temperatures of about .... After this the ...... structure of the enzyme is affected and the ...... of the .... is changed. The enzyme has been ....

2 Look at Figure 3.

a) At which pH does pepsin work best?
b) At which pH does amylase work best?
c) What happens to the activity of the enzymes as the pH increases?
d) Explain why this change in activity happens.

GET IT RIGHT!

The rate of enzyme-controlled reactions increases as the temperature goes up to about 40°C because the particles are moving faster. So substrate molecules collide with enzymes more often. Once the temperature goes much over 40°C most enzymes are denatured and no longer work as catalysts.

Enzymes aren't killed (they are molecules, not living things themselves) – use the term **denatured**.

Figure 4 The magical light display of a firefly is caused by the action of a very special enzyme called luciferase

KEY POINTS

1 Enzyme activity is affected by temperature and pH.
2 High temperatures and the wrong pH can affect the shape of the active site of an enzyme and stop it working.
3 Enzymes catalyse processes such as respiration, photosynthesis and protein synthesis in living cells.

B2 4.3

Aerobic respiration

LEARNING OBJECTIVES

1 What is aerobic respiration?
2 Where in your cells does respiration take place?

DID YOU KNOW?

The average energy needs of a teenage boy are 11 510 kJ of energy every day – but teenage girls only need 8830 kJ a day. This is partly because on average girls are smaller than boys but also because boys have more muscle cells, which means more mitochondria demanding fuel for aerobic respiration.

One of the most important enzyme-controlled processes in living things is **aerobic respiration**.

Your digestive system, lungs and circulation all work to provide your cells with what they need for respiration to take place.

During aerobic respiration glucose (a sugar produced as a result of digestion) reacts with oxygen. This reaction releases energy which your cells can use. This energy is vital for everything else that goes on in your body.

Carbon dioxide and water are produced as waste products of the reaction.

We call the process **aerobic** respiration because it uses oxygen from the air.

Aerobic respiration can be summed up by the equation:

$$\text{glucose} + \text{oxygen} \rightarrow \text{carbon dioxide} + \text{water } \textbf{(+ energy)}$$

a) Why is aerobic respiration so important?

PRACTICAL

Investigating respiration

Animals and plants – even bacteria – all respire. To show that cellular respiration is taking place, you can either deprive a living organism of the things it needs to respire, or show that waste products are produced from the reaction.

Depriving a living thing of food and/or oxygen would kill it – so this would be an unethical investigation. So we concentrate on the waste products of respiration. Carbon dioxide and energy in the form of heat are the easiest to identify.

Lime water goes cloudy when carbon dioxide bubbles through it. The higher the concentration of carbon dioxide, the quicker the lime water goes cloudy. This gives us an easy way of demonstrating that carbon dioxide has been produced. We can also look for a rise in temperature to show that energy is being produced during respiration.

● Plan an ethical investigation into aerobic respiration in living organisms.

Mitochondria – the site of respiration

Aerobic respiration involves lots of chemical reactions, each one controlled by a different enzyme. Most of these reactions take place in the **mitochondria** of your cells.

Mitochondria are tiny rod-shaped bodies (**organelles**) which are found in almost all plant and animal cells. They have a folded inner membrane which provides a large surface area for the enzymes involved in aerobic respiration.

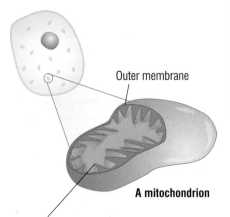

Outer membrane

A mitochondrion

Folded inner membrane gives a large surface area where the enzymes which control cellular respiration are found

Figure 1 Mitochondria are the powerhouses which provide energy for all the functions of your cells

Cells which need a lot of energy – like muscle cells and sperm – have lots of mitochondria. Cells which use very little energy – like fat cells – have very few mitochondria.

b) Why do mitochondria have folded inner membranes?

Reasons for respiration

- Respiration releases energy from the food we eat so that the cells of the body can use it.

- Both plant and animal cells need energy to carry out the basic functions of life. They build up large molecules from smaller ones to make new cell material. Much of the energy released in respiration is used for these 'building' activities (synthesis reactions). For example in plants, the sugars, nitrates and other nutrients are built up into amino acids which are then built up into proteins.

- Another important use of the energy from respiration in animals is in making muscles contract. Muscles are working all the time in our body, whether we are aware of them or not. Even when you sleep your heart beats, you breathe and your gut churns – and these muscular activities use energy.

- Finally, mammals and birds are 'warm-blooded'. This means that our bodies are the same temperature inside almost regardless of the temperature around us. On cold days we use energy to keep our body warm, while on hot days we use energy to sweat and keep our body cool.

GET IT RIGHT!

Make sure you know the equation for respiration. Remember that aerobic respiration takes place in the mitochondria.

Figure 2 Warm-blooded animals like this bird use up some of the energy they produce by aerobic respiration just to keep a steady body temperature. When the weather is cold, they use up a lot more energy to keep warm. Giving them extra food supplies can mean the difference between life and death.

SUMMARY QUESTIONS

1 Copy and complete these sentences, matching the pairs.

a) Energy is released from glucose	A ...... energy is released.
b) During respiration chemical reactions take place	B ...... because it uses oxygen from the air.
c) When glucose reacts with oxygen	C ...... are formed as waste products.
d) Carbon dioxide and water	D ...... by a process known as respiration.
e) The process is known as aerobic respiration	E ...... inside the mitochondria in the cells of your body.

2 Why are mitochondria so important and how is their structure adapted for the job that they do?

3 You need a regular supply of food to provide energy for your cells. If you don't get enough to eat you become thin and stop growing. You don't want to move around and you start to feel cold. There are three main uses of the energy released in your body during aerobic respiration. What are they and how does this explain the symptoms of starvation described above?

4 Suggest an experiment to show that a) oxygen is taken up, and b) carbon dioxide is released, during aerobic respiration.

KEY POINTS

1 Aerobic respiration involves chemical reactions which use oxygen and sugar and release energy. The reaction is summed up as:

glucose + oxygen →
carbon dioxide + water
(+ **energy**).

2 Most of the reactions in aerobic respiration take place inside the mitochondria.

B2 4.4

Enzymes in digestion

LEARNING OBJECTIVES

1 How are enzymes involved in the digestion of your food?
2 What happens to the digested food?

GET IT RIGHT!

Learn the different types of digestive enzymes and the end products of the breakdown of your food. Make sure you know where the different digestive enzymes are made.

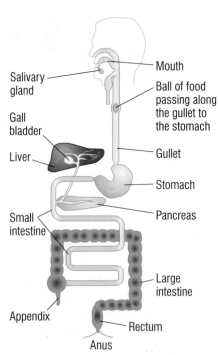

Salivary gland

Mouth

Ball of food passing along the gullet to the stomach

Gall bladder

Liver

Gullet

Stomach

Small intestine

Pancreas

Large intestine

Appendix

Rectum

Anus

Figure 1 The human digestive system

The food you eat is made up of large insoluble molecules which your body cannot absorb. They need to be broken down or *digested* to form smaller, soluble molecules. These can then be absorbed and used by your cells. This chemical breakdown is controlled by your digestive enzymes.

Most of your enzymes work *inside* the cells of your body. Your digestive enzymes are different – they work *outside* of your cells. They are produced by specialised cells which are found in glands (like your salivary glands and your pancreas), and in the lining of your gut.

The enzymes then pass out of these cells into the gut itself. It is here that they get mixed up with your food molecules and break them down.

Your gut is a hollow muscular tube which squeezes your food. The gut:

- helps to break up your food into small pieces with a large surface area for your enzymes to work on,
- mixes your food with your digestive juices so that the enzymes come into contact with as much of the food as possible, and
- uses its muscles to move your food along its length from one area to the next.

a) How do your digestive enzymes differ from most of your other enzymes?

Digesting carbohydrates

Enzymes which break down carbohydrates are known as **carbohydrases**. Starch is one of the most common carbohydrates that you eat. It is broken down into **sugars** like glucose. This reaction is catalysed by the carbohydrase called *amylase*.

Amylase is produced in your salivary glands, so the digestion of starch starts in your mouth. Amylase is also made in your pancreas and your small intestine. No digestion takes place in the pancreas. All the enzymes made there flow into your small intestine, which is where most of the starch you eat is digested.

b) What is the name of the enzyme which breaks down starch in your gut?

Digesting proteins

The breakdown of protein food like meat, fish and cheese into amino acids is catalysed by **protease** enzymes. Proteases are produced by your stomach, your pancreas and your small intestine. The breakdown of proteins into **amino acids** takes place in your stomach and small intestine.

c) Which enzymes breaks down protein in your gut?

Digesting fats

The **lipids** (fats and oils) that you eat are broken down into **fatty acids** and **glycerol** in your small intestine. The reaction is catalysed by *lipase* enzymes which are made in your pancreas and your small intestine. Yet again the enzymes made in the pancreas are passed into the small intestine.

Once your food molecules have been completely digested into soluble glucose, amino acids, fatty acids and glycerol, they leave your small intestine. They pass into your blood supply to be carried around the body to the cells which need them.

d) Which enzymes break down fats in your gut?

PRACTICAL

Investigating digestion

You can make a model gut using a bag of special membrane containing starch and amylase enzymes. When the enzyme has catalysed the breakdown of the starch, you can detect the presence of sugar on the outside of the 'gut'!

● How can you test for sugars?

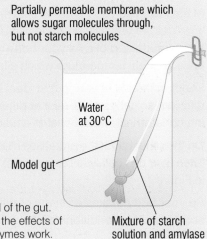

Partially permeable membrane which allows sugar molecules through, but not starch molecules

Water at 30°C

Model gut

Mixture of starch solution and amylase

Figure 2 This apparatus provides you with a model of the gut. You can use it to investigate how the gut works and the effects of factors like temperature and pH on how the gut enzymes work.

Using the digested food

● The glucose produced by the action of amylase and other carbohydrases is used by the cells of your body in respiration.
● Fatty acids and glycerol may be used as a source of energy or to build cell membranes, make hormones and as fat stores.
● The amino acids produced when you digest protein are not used as fuel. Once inside your cells, amino acids are built up into all the proteins you need. These synthesis reactions are catalysed by enzymes. In other words, your enzymes make new enzymes as well as all the other proteins you need in your cells. This **protein synthesis** takes place in the **ribosomes**.

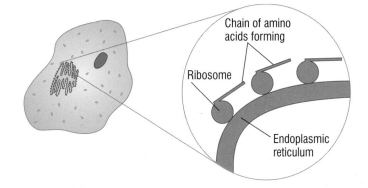

Chain of amino acids forming

Ribosome

Endoplasmic reticulum

Figure 3 Ribosomes are very small. They can only be seen using the most powerful microscopes. However their role in protein synthesis means they are vital to the working of your cells and your whole body!

SUMMARY QUESTIONS

1 Copy and complete using the words below:

   absorbed   broken down   cells   digestive enzymes
   food   insoluble   soluble

   The ...... you eat is made up of large ...... molecules which need to be .... to form smaller, ... molecules. These can be ...... by your body and used by your .... This chemical breakdown is controlled by your ....

2 Make a table which shows amylase, protease and lipase. For each enzyme show where it is made, which reaction it catalyses and where it works in the gut.

3 Why is digestion of your food so important? Explain in terms of the molecules involved.

KEY POINTS

1 Enzymes catalyse the breakdown of large food molecules into smaller molecules during digestion.
2 Digestive enzymes are produced inside cells but they work outside of cells in the gut.
3 Enzymes in the ribosomes catalyse the build up of proteins from amino acids.

B2 4.5

Speeding up digestion

1 Why does your stomach contain hydrochloric acid?
2 What is bile and why is it so important in digestion?

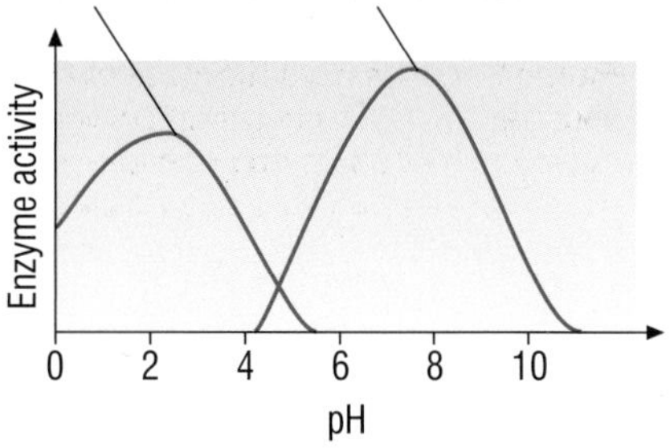

Pepsin – protease from the stomach   Trypsin – protease from the small intestine

Enzyme activity

pH
0 2 4 6 8 10

Figure 1 Both of these enzymes catalyse the breakdown of proteins. But as these graphs show, the enzyme found in the stomach works best at a very different pH to the one made in the pancreas and used in the small intestine.

Your digestive system produces many enzymes which speed up the breakdown of the food you eat. However enzymes aren't the only important chemicals in your gut. As you saw on pages 44 and 45, enzymes are very sensitive to temperature and pH. As your body is kept at a fairly steady 37°C, your enzymes have an ideal temperature which allows them to work as fast as possible.

Keeping the pH in your gut at ideal levels isn't quite so easy. That's because different enzymes work best at different pH levels. The protease enzyme found in your stomach works best in acidic conditions.

On the other hand, the proteases made in your pancreas need alkaline conditions. Then they can catalyse protein breakdown as fast as they can. Look at the graph in Figure 1.

So your body makes a variety of different chemicals which help to give your enzymes ideal conditions all the way through your gut.

a) Why do your enzymes almost always have the right temperature to work at their best?

Changing pH in the gut

You have around 35 million glands in the lining of your stomach secreting protease enzymes to digest the protein you eat. These enzymes work best in an acid pH. So your stomach also produces a concentrated solution of hydrochloric acid from the same glands. In fact your stomach produces around 3 litres of acid a day!

This acid allows your stomach protease enzymes to work very effectively. It also kills most of the bacteria which you take in with your food.

Finally, your stomach also produces a thick layer of mucus which coats your stomach walls and protects them from being digested by the acid and the enzymes!

b) How does your stomach avoid digesting itself?

Pigments from your bile are largely responsible for the brown colour of your faeces. If you have a disease which stops bile getting into your gut, your faeces will be white or silvery grey!

Breaking down protein

You can see the effect of acid on pepsin, the protease found in the stomach, quite simply. Set up three test tubes, one containing pepsin only, one containing only hydrochloric acid and one containing a mixture of the two. Keep them at body temperature in a water bath. Add a similar sized chunk of meat to all three of them. Set up a web cam and watch for a few hours to see what happens!

- What conclusions can you make?

Figure 2 These test tubes show clearly the importance of protein-digesting enzymes *and* hydrochloric acid in your stomach. Meat was added to each tube at the same time.

After a few hours – depending on the size and type of the meal you have eaten – your food leaves your stomach and moves on into your small intestine. Some of the enzymes which catalyse digestion in your small intestine are made in your pancreas. Some are also made in the small intestine itself. They all work best in an alkaline environment.

The acidic liquid coming from your stomach needs to become an alkaline mix in your small intestine! So how does it happen?

Your liver carries out many important jobs in your body and one of them is producing bile. Bile is a greenish-yellow alkaline liquid which is stored in your gall bladder until it is needed.

As food comes into the small intestine from the stomach, bile is squirted onto it. The bile neutralises the acid from the stomach and then makes the semi-digested food alkaline. This provides the ideal conditions for the enzymes in the small intestine.

c) Why does the food coming into your small intestine need neutralising?

Altering the surface area

It is very important for the enzymes of the gut to have the largest possible surface area of food to work on. This is not a problem with carbohydrates and proteins. However, the fats that you eat do not mix with all the watery liquids in your gut. They stay as large globules – think of oil in water – which makes it difficult for the lipase enzymes to act.

This is the second important function of the bile. It **emulsifies** the fats in your food. This means it physically breaks up large drops of fat into smaller droplets. This provides a much bigger surface area for the lipase enzymes to act on. The larger surface area helps them chemically break down the fats more quickly into fatty acids and glycerol.

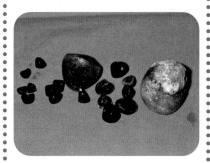
GET IT RIGHT!

Remember food is not digested in the liver or the pancreas.

Bile is **not** an enzyme and it does **not** break down fat molecules.

Bile emulsifies fat droplets to increase the surface area, which in turn increases the rate of fat digestion by lipase.

SUMMARY QUESTIONS

1 Copy and complete using the words below:

alkaline   emulsifies   gall bladder   liver   neutralises   small intestine

Bile is an...... liquid produced by your .... It is stored in the .... and released onto food as it comes into the .... It ...... the acid food from the stomach and makes it alkaline. It also ...... fats.

2 Look at Figure 1.

a) At what pH does the protease from the stomach work best?
b) How does your body create the right pH in the stomach for this enzyme?
c) At what pH does the protease from the intestine work best?
d) How does your body create the right pH in the small intestine for this enzyme?

3 Draw a diagram to explain how bile produces a big surface area for lipase to work on and explain why this is important.

KEY POINTS

1 The enzymes of the stomach work best in acid conditions.
2 The enzymes made in the pancreas and the small intestine work best in alkaline conditions.
3 Bile produced by the liver neutralises acid and emulsifies fats.

B2 4.6

Making use of enzymes

1 How do biological detergents work?
2 How are enzymes used in the food industry?

Figure 1 More and more homes now have a dishwasher – and dishwasher powders contain enzymes. They digest the cooked-on proteins like eggs which are often hard to remove even in a dishwasher.

Figure 2 Learning to eat solid food isn't easy. Having some of it pre-digested by protease enzymes can make it easier to get the goodness you need to grow!

Enzymes were first isolated from living cells in the 19th century, and ever since we have found more and more ways of using them in industry. Some microorganisms produce enzymes which pass out of the cells and are easy for us to use. In other cases we use the whole microorganism.

Enzymes in the home

In the past, people boiled and scrubbed their clothes to get them clean – and did it all by hand. Now we not only have washing machines to do the washing for us, we also have enzymes ready and waiting to digest the stains.

Many people use **biological detergents** to remove stains from their clothes from substances such as grass, sweat, food and blood. Biological washing powders contain proteases and lipases which break down the proteins and fats in the stains. They help provide us with a cleaner wash. We also use them at the lower temperatures that enzymes need to work best, so we use less electricity too.

a) What is a biological washing powder?

PRACTICAL

Investigating biological washing powder

Weigh a chunk of cooked egg white and leave it in a strong solution of biological washing powder.

● What do you think will happen to the egg white?
● How can you measure just how effective the protease enzymes are?
● How could you investigate the effect of surface area in enzyme action?

Enzymes in industry

Pure enzymes have many uses in industry.

Proteases are used in the manufacture of baby foods. They 'pre-digest' some of the protein in the food. When babies first begin to eat solid foods they are not very good at it. Treating the food with protease enzymes makes it easier for a baby's digestive system to cope with. It is easier for them to get the amino acids they need from their food.

Carbohydrases (carbohydrate digesting enzymes) are used to convert starch into sugar (glucose) syrup. We use huge quantities of sugar syrup in food production – just have a look at the ingredients labels on all sorts of foods.

Starch is made by plants like corn, and it is very cheap. Using enzymes to convert this plant starch into sweet sugar provides a cheap source of sweetness for food manufacturers.

It is also important for the process of making fuel (ethanol) from plants.

b) Why does the starch need to be converted to sugar before it is used to make ethanol?

Sometimes the glucose syrup made from starch is passed into another process which uses a different set of enzymes. **Isomerase** enzyme is used to convert glucose syrup into **fructose syrup** by rearranging the atoms in the glucose molecule.

Glucose and fructose contain exactly the same amount of energy (1700 kJ or 400 kcal per 100 g) but fructose is much sweeter than glucose. This means much smaller amounts of it are needed to make food taste sweet. So fructose is widely used in 'slimming' foods. The food tastes sweet but contains fewer calories.

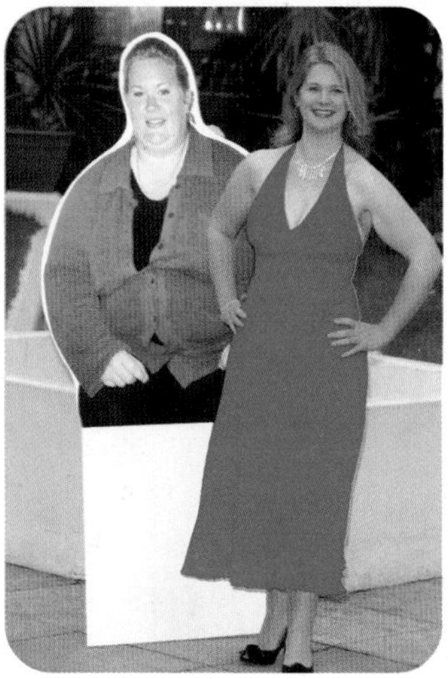

Figure 3 The market for slimming foods is enormous and growing all the time. Enzyme technology is being used to convert more and more glucose syrup into fructose syrup to make so-called 'slimming' foods.

The advantages and disadvantages of using enzymes

In an industrial process, many of the reactions need high temperatures and pressures to make them fast enough to produce the products needed. Supplying heat and building chemical plants which can stand high pressures costs a lot of money.

However, enzymes can provide the perfect answer to industrial problems like these. They catalyse reactions at relatively low temperatures and normal pressures. Enzyme-based processes are therefore often fairly cheap to run.

The main problem with enzymes is that they are very sensitive to their surroundings. For enzymes to function properly the temperature must be kept down (usually below 45°C). The pH also needs to be kept within carefully monitored limits which suit the enzyme. It costs money to control these conditions.

Whole microbes are relatively cheap, but need to be supplied with food and oxygen and their waste products removed. What's more, they use some of the substrate to grow more microbes. Pure enzymes use the substrate more efficiently, but they are also more expensive to produce.

GET IT RIGHT!

Remember that most enzyme names end in -ase. Some enzymes used in industry work at quite high temperatures – so don't be put off if a graph shows an optimum temperature well above 45°C!

SUMMARY QUESTIONS

1 List three enzymes and the ways in which we use them in the food industry.

2 Biological washing powders contain enzymes in tiny capsules. Explain why:

a) they are more effective than non-biological powders at lower temperatures,

b) they are not more effective at high temperatures.

3 Make a table to show the advantages and disadvantages of using enzymes in industry.

KEY POINTS

1 Some microorganisms produce enzymes which pass out of the cells and can be used in different ways.

2 Biological detergents may contain proteases and lipases.

3 Proteases, carbohydrases and isomerase are all used in the food industry.

B2 4.7   High-tech enzymes

The washing powder debate

I've got three children and they are all messy eaters! Their clothes get lots of mud and grass stains as well. I always use biological detergents because they get my washing really clean.

I've got very sensitive skin. When my mum changed to a biological detergent I got dermatitis so we never use biological detergents now.

When we first started manufacturing our biological detergent we found a lot of our factory staff developed allergies. We realised they were reacting to enzyme dust in the air – proteins often trigger allergies. But once we put the enzymes in tiny capsules all the allergy problems stopped. Unfortunately it got some bad publicity and lots of people still seem to think biological detergents cause allergies.

I try to be as green as possible in my lifestyle but I'm not sure about biological detergents. Enzymes are natural, after all – but I've heard they can cause allergies. On the other hand, biological powders use a lot less electricity because they clean at lower temperatures. That's good for the environment and cheaper for me!

ACTIVITY

You are part of a team producing an article for a lifestyle magazine about biological washing powders. Create a double-page article – make it lively, interesting to look at, scientifically accurate and informative!

Allergies aren't really a problem with biological detergents. However, if the clothes aren't rinsed really thoroughly, protein-digesting enzymes can get left in the fabric. Then the enzymes may digest some of the protein in your skin and set up dermatitis. But if the detergent is used properly, there shouldn't be a problem.

Enzymes and medicine

Here are just some of the ways in which enzymes are used in medicine:

To diagnose disease

If your liver is damaged or diseased, some of your liver enzymes may leak out into your blood. If your symptoms suggest your liver isn't working properly, doctors can test your blood for these enzymes to find out if your liver really is damaged.

To diagnose and control disease

People who have diabetes have too much sugar in their blood. As a result, they also get sugar in their urine. One common test for sugar in the urine relies on a colour change on a test strip.

The test strip contains a chemical indicator and an enzyme. It is placed in a urine sample. The enzyme catalyses the breakdown of any glucose found in the urine. The products of the reaction then make the indicator change colour if glucose is present.

To cure disease

- If your pancreas is damaged or diseased it cannot make enzymes. So you have to take extra enzymes – particularly lipase – to allow you to digest your food. The enzymes are in special capsules to stop them being digested in your stomach!
- If you have a heart attack, an enzyme called streptokinase will be injected into your blood as soon as possible. It dissolves clots in the arteries of the heart wall and reduces the amount of damage done to your heart muscle.
- An enzyme from certain bacteria is being used to treat a type of leukaemia in children. The cancer cells cannot make one particular amino acid, so they need to take it from your body fluids. The enzyme catalyses the breakdown of this amino acid, so the cancer cells cannot get any and they die. Your normal cells can make the amino acid so they are not affected. Doctors hope something similar may work against other types of cancer.

Health special

IN THE RAW!

WILL YOU TRY THE NEW DIET SENSATION?

The latest food craze to sweep the US is to eat your food – including meat – completely raw. And now it's coming to the UK!

It has been reported that there are lots of health benefits to this new way of eating. It is claimed that raw food contains live enzymes which will help to give you more energy. Apparently when food is cooked these enzymes die.

One of the owners of a new raw food restaurant has been quoted as saying,

'It is an amazingly interesting way of preparing food, it is good to have live enzymes in your system and, most importantly, it is yummy.'

Dodgy science

SUMMARY QUESTIONS

1 a) Copy and complete the following sentences, matching the parts of the sentences.

i)	A catalyst will speed up or slow down a reaction	1	...... could not occur without enzymes.
ii)	Living organisms make very efficient catalysts	2	...... made of protein.
iii)	All enzymes are	3	...... binds to the active site.
iv)	The reactions which keep you alive	4	...... known as enzymes.
v)	The substrate of an enzyme	5	...... a specific type of molecule.
vi)	Each type of enzyme affects	6	...... but is not changed itself.

b) Explain how an enzyme catalyses a reaction. Use diagrams if they make your explanation clearer.

2 Use Figure 2 on page 178 to help you answer this question.

a) What effect does the enzyme amylase have on starch?

b) What do these results tell you about the effect of temperature on the action of the amylase?

c) Why is one tube of starch solution kept at each temperature without the addition of the enzyme?

d) How could you improve this investigation?

e) What do you predict would happen to the activity of the enzyme if acid from the stomach were added to the mixture?

3 The table gives some data about the relative activity levels of an enzyme at different pH levels.

pH	Relative activity
4	0
6	3
8	10
10	1

a) Plot a graph of this data.

b) Does this enzyme work best in an acid or an alkaline environment?

c) This is a protein-digesting enzyme. Where in the gut do you think it might be found? Explain your answer.

EXAM-STYLE QUESTIONS

1 (a) In the summary of aerobic respiration shown below, choose a word from each of the boxes that best completes the equation. **(2)**

Glucose + BOX A → carbon dioxide + BOX B (+ energy)

BOX A: water / oxygen / nitrogen

BOX B: water / oxygen / nitrogen

(b) (i) State two ways in which the energy released during respiration is used in **all** animals. **(2)**

(ii) How else might the energy released be used in mammals and birds only? **(1)**

(iii) Give a further use of the energy released that applies to plants rather than animals. **(1)**

2 **A**, **B**, **C**, **D** and **E** are the names of enzymes or groups of enzymes. The numbers **1**, **2**, **3**, **4** and **5** refer to the functions or uses of each of these enzymes.

Match each letter with the appropriate number. **(5)**

A	Lipase	**1**	Used in the manufacture of baby foods
B	Amylase	**2**	Group of enzymes that act on carbohydrates
C	Proteases	**3**	Its substrate is starch
D	Isomerase	**4**	Used in the production of slimming foods
E	Carbohydrases	**5**	The products of its catalytic action are glycerol and fatty acids

3 Amylase is an enzyme that catalyses the conversion of starch into sugar.

(a) To which of the following groups of food does starch belong? **(1)**

carbohydrates   fats   protein   vitamins

(b) Give the names of the **three** organs in the human body that secrete the enzyme amylase. **(3)**

The graph on the next page shows the effect of temperature on the activity of amylase.

(c) (i) At what temperature did the amylase work fastest? **(1)**

(ii) Why did the amylase not work above 56°C? **(1)**

(iii) State one other factor apart from temperature that will affect the rate of reaction of amylase. **(1)**

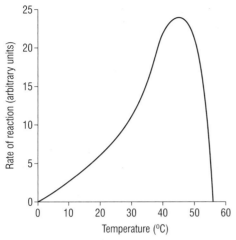

Rate of reaction (arbitrary units) vs Temperature (°C)

4 In the making of cheese, a commercially prepared form of an enzyme called rennin is used to make the protein in milk more solid. Rennin is an enzyme that is produced naturally in the stomachs of young mammals. The owner of a cheese making factory wanted to use a different source of rennin. She needed to find out the best temperature to use for the new rennin. She planned to set up 20 test tubes. All would have 20 cm³ of milk in them: half with the rennin added (A) and half to be left without rennin (B). One tube of each type would be left in a water bath until one of them clotted. When this happened, the time taken would be recorded.

(a) Construct a table that could be used by the owner. (3)

(b) Fill in the table to show the range of temperatures she might use. (1)

(c) Fill in the table to show the interval for the independent variable. (1)

(d) Suggest how the owner might know when the milk is clotted. (1)

(e) Would you suggest that she repeats her results? Explain your answer. (1)

(f) Why do you think she used tubes A and B at each temperature? (1)

5 Bile is a greenish liquid that plays an important role in the digestion of food.

(a) In which organ is bile produced? (1)

(b) Where is bile stored in the body? (1)

(c) Into which region of the digestive system is bile released? (1)

(d) Describe how bile is involved in the digestion of fats. (3)

(e) What is the name of the enzyme that digests fats? (1)

(f) Name two places where this enzyme is produced in the body. (2)

HOW SCIENCE WORKS QUESTIONS

Najma had carried out a 'rates of reaction' investigation in chemistry. Her results are in Table 1 below.

Table 1 Chemistry investigation

Temperature (°C)	Time taken (secs)
20	106
30	51
40	26
50	12
60	5

When asked in biology to do a 'rates of reaction' investigation she expected to get the same results. She reasoned that in both cases she was collecting the oxygen produced from hydrogen peroxide. The only difference was that she used manganese(IV) oxide in chemistry and she was using mashed up plant cells in biology! Her results from biology are in Table 2.

Table 2 Biology investigation

Temperature (°C)	Time taken (secs)
20	114
30	96
40	80
50	120
60	No reaction

a) What was Najma's prediction for the biology investigation? (1)

b) Was her prediction supported, refuted or should she rethink the prediction? (2)

c) Najma checked her results against some results in a textbook. Why was this a good idea? (1)

d) Najma was feeling happier now that she had been supported by other scientists' results. She had also learned that enzymes had a temperature at which they worked best. How could she change her investigation so that she could find the best temperature for this enzyme? (1)

e) Najma also learned that this enzyme was called catalase and that it occurs in nearly all organisms, even those living in hot water springs. How could she change her investigation to find the best temperature for catalase in hot water spring organisms? (2)

191

B2 5.1 Controlling internal conditions

Figure 1 Whatever you choose to do in life, the conditions inside your body will stay more-or-less exactly the same. When you think of the range of things you can do, it is amazing how the balance is maintained.

For your body to work properly the conditions surrounding your millions of cells must stay as constant as possible. On the other hand, almost everything you do tends to change things.

As you move you produce heat, as you respire you produce waste, when you digest food you take millions of molecules into your body. Yet you somehow keep your internal conditions constant within a very narrow range. How do you manage this?

The answer is through **homeostasis**. As you saw on page 32, many of the functions in your body help to keep your internal environment as constant as possible. Now you are going to find out more about some of them.

a) What is homeostasis?

Removing waste products

No matter what you are doing, even sleeping, the cells of your body are constantly producing waste products as a result of the chemical reactions which are taking place. The more extreme the conditions you put yourself in, the more waste products your cells will make. There are two main poisonous waste products which would cause major problems for your body if the levels built up. These are carbon dioxide and urea.

Carbon dioxide

Carbon dioxide is produced during cellular respiration.

Every cell in your body respires, and so every cell produces carbon dioxide. It is vital that you remove this carbon dioxide. That's because if it all remained dissolved in the cytoplasm of your cells it would affect the pH. Dissolved carbon dioxide produces an acidic solution – and a lower pH would affect the working of all the enzymes in your cells!

PRACTICAL

Investigating breathing

Find out the capacity of your lungs or the effect of exercise on breathing.

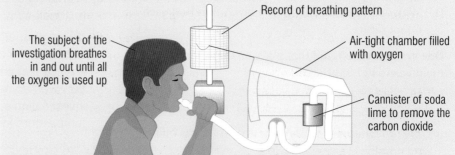

Record of breathing pattern

The subject of the investigation breathes in and out until all the oxygen is used up

Air-tight chamber filled with oxygen

Cannister of soda lime to remove the carbon dioxide

Figure 2 Because you breathe in and out of the machine all the time, you can't get rid of your waste carbon dioxide in the normal way. There has to be a special filter to remove the carbon dioxide so it doesn't poison you!

● How can we improve the reliability of investigations involving living organisms?

The carbon dioxide moves out of the cells into your blood. Your blood stream carries it back to your lungs. Almost all of the carbon dioxide you produce is removed from your body via your lungs when you breathe out. The air you breathe in contains only 0.04% carbon dioxide, but the air you breathe out contains about 4% carbon dioxide!

b) How do you remove carbon dioxide from your body?

Urea

The other main waste product of your body is **urea**.

Urea is produced in your **liver** when excess amino acids are broken down. When you eat more protein than you need, or when body tissues are worn out, the extra protein has to be broken down. Amino acids cannot be used as fuel for your body. But in your liver the amino group is removed and converted into urea.

The rest of the amino acid molecule can then be used in respiration or to make other molecules. The urea passes from the liver cells into your blood.

Urea is poisonous and if the levels build up in your blood it will cause a lot of damage. Fortunately the urea is filtered out of your blood by your **kidneys**. It is then removed in your **urine**, along with any excess water and salt.

Urine is produced all the time by your kidneys. It leaves your kidneys and is stored in your **bladder** which you then empty from time to time!

c) Where is urea made?

Maintaining body balance

Water and ions enter your body when you eat or drink. The water and ion content of your body are carefully controlled to prevent damage to your cells. Water is lost through breathing, through sweating and in the urine, while ions are lost in the sweat and in the urine.

If the water or ion content of your body is wrong, too much water may move into or out of your cells. That's why control is vital.

It is also very important to control your body temperature and the levels of sugar in your blood. So homeostasis plays a very important role in your body.

> ### SUMMARY QUESTIONS
>
> 1 Copy and complete using the words below:
>
> > **blood   carbon dioxide   constant   controlled   environment**
> > **enzymes   homeostasis   sugar   temperature   urea   water**
>
> The internal ...... of your body is kept relatively ...... by a whole range of processes which make up .... Waste products such as .... and ...... have to be removed from your ...... all the time. The ...... and ion concentration of your blood are constantly ...... and so is your blood ...... level. Your body ...... is kept the same so your ...... work effectively.
>
> 2 There are two main waste products which have to be removed from the human body – carbon dioxide and urea. For each waste product, describe:
>
> a) how it is formed, b) why it has to be removed, c) how it is removed from the body.
>
> 3 Explain briefly a) how a period of exercise would affect the internal conditions of your body, and b) how the conditions would be returned to normal.

FOUL FACTS

The average person produces between 1.5 and 2.5 litres of urine a day – that's up to 900 litres of urine a year!

GET IT RIGHT!

Don't confuse urea and urine. Urea is made in the liver; urine is produced by the kidney. Urine contains urea.

KEY POINTS

1 The internal conditions of your body have to be controlled to maintain a constant internal environment.

2 Poisonous waste products are made all the time and need to be removed.

3 Carbon dioxide is produced during respiration and leaves the body via the lungs when you breathe out.

4 Urea is produced by your liver as excess amino acids are broken down, and it is removed by your kidneys in the urine.

B2 5.2 | Controlling body temperature

LEARNING OBJECTIVES

1 How does your body monitor its temperature?
2 How does your body stop you getting too hot? [Higher]
3 How does your body keep you warm? [Higher]

Figure 1 People in different parts of the world live in conditions of extreme heat and extreme cold and still maintain a constant internal body temperature

Wherever you go and whatever you do it is vital that your body temperature is maintained at around 37°C. This is the temperature at which your enzymes work best. Your skin temperature can vary enormously without causing harm. It is the temperature deep inside your body, known as the core body temperature, which must be kept stable.

At only a few degrees above or below normal body temperature your enzymes cannot function properly. All sorts of things can affect your internal body temperature, including:

● heat produced in your muscles during exercise,
● fevers caused by disease, and
● the external temperature rising or falling.

People can control some aspects of their own temperature. We can change our clothing, light a fire, and turn on the heating or air-conditioning. But it is our internal control mechanisms which are most important in controlling our body temperature.

a) Why is control of your body temperature so important?

Control of the temperature relies on the **thermoregulatory centre** in the brain. This centre contains receptors which are sensitive to temperature changes. They monitor the temperature of the blood flowing through the brain itself.

Extra information comes from the temperature receptors in the skin. These send impulses to the thermoregulatory centre giving information about the skin temperature. The receptors are so sensitive they can detect a difference of as little as 0.5°C!

Sweating helps to cool your body down. So the loss of salt and water when you sweat can affect your water and ion balance. If you are sweating a lot you need to take in more drink or food to replace the water and ions you have lost – just watch a marathon runner!

> ### Cooling the body down
>
> If you get too hot, your enzymes denature and can no longer catalyse the reactions in your cells. When your core body temperature begins to rise, impulses are sent from the thermoregulatory centre to the body so more heat is lost:
>
> ● The blood vessels, which supply your skin capillaries, **dilate** (open wider). This lets more blood flow through the capillaries. Your skin flushes, so you lose more heat by radiation.
> ● Your rate of sweating goes up. Sweat (made up mainly of water, salt and a little protein) oozes out of your sweat glands and spreads over your skin. As the water evaporates it cools the skin, taking heat from your body. In very humid conditions, when the sweat doesn't evaporate very easily, it is very difficult to cool down.

HIGHER

HIGHER

Reducing heat loss

It is just as dangerous for your core temperature to drop as it is to rise. If you get very cold, the rate of the enzyme-controlled reactions in your cells falls too low. You don't make enough energy and your cells begin to die. If your core body temperature starts to get too low, impulses are sent from your thermoregulatory centre to the body to conserve and even generate more heat.

● The blood vessels which supply your skin capillaries **constrict** (close up) to reduce the flow of blood through the capillaries. This reduces the heat lost through the surface of the skin, and makes you look pale.
● Shivering begins – your muscles contract and relax rapidly which involves lots of cellular respiration. This releases some energy as heat which you use to raise your body temperature. As you warm up, shivering stops.
● Sweat production is reduced.

b) Why is a fall in your core body temperature so dangerous?

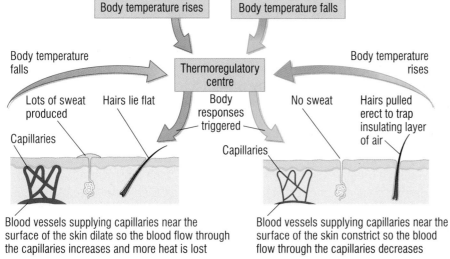

Figure 2 Changes in your core body temperature set off automatic responses to oppose the changes and maintain a steady internal environment.

SUMMARY QUESTIONS

1 Here is a jumbled list of some of the events by which your body temperature is controlled when it starts to go up. Sort them out into the right order and then copy them out.

   A Her body temperature starts to rise.
   B Sally takes a long, cool drink to replace the liquid she has lost through sweating.
   C Her temperature returns to normal.
   D Her skin goes red and her rate of sweating increases so the amount of heat lost through her skin goes up.
   E Sally exercises hard.

2 a) Why is it so important to maintain a body temperature of about 37°C?
   b) Explain the role of i) the thermoregulatory centre in the brain and
      ii) the temperature sensors in the skin in maintaining a constant core body temperature.

3 Explain how the body responds to both an increase and a decrease in core temperature to return its temperature to normal levels. [Higher]

DID YOU KNOW?

Birds and mammals can help reduce heat loss from their bodies by pulling the hairs or feathers on their skin upright to trap an insulating layer of air. Our bodies try to do this, but we just get goose-pimples. The tiny muscles pulling on our hairs show up more than the hairs themselves!

PRACTICAL

Body temperature

Use a temperature sensor and data logger to record your skin and core body temperature on one hand as you plunge the other into icy water.

● Explain your observations.

GET IT RIGHT!

Use the terms dilate and constrict for the changes which take place in the blood vessels in your skin. Remember sweating only cools your body when the sweat actually evaporates.

KEY POINTS

1 Your body temperature must be maintained at the level at which enzymes work best.
2 Your body temperature is monitored and controlled by the thermoregulatory centre in your brain.
3 Your body responds to cool you down if you are overheating and to warm you up if your core body temperature falls. [Higher]

B2 5.3

Controlling blood sugar

LEARNING OBJECTIVES

1 How is your blood sugar level controlled?
2 What is diabetes and how is it treated?

It is very important that your cells have a constant supply of the glucose they need for cellular respiration. Glucose is transported around your body to all the cells by your blood. However you don't spend all of your time eating to keep your blood sugar levels high. Instead the level of sugar in your blood is controlled by hormones produced in your pancreas.

a) Why are the levels of glucose in your blood so important?

The pancreas and the control of blood sugar levels

When you digest a meal, large amounts of glucose pass into your blood. Without a control mechanism your blood glucose levels would vary wildly. After a meal they would soar to a point where glucose would be removed from the body in the urine. A few hours later the levels would plummet and cells would not have enough glucose to respire.

This internal chaos is prevented by your **pancreas**. The pancreas is a small pink organ found under your stomach. It constantly monitors your blood glucose concentration and controls it using two hormones known as **insulin** and **glucagon**.

When your blood glucose concentration rises above the ideal range after you have eaten a meal, insulin is released. Insulin causes your liver to remove any glucose which is not needed at the time from the blood. The soluble glucose is converted to an insoluble carbohydrate called **glycogen** which is stored in your liver.

When your blood glucose concentration falls below the ideal range, the pancreas secretes glucagon. Glucagon makes your liver break down glycogen, converting it back into glucose. In this way the stored sugar is released back into the blood.

By using these two hormones and the glycogen store in your liver, your pancreas keeps your blood glucose concentration fairly constant. Its normal concentration is usually about 90 mg glucose per 100 cm^3 of blood.

b) Which two hormones are involved in the control of your blood sugar levels?

SCIENCE @ WORK

In 2005, research scientists produced insulin-secreting cells from stem cells which cured diabetes in mice. More research is needed but the scientists hope that before long diabetes will be a disease that we can cure instead of just treating the symptoms.

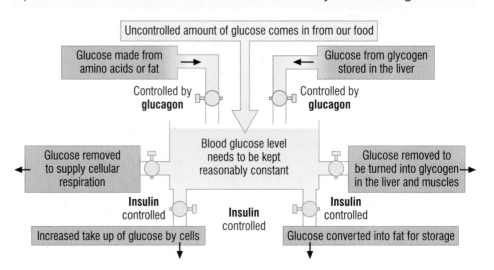

Figure 1 This model of your blood glucose control system shows the blood glucose as a tank. It has both controlled and uncontrolled inlets and outlets. In every case the control is given by the hormones insulin and glucagon.

What causes diabetes?

Most of us never think about our blood sugar levels because they are perfectly controlled by our pancreas. But for some people life isn't quite this simple. Unfortunately, their pancreas does not make enough – or any – insulin.

Without insulin your blood sugar levels get higher and higher after you eat food. Eventually your kidneys produce glucose in your urine. You produce lots of urine and feel thirsty all the time.

Without insulin, glucose cannot get into the cells of your body, so you lack energy and feel tired. You break down fat and protein to use as fuel instead, so you lose weight.

Before there was any treatment for diabetes, people would waste away. Eventually they would fall into a coma and die. Fortunately there are now some very effective ways of treating diabetes!

c) Why do people with untreated diabetes feel very tired and lack energy?

Treating diabetes

If you have a mild form of diabetes, managing your diet is enough to keep you healthy. Avoiding carbohydrate-rich foods keeps the blood sugar levels relatively low. So your reduced amount of insulin can cope with small amounts of glucose.

However, other people with diabetes need replacement insulin before meals. Insulin is a protein which would be digested in your stomach. So it is usually given as an injection to get it into your blood.

This injected insulin allows glucose to be taken into your body cells and converted into glycogen in the liver. This stops the concentration of glucose in your blood from getting too high.

Then as the blood glucose levels fall, natural glucagon makes sure glycogen is converted back to glucose. As a result your blood glucose levels are kept as stable as possible. (See graphs on page 199.)

Insulin injections treat diabetes successfully but they do not cure it. Until a cure is developed, someone with diabetes has to inject insulin several times every day of their life.

d) How can people with mild diabetes control the disease?

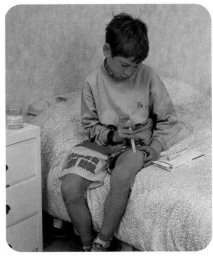

Figure 2 The treatment of diabetes involves regular blood sugar tests and insulin injections. These could become a thing of the past if some of the new treatments being developed work as well as scientists hope!

KEY POINTS

1 Your blood glucose concentration is monitored and controlled by your pancreas.
2 Insulin and glucagon are the hormones involved in controlling blood sugar concentration. Insulin converts glucose to glycogen; glucagon converts glycogen to glucose.
3 In diabetes, the blood glucose may rise to fatally high levels because the pancreas does not secrete enough insulin. It can be treated by injections of insulin before meals.

SUMMARY QUESTIONS

1 Define the following words:
   hormone;
   insulin;
   diabetes;
   glycogen.

2 a) Explain how your pancreas keeps the blood glucose levels of your body constant.
   b) Why is it so important to control the level of glucose in your blood?

3 What is diabetes and how can it be treated?

B2 5.4 Homeostasis matters!

HYPOTHERMIA = THE SILENT KILLER

If your core body temperature falls too low you suffer from hypothermia. About 30 000 people die of it every year in the UK alone. Here is some more information about hypothermia:

- Hypothermia is when your body temperature drops below 35°C and the normal working of your body is affected.
- Old people, small children and people exposed in bad weather conditions are most at risk.
- Young people on outdoor expeditions are often at risk if they do not wear the right clothing. Wet weather and wind make you lose heat faster.
- The first signs of hypothermia are extreme tiredness and not wanting to move – you may not realise how cold you are.
- Up to 20% of your body heat is lost through your head.
- Warm clothing, adequate heating, regular food and warm drinks, together with exercise all help to prevent hypothermia.
- People with hypothermia have greyish-blue, puffy faces and blue lips. Their skin feels very cold to the touch. They will be drowsy, with slurred speech. As it gets worse, they will stop shivering. If the body temperature falls too low the sufferer will become unconscious and may die.
- It has been estimated that every time the temperature drops one degree Celsius below average in the winter, 8000 more elderly people will die of hypothermia.

HEAT WAVE KILLS SEVEN

The latest spell of very hot weather has led to seven deaths this week. As Britain sizzles in the latest heat-wave, with temperatures of over 33°C, people are dropping like flies.

Heat stroke and other heat-related illnesses are hitting the elderly, small babies and people with existing heart problems particularly hard.

The World Health Organisation along with the World Meteorological Organisation have suggested that a hot weather warning is added to our weather forecasts along with pollen levels, air pollution and flood warnings.

To reduce your risk of heat stroke as Britain continues to fry, stay in air-conditioned rooms where possible, drink plenty of water and take cool baths.

ACTIVITY

If more people were aware of the risks of hypothermia, fewer people would die from it. Use the information to help you design **either** a poster **or** a leaflet informing people about the dangers of hypothermia and ways to avoid it.

ACTIVITY

Climate change may well result in colder winters and hotter summers. Write an article for the lifestyle pages of a newspaper on:
- how your body copes with changes in temperature,
- the dangers to health of hot summers and cold winters, and
- the best ways to avoid any problems.

THE DIABETES DEBATE

The treatment of diabetes has changed a great deal over the years. For centuries nothing could be done. Then in the early 1920s Frederick Banting and Charles Best realised that extracts of animal pancreas could be used to keep people with diabetes alive. For many years insulin from pigs and cows was used to treat affected people. This saved millions of lives.

In recent years, bacteria have been developed using genetic engineering which produce pure human insulin. This is now injected by the majority of people affected by diabetes.

Scientists are trying to find easier ways – like nasal sprays – to get insulin into the body. Transplanting working pancreas cells from both dead and living donors has been shown to work for some people. And for the future, scientists are hoping to use embryonic stem cells to provide people affected by diabetes with new, functioning pancreas cells which can make their own insulin.

The difference these treatments have made to the lives of people with diabetes and their families is enormous. If a cure is found, it will be even better. But most of these developments have some ethical issues linked to them.

- Banting and Best did their experiments on dogs. They made some of the dogs diabetic by removing most of their pancreas, and they extracted insulin from the pancreases of other dogs. Many dogs died in the search for a successful treatment – but the scientists found a treatment to a disease which has killed millions of people over the centuries.
- Human insulin is now mass-produced using genetically engineered bacteria. The gene for human insulin is stuck into the bacterial DNA and the bacteria make pure human protein.
- There are not enough dead donors to give pancreas transplants to the people who need them. However, in living donor transplants there is a risk to the health of the donor as they have to undergo surgery.
- Stem cell research promises a possible cure – but the stem cells come from human embryos which have been specially created for the process.

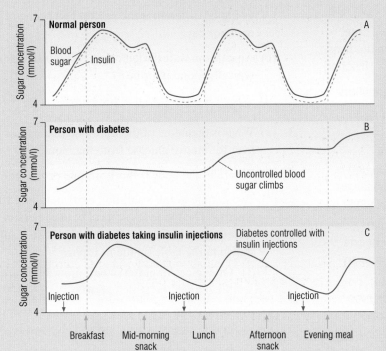

ACTIVITY

a) You are to plan a three-minute speech for a debate. The title of the debate is:

  'Ethical concerns are less important than a cure for diabetes.'

You can argue for or against the motion, but your arguments must be clear and sensible and backed up by scientific evidence.

b) Work in groups of 9 and set up a role play involving the following characters:

- Frederick Banting who first showed that animal insulin could be used to treat humans with diabetes.
- A spokesperson from a pharmaceutical company manufacturing human insulin.
- A daughter who has been cured of diabetes by receiving pancreas tissue from her mother, and her mother who donated the tissue.
- A scientist working on the development of insulin-producing cells from embryonic stem cells.
- Someone who has had diabetes since they were 10 years old.
- An animal rights activist.
- A 'pro-life' activist who is against any use of stem cells.
- A representative of a group opposed to genetic engineering.
- The chair of the discussion.

Each character must explain to the chair why research into diabetes should – or should not – continue.

Figure 1 These graphs show the impact insulin injections have on people affected by diabetes. The injections keep the blood sugar level within safe limits. They cannot mimic the total control given by the natural production of the pancreas – but they work well enough to let people lead a full and active life.

SUMMARY QUESTIONS

1 a) Draw and annotate a diagram explaining the basic principles of homeostasis.

   b) Write a paragraph explaining why control of the conditions inside your body is so important.

2 We humans maintain our body temperature at a constant level over a wide range of environmental temperatures. Many other animals – fish, amphibians and reptiles as well as the invertebrates – cannot do this. Their body temperature is always very close to the environmental temperature.

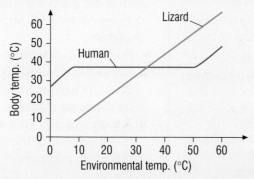

   a) What is the body temperature of a person and a lizard at an atmospheric temperature of 20°C?

   b) From the graph, at what external temperature does the human core temperature become dangerously low? Why is it dangerous?

   c) At what external temperature does the human core temperature become dangerously high? Why is it dangerous?

   d) Explain how a person maintains a constant core body temperature as the external temperature falls.

   e) Explain how a person maintains a constant core body temperature as the external temperature rises.
   [Higher]

3 Use Figure 1 on page 199 to answer this question.

   a) Look at graph A. Why does the level of insulin increase after a meal?

   b) Graph B shows the blood sugar pattern of someone who has just developed diabetes and is not yet using injected insulin. What differences are there between this pattern and the one shown in A?

   c) Graph C shows the effect of regular insulin injections on the blood sugar level of someone with diabetes. Why are the insulin injections so important to their health?

   d) People who are mildly diabetic and those who inject insulin all have to watch the amount of carbohydrate in their diet. Explain why.

EXAM-STYLE QUESTIONS

1 Complete the passage below by choosing the correct terms from the box and matching them with the numbers in the passage.

sweating	dilate	shivering
thermoregulatory	radiation	constrict

Body temperature is controlled by the ....**1**.... centre in the brain. On a hot day it causes blood vessels in the skin to ...**2**... and so lose heat by ....**3**.... Heat may also be lost by ...**4**... . On a cold day the blood vessels ....**5**.... to conserve heat. When cold, ...**6**.... may also occur to create some heat. [Higher] (6)

2 The table shows the daily water loss from a typical human being.

Water lost in	Volume of water (cm³ per day)
Urine	1500
X	400
Evaporation from the skin	350
Faeces	150
Sweat	100

   (a) One way in which water is lost from the body has been missed out and replaced by the letter **X**. What does **X** represent? (1)

   (b) These figures were taken on a cool day with the person at rest. State two ways in which the figures would be different if the person had been exercising on a hot day. (2)

   (c) Apart from water, what other two substances are typically found in urine? (2)

   (d) Where is urine stored in the body? (1)

3 (a) What is the name of the hormone that causes the liver to remove glucose from the blood? (1)

   (b) Where in the body is this hormone produced? (1)

   (c) Two people drank a solution that contained 100 g of glucose. The blood sugar level of each person was measured over the next three hours. The results are shown in the table on the next page.

      (i) On a piece of graph paper, draw a line graph of the data in the table opposite. (5)

      (ii) One of the two persons is diabetic. From the graph suggest which one and give two reasons for your answer. (2)

Time in minutes	Blood sugar level (mg/100 cm³ blood)	
	Person X	Person Y
0 (glucose drunk)	90	90
30	160	140
60	220	90
90	200	80
120	150	70
150	130	80
180	110	90

4 Read the following passage about diabetes.

Diabetes is a metabolic disorder in which there is an inability to control blood glucose levels due to the lack of the hormone insulin. Diabetes was a fatal disease until in 1921 Banting and Best succeeded in isolating insulin from the pancreases of pigs and cows, having first carried out experiments on dogs. Insulin is a small protein of 51 amino acids, the sequence of which was determined in the 1950s by Sanger. More recently the gene for human insulin has been isolated and the hormone can now be produced by bacteria as a result of genetic engineering. Diabetics must test their blood sugar levels regularly and inject insulin if they are to lead normal lives.

(a) Why do diabetics inject insulin rather than taking it by mouth? (2)

(b) What would happen to the blood sugar level of a diabetic who failed to inject insulin? (1)

(c) Suggest one other symptom of diabetes other than changes to blood sugar. (1)

(d) Give three advantages of using genetically engineered insulin rather than extracting the hormone from animal pancreases. (3)

(e) Injecting insulin only *treats* diabetes. In future it may be possible to replace the damaged pancreas by transplantation.
  (i) What would be the benefits to the person with diabetes of such treatment?
  (ii) State the drawbacks of this treatment. (4)

HOW SCIENCE WORKS QUESTIONS

You have probably heard the weatherman, during winter, tell you about the 'wind chill factor'. This is to give you a better idea of how cold your skin might feel if you were to go out whilst the temperature was low and it was windy. Remember that wind will cool the skin by evaporating moisture from it and therefore make it feel colder than the actual air temperature.

Until recently the wind chill factor was calculated by measuring temperatures of some water in a container in the Arctic. The tank of water was 10 metres above the ground.

a) Explain why this was a poor way to calculate the effect of wind chill on humans. (2)

Recently some investigations were carried out to get a better measure of the effect of wind chill on humans. The tests were carried out on humans dressed in protective clothing, except their cheeks were left exposed, so that their cheek skin temperature could be measured.

b) Why do you think the cheeks were chosen? (1)

c) The people were tested at different temperatures and wind speeds. What would have been a suitable sensitivity for the thermometer? (1)

d) How many people would have been chosen? (1)

e) How would these people have been chosen? (1)

f) Imagine you were carrying out these tests. Draw up a table that would let you fill in the results as you did the tests on just one person. (2)

g) Now fill in the table with some temperatures and some wind speeds that you think might be useful. (5)

B2 6.1   Cell division and growth

1 What is mitosis?
2 Why do plants grow throughout their lives while most animals stop growing once they are adults?

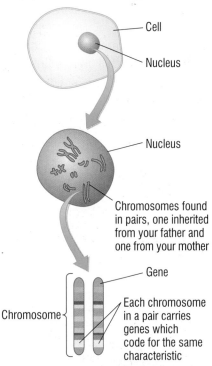

Figure 1 The nucleus of your cell contains the chromosomes that carry the genes which control the characteristics of your whole body

Cell
Nucleus

Nucleus

Chromosomes found in pairs, one inherited from your father and one from your mother

Gene

Chromosome

Each chromosome in a pair carries genes which code for the same characteristic

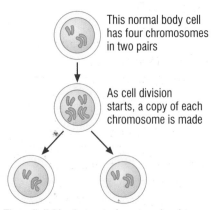

This normal body cell has four chromosomes in two pairs

As cell division starts, a copy of each chromosome is made

The cell divides in two to form two daughter cells. Each daughter cell has a nucleus containing four chromosomes identical to the ones in the original parent cell.

New cells are needed for an organism, or part of an organism, to grow. They are also needed to replace cells which become worn out and repair damaged tissue. However the new cells must have the same genetic information in them as the originals, so they can do the same job.

Each of your cells has a nucleus containing the instructions for making whole new cells and even an entire new you! These instructions are carried in the form of genes.

A gene is a small packet of information which controls a characteristic, or part of a characteristic, of your body. The genes are grouped together on chromosomes. A chromosome may carry several hundred or even thousands of genes.

You have 46 chromosomes in the nucleus of your cells (except your gametes – sperm or ova). They come in 23 pairs. One of each pair is inherited from your father, and one from your mother.

a) Why are new cells needed?

Mitosis

Body cells divide to make new cells. The cell division which takes place in the normal body cells and produces identical daughter cells is called **mitosis**. As a result of mitosis all your body cells have the same genetic information.

In asexual reproduction, the cells of the offspring are produced by mitosis from cells of their parent. This is why they contain exactly the same genes with no variety.

How does mitosis work? Before a cell divides it produces new copies of the chromosomes in the nucleus. This means that when division takes place two genetically identical **daughter cells** are formed.

In some areas of the body of an animal or plant, cell division like this carries on rapidly all of the time. Your skin is a good example – cells are constantly being lost from the surface and new cells are constantly being formed by cell division to replace them.

b) What is mitosis?

Differentiation

In the early development of animal and plant embryos the cells are very unspecialised. Each one of them (known as **stem cells**) can become any type of cell which is needed.

In many animals, the cells become specialised very early in life. By the time a human baby is born most of its cells have become specialised for a particular job, such as liver cells, skin cells and muscle cells. They have **differentiated**. Some of their genes have been switched on and others have been switched off.

Figure 2 Identical daughter cells are formed by the simple division that takes place during mitosis. It supplies all the new cells needed in your body for growth, replacement and repair. Your cells really have 23 pairs of chromosomes – but for simplicity this cell is shown with only two pairs!

This means that when a muscle cell divides by mitosis it can only form more muscle cells. Liver cells can only produce more liver cells. So in adult animals, cell division is restricted because differentiation has occurred. Specialised cells can divide by mitosis, but this can only be used to repair damaged tissue and replace worn out cells. Each cell can only produce identical copies of itself.

In contrast, most plant cells can differentiate all through their life. Undifferentiated cells are formed at active regions of the stems and roots. In these areas mitosis takes place almost continuously.

Plants keep growing all through their lives at these 'growing points'. The plant cells produced don't differentiate until they are in their final position in the plant. What's more, the differentiation isn't permanent. If you move a plant cell from one part of a plant to another, it can re-differentiate and become a completely different type of cell. You just can't do that with animal cells – once a muscle cell, always a muscle cell!

PRACTICAL

Observing mitosis

Make a special preparation of a growing root tip to view under a microscope. Then you can see the actively dividing cells and the different stages of mitosis as it is taking place.

● Describe your observations of mitosis.

We can produce huge numbers of identical plant clones from a tiny piece of leaf tissue. Now you can see why this is possible. In the right conditions a plant cell will become unspecialised and undergo mitosis many times. In different conditions, each of these undifferentiated cells will produce more cells by mitosis. These will then differentiate to form a tiny new plant identical to the original parent.

The reason animal clones cannot be made easily is because animal cells differentiate permanently early in embryo development – and can't change back! Animal clones can only be made by cloning embryos in one way or another.

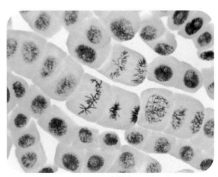

Figure 3 The undifferentiated cells in this onion root tip are dividing rapidly. You can see mitosis taking place, with the chromosomes in different positions as the cells divide.

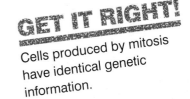

GET IT RIGHT!

Cells produced by mitosis have identical genetic information.

SUMMARY QUESTIONS

1 Copy and complete using the words below:

    chromosomes   genetic information   genes   growth
    mitosis   nucleus   replace

   New cells are needed for …… and to …… worn out cells. The new cells must have the same …… …… in them as the originals. Each cell has a …… containing the …… grouped together on ……. . The type of cell division which produces identical cells is known as …… .

2 Division of the body cells is taking place all the time in living organisms.

   a) Why is it so important?

   b) Explain why the chromosome number must stay the same when the cells divide to make other normal body cells.

3 The process of growth and differentiation is very different in plants and animals.

   a) What is differentiation?

   b) How is differentiation in animal and plant cells so different?

   c) How does this difference affect the cloning of plants and animals?

KEY POINTS

1 In body cells, chromosomes are found in pairs.

2 Body cells divide by mitosis to produce more identical cells for growth, repair, replacement or in some cases asexual reproduction.

3 Most types of animal cells differentiate at an early stage of development. Many plant cells can differentiate throughout their life.

B2 6.2   Stem cells

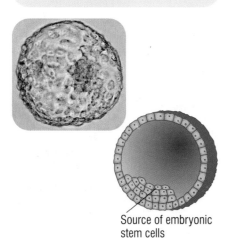

Source of embryonic stem cells

Figure 1 This ball of cells is an early human embryo. In the right conditions these few cells can form all the organs of the human body.

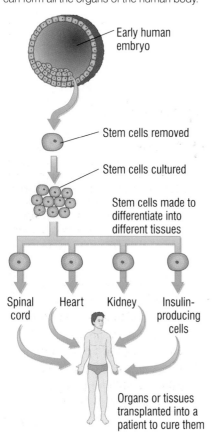

Early human embryo

Stem cells removed

Stem cells cultured

Stem cells made to differentiate into different tissues

Spinal cord   Heart   Kidney   Insulin-producing cells

Organs or tissues transplanted into a patient to cure them

Most of the cells in your body are differentiated. They are specialised and carry out particular jobs. But some of your most important cells are the completely unspecialised **stem cells**. They can differentiate (divide and change) into many different types of cell when they are needed. Human stem cells are found in human embryos and in some adult tissue including bone marrow.

The function of stem cells

Stem cells divide and form the specialised cells of your body which make up your various tissues and organs. When an egg and sperm fuse to form an embryo, they form a single new cell. That cell divides and the embryo is soon a hollow ball of cells. The inner cells of this ball are the stem cells which will eventually give rise to every type of cell in your body.

Even when you are an adult some of your more specialised stem cells remain. Your bone marrow is a good source of stem cells. What's more, scientists now think there may be a tiny number of stem cells in most of the different tissues in your body. This includes your blood, brain, muscle and liver.

The stem cells can stay there for many years until your tissues are injured or affected by disease. Then they start dividing to replace the different types of damaged cells.

a) What are stem cells?

Using stem cells

Many people suffer and even die because various parts of their body stop working properly. For example, spinal injuries can cause paralysis. That's because the spinal nerves do not repair themselves. Millions of people would benefit if we could replace damaged body parts.

In 1998, there was a breakthrough. Two American scientists managed to culture human embryonic stem cells that were capable of forming other types of cells.

Scientists hope that these embryonic stem cells can be encouraged to grow into almost any different type of cell needed in the body. For example, we may be able to grow new nerve cells. If new nerves grown from stem cells could be used to reconnect the spinal nerves, people who have been paralysed could walk again.

With stem cells we might also be able to grow whole new organs which could be used in transplant surgery. These new organs would not be rejected by the body. Conditions from infertility to dementia could eventually be treated using stem cells.

Unfortunately, at the moment no-one is quite sure just how the cells in an embryo are switched on or off. We don't yet know how to form particular types of tissue. Once we know how to do this, we can really start to use stem cells effectively.

b) What was the big scientific breakthrough by American scientists in 1998?

Figure 2 Some of the embryonic stem cells which scientists have produced and grown have formed into adult cells. Unfortunately no-one is quite sure how to control this process at the moment. Hopefully one day the technique shown in this diagram will be used to treat people.

Problems with stem cells

Many embryonic stem cells come from aborted embryos or from spare embryos in fertility treatment. This raises ethical problems. There are people, including many religious groups, who feel it is wrong to use a potential human being as a source of cells, even to cure others.

Some people feel that as the embryo cannot give permission, using it is a violation of its human rights. On top of this, progress with stem cells is slow. There is some concern that embryonic stem cells might cause cancer if they are used to treat sick people. This has certainly been seen in mice. Making stem cells is slow, difficult, expensive and hard to control.

c) What is the biggest ethical concern with the use of embryonic stem cells?

The future of stem cell research

We have found embryonic stem cells in the umbilical cord blood of newborn babies. These may help to overcome some of the ethical concerns.

Scientists are also finding ways of growing adult stem cells. Unfortunately the adult stem cells found so far can only develop into a limited range of cell types. However this is another possible way of avoiding the controversial use of embryonic tissue.

The area of stem cell research known as **_therapeutic cloning_** could be very useful – but it is proving very difficult.

Therapeutic cloning involves using cells from an adult person to produce a cloned early embryo of themselves as a source of perfectly matched embryonic stem cells. In theory these could then be used to heal the original donor and maybe many others as well.

Most people remain excited by the possibilities of embryonic stem cells in treating many diseases. Just how many of these early hopes will be fulfilled only time will tell!

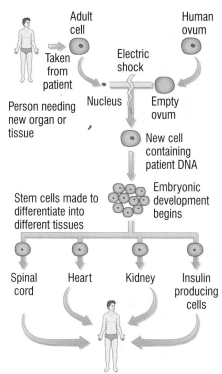

Organs or tissues transplanted into the patient with no risk of rejection

Figure 3 In 2005, a team led by Professor Woo Suk Hwang in South Korea claimed to have produced human embryos from adult cells and developed cloned stem cells from them. This seemed a huge step forward in stem cell research. But sadly, in 2006 the work was shown to be a massive scientific fraud. This was a massive blow to everyone working in stem cell research.

SUMMARY QUESTIONS

1 Copy and complete using the words below:

> **bone marrow   differentiate   embryos   hollow**
> **inner   stem cells**

Unspecialised cells known as …… …… can …… (divide and change) into many different types of cell when they are needed. Human stem cells are found in human …… and in adult …… …… . The embryo forms a …… ball of cells and the …… cells of this ball are the stem cells.

2 a) Why was the work of the American scientists in 1998 such a breakthrough in stem cell research?

b) How might stem cells be used to treat patients who are paralysed after a spinal injury?

3 a) What are the advantages of using stem cells to treat a wide range of diseases?

b) What are the difficulties with stem cell research?

c) How are scientists hoping to overcome the difficulties of using embryonic stem cells in their research?

GET IT RIGHT!

Make sure you refer to both pros and cons when you are giving information about the possible use of stem cells.

KEY POINT

1 Embryonic stem cells (from human embryos) and adult stem cells (from adult bone marrow) can be made to differentiate into many different types of cells.

B2 6.3 Cell division in sexual reproduction

GET IT RIGHT!

Be careful with the spelling of mitosis and meiosis. Make sure you know the differences between the two processes.

Mitosis is taking place all the time, in tissues all over your body. But mitosis is not the only type of cell division. There is another type which takes place only in the reproductive organs of animals and plants. **Meiosis** results in sex cells with only half the original number of chromosomes.

Meiosis

The reproductive organs in people, like most animals, are the **ovaries** and the **testes**. This is where the sex cells (the gametes) are made. The female gametes or **ova** are made in the ovaries. The male gametes or **sperm** are made in the testes.

The gametes are formed by meiosis, which is a special form of cell division where the chromosome number is reduced by half. When a cell divides to form gametes, the first stage is very similar to normal body cell division. The chromosomes are copied so there are four sets of chromosomes. The cell then divides twice in quick succession to form four gametes, each with a single set of chromosomes.

Why is meiosis so important?

Your normal body cells have 46 chromosomes in two matching sets – 23 come from your mother and 23 from your father. If two 'normal' body cells joined together in sexual reproduction, the new cell would have 92 chromosomes, which simply wouldn't work!

Fortunately, as a result of meiosis, your sex cells contain only one set of chromosomes, exactly half of the full chromosome number. So when the gametes join together at fertilisation, the new cell formed contains the right number of 46 chromosomes.

a) What are the names of the male and female gametes and how do they differ from normal body cells?

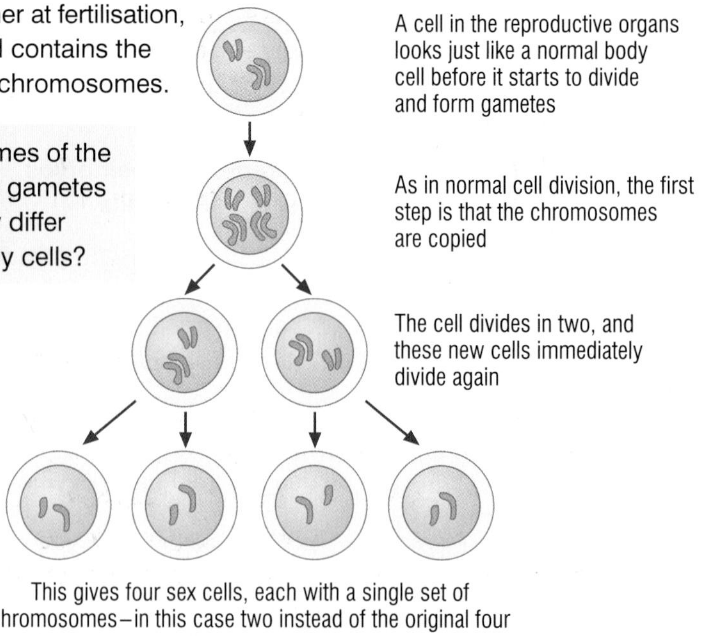

A cell in the reproductive organs looks just like a normal body cell before it starts to divide and form gametes

As in normal cell division, the first step is that the chromosomes are copied

The cell divides in two, and these new cells immediately divide again

This gives four sex cells, each with a single set of chromosomes – in this case two instead of the original four

Figure 1 The formation of sex cells in the ovaries and testes involves a special kind of cell division to halve the chromosome number. The original cell is shown with only two pairs of chromosomes to make it easier to follow what is happening.

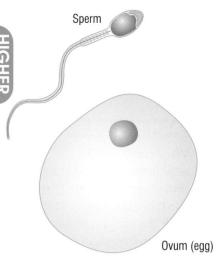

In girls, the first stage of meiosis is completed before they are even born. The tiny ovaries of a baby girl contain all the ova she will ever have.

In boys, meiosis doesn't start until puberty when the testes start to produce sperm. It then carries on for the rest of their lives.

Each gamete you produce is slightly different from all the others. The combination of chromosomes will be different. What's more, there is some exchange of genes between the chromosomes during the process of meiosis. This means that no two eggs or sperm are the same. This introduces lots of variety into the genetic mix of the offspring.

b) What type of cell division is needed to produce the gametes?

HIGHER

Sperm

Ovum (egg)

Figure 2 Once meiosis has taken place, the male and female gametes develop very differently – they are adapted for very different jobs

Fertilisation

More variety is added when fertilisation takes place. Each sex cell has a single set of chromosomes. When two sex cells join during fertilisation the new cell formed has a full set of chromosomes. In humans, the egg cell has 23 chromosomes and so does the sperm. When they join together they produce a new normal cell with the full human complement of 46 chromosomes.

The combination of genes on the chromosomes of every newly fertilised ovum is completely unique. Once fertilisation is complete, the unique new cell begins to divide by mitosis. This will continue long after the fetus is fully developed and the baby is born.

Variation

The differences between asexual and sexual reproduction are a reflection of the different types of cell division involved in the two processes.

In asexual reproduction the offspring are produced as a result of mitosis from the parent cells. (See the start of this chapter.) So they contain exactly the same chromosomes and the same genes as their parents. There is no variation in the genetic material.

In sexual reproduction the gametes are produced by meiosis in the sex organs of the parents. This introduces variety as each gamete is different. Then when the gametes fuse, one of each pair of chromosomes, and so one of each pair of genes, comes from each parent.

The combination of genes in the new pair will contain **alleles** (different forms of the gene) from each parent. This also helps to produce different characteristics in the offspring.

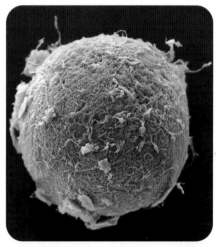

Figure 3 At the moment of fertilisation the chromosomes in the two gametes are combined so the new cell has a complete set, like any other body cell. This cell will then grow and reproduce by mitosis to form a new individual.

SUMMARY QUESTIONS

1 a) How many pairs of chromosomes are there in a normal human body cell?
   b) How many chromosomes are there in a human egg cell?
   c) How many chromosomes are there in a fertilised human egg cell?

2 Sexual reproduction results in variety. Explain how.

3 a) What is the name of the special type of cell division which produces gametes from ordinary body cells? Describe what happens to the chromosomes in this process.
   b) Where in your body would this type of cell division take place?
   c) Why is this type of cell division so important in sexual reproduction? [Higher]

KEY POINTS

1 Cells in the reproductive organs divide to form the gametes (sex cells).
2 Body cells have two sets of chromosomes; gametes have only one set.
3 Gametes are formed from body cells by meiosis. [Higher]
4 Sexual reproduction gives rise to variety because genetic information from two parents is combined.

B2 6.4

From Mendel to DNA

GET IT RIGHT!

Mendel knew nothing of chromosomes and genes. Make sure you don't confuse modern knowledge with what Mendel knew when he did his experiments.

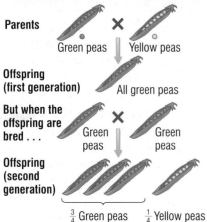

Parents — Green peas × Yellow peas

Offspring (first generation) — All green peas

But when the offspring are bred... Green peas × Green peas

Offspring (second generation)

$\frac{3}{4}$ Green peas $\frac{1}{4}$ Yellow peas

Figure 1 Gregor Mendel, the father of modern genetics. When he died in 1884 he was still hoping that eventually other people would acknowledge his discoveries. In the 21st century, we know just how right he was!

For hundreds of years people had no idea about how information moved from one generation to the next. Yet now we can identify people by the genetic information in their cells!

Mendel's discoveries

Gregor Mendel was born in 1822 in Brunn, Czechoslovakia. Clever but poor, he became a monk to get an education.

He worked in the monastery gardens and became fascinated by the peas growing there. He decided to carry out some breeding experiments, using pure strains of round peas, wrinkled peas, green peas and yellow peas for his work. Mendel cross-bred the peas and counted the different offspring carefully. He found that characteristics were inherited in clear and predictable patterns.

Mendel explained his results by suggesting there were separate units of inherited material. He realised some characteristics were dominant over others and that they never mixed together. This was an amazing idea for the time.

a) Why did Gregor Mendel become a monk?

Mendel kept records of everything he did, and analysed his results. This was almost unheard of in those days! Finally in 1866, when he was 44 years old, Mendel published his findings.

He never saw chromosomes and never heard of genes. Yet he explained some of the basic laws of genetics in a way we still use today.

Sadly Mendel's genius was ahead of his time. As no-one knew about genes or chromosomes, people simply didn't understand his theories. He died twenty years later with his ideas still ignored – but convinced that he was right!

b) What was unusual about Mendel's scientific technique at the time?

Sixteen years after his death, Gregor Mendel's work was finally recognised. By 1900, people had seen chromosomes through a microscope. Three scientists, discovered Mendel's papers and repeated his experiments. When they published their results, they gave Mendel the credit for what they observed! From then on ideas about genetics developed fast. It was suggested that Mendel's units of inheritance might be carried on the chromosomes seen beneath the microscope. And so the science of genetics as we know it today was born.

DNA – the molecule of inheritance

The work of Gregor Mendel was just the start of our understanding of inheritance. Today, we know that our features are inherited on genes carried on our chromosomes. We also know what those chromosomes are made of.

Your chromosomes are made up of long molecules of a chemical known as DNA (**d**eoxyribose **n**ucleic **a**cid). Your genes are small sections of this DNA. The DNA carries the instructions to make the proteins which form most of your cell structures. These proteins also include the enzymes which control your cell chemistry.

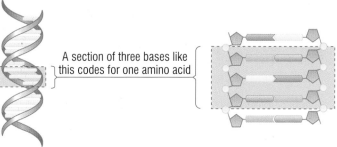

A section of three bases like this codes for one amino acid

Figure 2 It is at this fundamental level of chemistry that your characteristics are determined. A small quirk of chemistry would have resulted in a very different you – a very strange thought.

The long strands of your DNA are made up of combinations of four different chemical bases. (See Figure 2.) These are grouped into threes and each group of three codes form an amino acid.

Each gene is made up of hundreds or thousands of these bases. The order of the bases controls the order in which the amino acids are put together so that they make a particular protein for use in your body cells. Each gene codes for a particular combination of amino acids which make a specific protein.

A change or mutation in a single group of bases can be enough to change or disrupt the whole protein structure and the way it works.

DNA fingerprinting

Unless you have an identical twin, your DNA is unique to you. Other members of your family will have strong similarities in their DNA, but each individual has their own unique blueprint. Only identical twins have the same DNA. That's because they have both developed from the same original cell.

The unique patterns in your DNA can be used to identify you. A technique known as 'DNA fingerprinting' can be applied.

Certain areas of your DNA produce very variable patterns under the microscope. These patterns are more similar between people who are related than between total strangers. The patterns are known as **DNA fingerprints**. They can be produced from very tiny samples of DNA from body fluids such as blood, saliva and semen.

The likelihood of two identical samples coming from different people (apart from identical twins) is millions to one. As a result DNA fingerprinting is enormously useful in solving crimes. It is also used to show who is the biological father of a child when there is doubt.

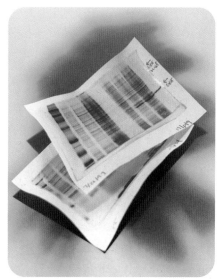

Figure 3 DNA fingerprints like these can be used to identify the guilty – and the innocent – in a crime investigation

SUMMARY QUESTIONS

1. a) How did Mendel's experiments with peas convince him that there were distinct 'units of inheritance' which were not blended together in offspring?
   b) Why didn't people accept his ideas?
   c) The development of the microscope played an important part in helping to convince people that Mendel was right. How?

2. Two men claim to be the father of the same child. Explain how DNA fingerprinting could be used to find out which one is the real father.

3. Explain the saying 'One gene, one protein'. [Higher]

KEY POINTS

1. Gregor Mendel was the first person to suggest separately inherited factors which we now call genes.
2. Chromosomes are made up of large molecules of DNA.
3. A gene is a small section of DNA which codes for a particular combination of amino acids which make a specific protein. [Higher]
4. Everyone (except identical twins) has unique DNA which can be used to identify them using DNA fingerprinting.

B2 6.5 Inheritance in action

1 How is sex determined in humans?
2 Can you predict what features a child might inherit? [Higher]

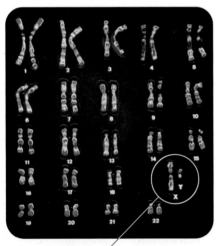

Sex chromosomes

Figure 1 The chromosomes of the human male. The X chromosome carries genes controlling lots of different features. The Y chromosome is much smaller than the X chromosome and carries information mainly about maleness!

Ideas about genetics, chromosomes and genes are everywhere in the 21st century. We read about them in the papers, see them on TV and learn about them in science lessons. The way features are passed from one generation to another follow some clear patterns. We can use these to predict what may be passed on.

How inheritance works

Scientiest have built on the work of Gregor Mendel. We now understand how genetic information is passed from parent to offspring.

Human beings have 23 pairs of chromosomes. In 22 cases, each chromosome in the pair is a similar shape and has genes carrying information about the same things. But one pair of chromosomes may be different – these are the **sex chromosomes**. Two X chromosomes mean you are female. However, one X chromosome and a much smaller one, known as the Y chromosome, give a male.

a) Twins are born. Twin A is XY and twin B is XX. What sex are the two babies?

The chromosomes we inherit carry our genetic information in the form of genes. Many of these genes have different forms, known as alleles. (See page 207.) A gene can be pictured as a position on a chromosome. An allele is the particular form of information in that position on an individual chromosome. For example, the gene for dimples may have the dimple or the no-dimple allele in place.

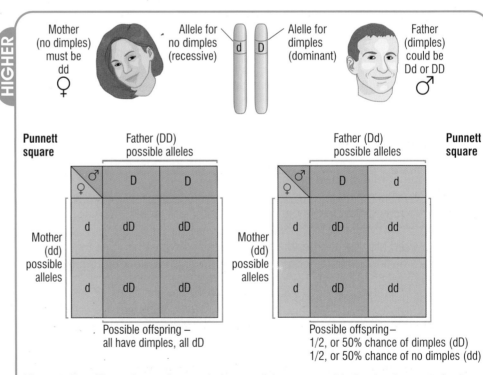

Figure 2 The different forms of genes, known as alleles, can result in the development of quite different characteristics. We can use diagrams like this Punnett square to explain what is happening or predict what the offspring might be like.

Most of your characteristics, like your eye colour and nose shape, are controlled by a number of genes. However, some characteristics, like dimples or having attached earlobes, are controlled by a single gene. Often there are only two possible alleles for a particular feature. However, sometimes you can inherit one from a number of different possibilities.

Some alleles control the development of a characteristic even when they are only present on one of your chromosomes. These alleles are **dominant**, e.g. dimples and dangly earlobes.

Some alleles only control the development of a characteristic if they are present on both alleles – in other words, no dominant allele is present. These alleles are **recessive**, e.g. no dimples and attached earlobes.

HIGHER

How does inheritance work?

We can use a simple model to help us understand how inheritance works. It explains how different features are passed on from one generation to another.

Imagine a bag containing marbles. If you put your hand in and – without looking – picked out two marbles at a time, what pairs might you get? If the bag contained only red marbles or only blue marbles, the pairs would all be the same. But if the bag held a mixture of red and blue marbles you could end up with three possible pairs – two blue marbles, two red marbles or one of each.

This is what happens when you inherit genes from your parents, depending on the different alleles they have. For example, if both of your parents have two alleles for dimples (like the red marbles) you will definitely inherit two dimple alleles – and you will have dimples! If both of your parents have two alleles for no dimples, you will inherit alleles for no dimples and you will be dimple free.

But if your parents both have one allele for dimples and one for no dimples, you could end up with two dimple alleles, two no dimple alleles – or one of each.

SUMMARY QUESTIONS

1 Copy and complete:

male   sex chromosomes   23   22   X   XX   Y

Human beings have ...... pairs of chromosomes. In ...... pairs the chromosomes are always the same. The final pair are known as .... If you inherit ...... you will be female, while an ...... and a ...... make you ....

2 a) What is meant by the term 'dominant allele'?
b) What is meant by the term 'recessive allele'?
c) Try and discover as many human characteristics as you can which are inherited on a single gene. Which alleles are dominant and which are recessive?

3 Use a Punnett square like the one in Figure 2 to show the possible offspring from a cross between two people who both have dimples and the genotype Dd. [Higher]

KEY POINTS

1 In human body cells the sex chromosomes determine whether you are female (XX) or male (XY).
2 Some features are controlled by a single gene.
3 Genes can have different forms called alleles.
4 Some alleles are dominant and some are recessive.
5 We can construct genetic diagrams to predict features. [Higher]

B2 6.6 Inherited conditions in humans

LEARNING OBJECTIVES

1 How are human genetic disorders inherited?
2 Can you predict if a child will inherit a genetic disorder? [Higher]

Not all diseases are infectious. Sometimes diseases are the result of a problem in your genes and can be passed on from parent to child. They are known as **genetic diseases** or **genetic disorders**.

We can use our knowledge of dominant and recessive alleles to work out the risk of inheriting a genetic disease.

a) How is a genetic disease different from an infectious disease?

Huntington's disease

One example of a very serious, although very rare, genetic disorder is Huntington's disease. This is a disorder of the nervous system. It is caused by a dominant allele and so it can be inherited from one parent who has the disease. If one of your parents is affected by Huntington's you have a 50% chance of inheriting the disease. That's because half of their gametes will contain the faulty allele.

The symptoms of this inherited disease usually appear when you are between 30 and 50 years old. Sadly, the condition is fatal. Because the disease does not appear until middle-age, many people have already had children and passed on the faulty allele before they realise they are affected.

b) You may inherit Huntington's disease even if only one of your parents is affected. Why?

DID YOU KNOW?

Around 7500 babies, children and young adults are affected by cystic fibrosis in the UK alone. Between 6800 and 8000 people are affected by Huntington's disease.

Cystic fibrosis

Another genetic disease which has been studied in great detail is **cystic fibrosis**. This is a disorder which affects many organs of the body, particularly the lungs and the pancreas.

The organs become clogged up by a very thick sticky mucus which stops them working properly. The reproductive system is affected so most people with cystic fibrosis are infertile.

Treatment for cystic fibrosis includes physiotherapy and antibiotics to help keep the lungs clear of mucus and infections. Enzymes are used to replace the ones the pancreas cannot produce and to thin the mucus.

However, although treatments are getting better all the time, there is still no cure.

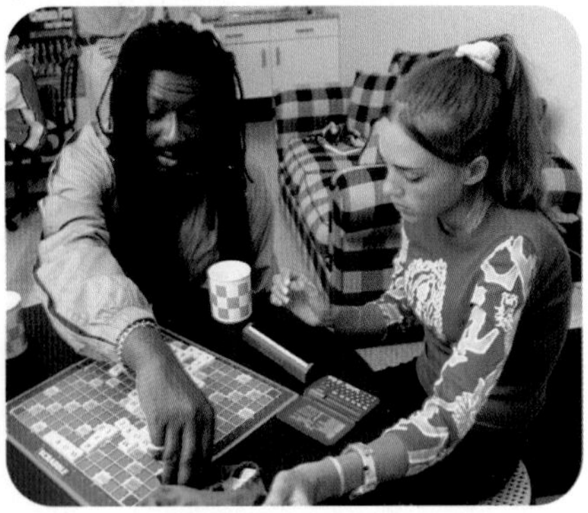

Figure 1 Modern medicine and determination mean that many sufferers from cystic fibrosis manage to lead full and active lives. However, the cells in their bodies are still carrying the faulty alleles and cannot function properly.

Cystic fibrosis is caused by a recessive allele so it must be inherited from both parents. Children affected by cystic fibrosis are born to parents who do not suffer from the disease. They have a dominant healthy allele which means their bodies work normally but they carry the cystic fibrosis allele. Because it gives them no symptoms, they have no idea it is there.

People who have a silent disease-causing allele like this are known as **carriers**. In the UK, one person in 25 carries the cystic fibrosis allele. Most of them will never be aware of it, unless they happen to have children with a partner who also carries the allele. Then there is a 25% (one in four) chance that any child they have will be affected.

c) You will only inherit cystic fibrosis if you get the allele from both parents. Why?

DID YOU KNOW?

Some genetic disorders cause such chaos in the cells that the fetus does not develop properly and miscarries. Yet other genetic disorders – such as colour-blindness – have little effect on people's lives!

The genetic lottery

When the genes from parents are combined, it is called a genetic cross. We can show this using a genetic diagram (see Figures 2 and 3). A genetic diagram shows us:

- the alleles for a characteristic carried by the parents,
- the possible gametes which can be formed from these, and
- how these could combine to form the characteristic in their offspring.

When looking at the possibility of inheriting genetic diseases, it is important to remember that every time an egg and a sperm meet it is down to chance which alleles combine. So if two parents who both carry the cystic fibrosis allele have four children, there is a 25% chance (one in four) that each child might have the disease.

But in fact all four children could have cystic fibrosis, or none of them might be affected. They might all be carriers, or none of them might inherit the faulty alleles at all. It's all down to chance!

Parent with Huntington's disease   Hh
Normal parent   hh

	H	h
h	Hh	hh
h	hH	hh

50% chance Huntington's disease, Hh or hH
50% chance normal, hh

Figure 2 A genetic diagram for Huntington's disease shows us how a dominant allele can affect offspring. It is important to realise that this shows that the chance of passing on the disease allele is 50%, but it cannot tell us which, if any, of the children will actually inherit the allele.

Both parents are carriers, so Cc

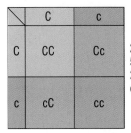

	C	c
C	CC	Cc
c	cC	cc

25% normal (CC)
50% carriers (Cc)
25% affected by cystic fibrosis (cc)

3/4, or 75% chance normal
1/4, or 25% chance cystic fibrosis

Figure 3 The arrival of a child with cystic fibrosis in a family often comes as a complete shock. The faulty alleles can be covered up by normal alleles for generations until two carriers have a child and by chance both of the cystic fibrosis alleles are passed on.

Curing genetic diseases

So far we have no way of curing genetic diseases. Scientists hope that genetic engineering will enable them to cut out faulty alleles and replace them with healthy ones. They have tried this in people affected by cystic fibrosis. But so far they have not managed to cure anyone.

There are genetic tests which can show people in affected families if they carry the faulty allele. This allows them to make choices such as whether to have a family. It is also possible to screen embryos for the alleles which cause these and other genetic disorders. These tests are very useful but raise many ethical issues. (See page 215.)

SUMMARY QUESTIONS

1 a) What is Huntington's disease?
   b) Why can one parent carrying the allele for Huntington's disease pass it on to their children even though the other parent is not affected?

2 At the moment, only people who have genetic diseases in their family are given genetic screening. What would be the pros and cons of screening everyone for diseases like cystic fibrosis and Huntington's disease?

3 a) Why are carriers of cystic fibrosis not affected by the disease themselves?
   b) A couple have a baby who has cystic fibrosis. Neither of the couple, nor their parents, have any signs of the disease.
   Draw genetic diagrams of the grandparents and the parents to show how this could happen. [Higher]

GET IT RIGHT!

If one parent has a characteristic caused by a single dominant allele (e.g. Huntington's disease, dangly earlobes) you have a 50% chance of inheriting it.
If one parent has two dominant alleles (e.g. for Huntington's disease, dangly earlobes) you have 100% chance of inheriting it.
If both parents have a recessive allele for a characteristic (e.g. cystic fibrosis, attached earlobes) you have a 25% chance of inheriting that characteristic.

KEY POINTS

1 Some disorders are inherited.
2 Huntington's disease is caused by a dominant allele of a gene and can be inherited from only one parent.
3 Cystic fibrosis is caused by a recessive allele of a gene and so must be inherited from both parents.

B2 6.7 Stem cells and embryos – an ethical minefield

The stem cell dilemma

Doctors have treated people with adult stem cells for many years by giving bone marrow transplants. Now scientists are moving ever closer to treating very ill people using embryonic stem cells. This area of medicine raises many issues. Here are just a few different opinions:

I think it is absolutely wrong to use human embryos in this way. Each life is precious to God. They may only be tiny balls of cells – but they could become people.

The accident happened so quickly. Now I'm stuck in this wheelchair for the rest of my life. I can't walk or even control when I go to the loo. It would be wonderful if they could develop cell stem therapy. I want them to heal my spinal cord so I can walk again!

It may become possible to take stem cells from the umbilical cord of every newborn baby. They could be frozen and stored ready for when the person might need them later in their life.

The embryos we use would all be destroyed anyway. Now we are even making our own embryos from adult cells. We could do so much good for people that we all feel it is very important for the research to continue.

It was terrible to see my husband suffer. By the time he died he didn't know who I was or any of the children. If these stem cells can cure Alzheimer's disease then we should do the research as fast as possible.

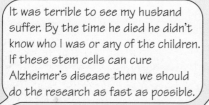

We need to be careful. There are some real problems with these stem cell treatments. We don't want to solve one problem and cause another.

I am going to volunteer to let them use some of my cells for therapeutic cloning. It is too late to help me now, but I'd like to think I could help other people.

ACTIVITY

Here is an opportunity to make your voice heard. Your class is going to produce a large wall display covered with articles both for and against stem cell research. Your display is aimed at students in Years 10 and 11, so make sure the level of content is right for your target group.

Try and carry out a survey or vote of your target group before the display is put up. Find out:

- how many people support stem cell research,
- what proportion are completely against it, and
- how many haven't made up their minds.

Record your findings

Work on your own or in a small group. Each group is to produce one piece of display material. Make sure that some of you give information in favour of stem cell research and others against. Use a variety of resources to help you – the material in this chapter is a good starting point. Make sure that your ideas are backed up with as much scientific evidence as possible.

Once the material has been displayed for a week or two, repeat your initial survey or vote. Analyse the data to see if easy access to information has changed people's views!

Can we know too much?

Today we not only understand the causes of many genetic disorders, we can also test for them. But being able to test for a genetic disorder doesn't necessarily mean we should always do it.

- People in families affected by Huntington's disease can take a genetic test which tells them if they have inherited the faulty gene. If they have, they know that they will develop the fatal disease as they get older and may pass on the gene to their children. Some people in affected families take the test and use it to help them decide whether to marry or have a family. Others prefer not to know.

- If a couple have a genetic disease in their family or already have a child with a genetic disorder, they can have a developing embryo tested during pregnancy. Cells from the embryo are checked. If it is affected, the parents have a choice. They may decide to keep the baby, knowing that it will have a genetic disorder when it is born. On the other hand, they may decide to have an abortion. This prevents the birth of a child with serious problems and allows them to try again to have a healthy baby.

- Some couples who have a genetic disease in the family or who already have a child affected by a genetic disease have their embryos screened before they are implanted in the mother. Embryos are produced by IVF (*in vitro* fertilisation). Doctors remove a single cell from each embryo and screen it for genetic diseases. Only healthy embryos free from genetic disease are implanted back into their mother. Using this method, only healthy babies are born.

ACTIVITY

Many couples who have a genetic disease in the family spend time with a genetic counsellor to help them understand what is happening and the choices they have. Plan a role-play of an interview with a genetic counsellor.

Either: Plan the role of the counsellor. Make sure you have all the information you need to talk to a couple who have already got one child with cystic fibrosis who would like to have another child. You need to be able to explain the chances of another child being affected and the choices that are open to them.

Or: Plan the role of a parent who already has one child with cystic fibrosis and who wants to have another child. Work in pairs to give the views of a couple if you like. Think carefully about the factors which will affect your decision such as: Can you cope with another sick child? Are you prepared to have an abortion? Do you have religious views on the matter? What is fairest to the unborn child – and the child you already have? Is it ethical to choose embryos to implant?

SUMMARY QUESTIONS

1. a) What is mitosis?

    b) Explain, using diagrams, what takes place when a cell divides by mitosis.

    c) Mitosis is very important during the development of a baby from a fertilised egg. It is also important all through life. Why?

2. a) What are stem cells?

    b) It is hoped that many different medical problems may be cured using stem cells. Explain how this might work.

    c) There are some ethical issues about the use of embryonic stem cells. Explain the arguments both for and against their use.

3. a) What is meiosis and where does it take place?

    b) Explain, using labelled diagrams, what takes place when a cell divides by meiosis.

    c) Why is meiosis so important?

    [Higher]

4. Hugo de Vries is one of the scientists who made the same discoveries as Mendel several years after his death. Write a letter from Hugo to one of his friends after he has found Mendel's writings. Explain what Mendel did, why no-one took any notice of him and how the situation is so different now for you if you were doing the same sort of experiments.

5. Whether you have a straight thumb or a curved one is decided by a single gene with two alleles. The straight allele **S** is dominant to the curved allele **s**. Use this information to help you answer these questions.

    Josh has straight thumbs but Sami has curved thumbs. They are expecting a baby.

    a) We know exactly what Sami's thumb alleles are. What are they and how do you know?

    b) If the baby has curved thumbs, what does this tell you about Josh's thumb alleles? Fill in a Punnett square to show the genetics of your explanation.

    c) If the baby has straight thumbs, what does this tell us about Josh's thumb alleles? Fill in a Punnett square to show the genetics of your explanation.

    [Higher]

EXAM-STYLE QUESTIONS

1. The diagram below is of stages in sexual reproduction in a mammal.

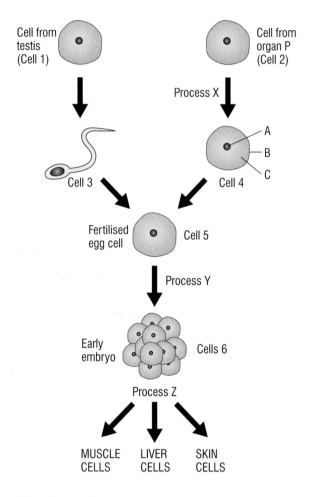

(a) What is the name of organ **P**? (1)

(b) Give the names of parts **A**, **B** and **C** in cell 4. (3)

(c) What is the name of cell 3? (1)

(d) What type of cell division takes place in processes **Y** and **Z**? (1)

(e) Which two of the cells labelled 1–6:

   (i) are genetically identical to one another? (1)

   (ii) are known as gametes? (1)

(f) Cells 6 will in due course change into a range of different cell types.

   (i) What name is given to the type of cell labelled as cells 6? (1)

   (ii) What is the process called by which these cells change into different cell types? (1)

2 Cystic fibrosis is a condition in which people suffer from the accumulation of thick and sticky mucus in their lungs. The chart shows part of a family tree in which some members have cystic fibrosis.

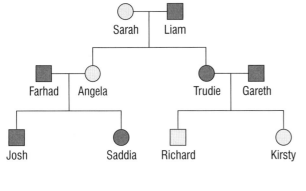

○ Female with cystic fibrosis   ☐ Male with cystic fibrosis

● Unaffected female   ■ Unaffected male

(a) Using two pieces of evidence from the family tree, explain why cystic fibrosis appears to be controlled by a recessive gene. (2)

(b) Trudie and Gareth want to have another child. What is the chance that this child will inherit cystic fibrosis? Explain, with the aid of a genetic diagram, how you reached your answer. (4)

(c) The letters **A**, **B** and **C** show the three different possible combinations of alleles possessed by the members of this family tree

   A   dominant and dominant

   B   dominant and recessive

   C   recessive and recessive

   For each of the individuals below, give the letter that represents the alleles they possess.

   (i)   Liam (1)

   (ii) Angela (1)

   (iii) Saddia (1)

(d) Explain how it is possible that Farhad and Angela could have a child with cystic fibrosis. (3)

[Higher]

HOW SCIENCE WORKS QUESTIONS

Amjid grew some purple flowering pea plants he had bought at the garden centre.

Here are his results.

Seeds planted	247
Purple-flowered plants	242
White-flowered plants	1
Seeds not growing	4

a) Is the white flowered plant an anomaly? (1)

b) Are the seeds that did not grow anomalies? (1)

c) What might Amjid do with the white-flowered plant? (1)

Amjid was interested in these plants, so he collected the seed from some of the purple-flowered plants and used them in the garden the following year. He made a careful note of what happened.

Here are his results:

Seeds planted	406
Purple-flowered plants	295
White-flowered plants	102
Seeds not growing	6

Amjid was slightly surprised. He did expect to find that a third of his flowers would be white.

d) Suggest how Amjid could display his results. (1)

e) Check the accuracy of Amjid's results. How accurate were they? (3)

f) How could Amjid have improved his method of growing the peas to make his results more valid? (1)

EXAMINATION-STYLE QUESTIONS

1   Each autumn, many trees lose their leaves.

See pages 168–71

(a)   Describe how carbon compounds in the leaves can be recycled so that they can be used again by the trees.   *(4 marks)*

To gain full marks in this question you should write your ideas in good English. Put them into a sensible order and use the correct scientific words.

(b)   Give **two** environmental conditions that speed up the processes that you have described in part (a).   *(2 marks)*

2   In an investigation, an enzyme was added to glucose syrup in test tube A. In another test tube (B) glucose was left without the enzyme. In a third test tube (C) the enzyme was left without the glucose. The concentrations of glucose, fructose and the enzyme were measured for thirty minutes. The results for test tube A are shown in the graph.

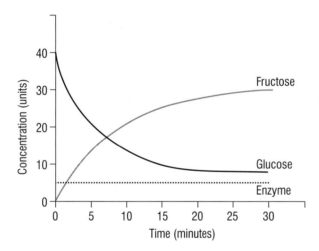

(a)   Describe the changes in the concentration of fructose.   *(2 marks)*

(b)   (i)   Explain why test tubes B and C were used.   *(1 mark)*

(ii)   How should tubes B and C have been treated.   *(1 mark)*

(c)   Fructose is often added to foods used by people on a slimming diet.

(i)   Give **one** advantage of this for the company making the slimming food.

(1 mark)

(ii)   Explain **one** advantage of this for a person on a slimming diet.   *(2 marks)*

See pages 136–9

See pages 186–7

GET IT RIGHT!

When giving a change in any environmental condition, remember to say in which direction the change takes place. Use terms such as 'higher', 'lower', 'more', 'less'.

3 In 1868 a German scientist, Wunderlich, took the mouth temperature of 25,000 people a total of 1 million times. He concluded that the normal range for temperatures recorded from the mouth using a mercury thermometer was 36.25°C to 37.5°C. Mean temperature was 37°C.

See pages 4–19

In 2005 scientists in Baltimore measured the mouth temperature of 148 men and women aged 18–40 years. Measurements were taken 4 times daily for 3 days using an electronic digital thermometer. They found that the normal range was 37.2°C to 37.7°C, with a mean of 36.8°C.

(a) Which control variable was the same for both investigations? *(1 mark)*

(b) How could supporters of Wunderlich argue that he had the most accurate technique. *(1 mark)*

(c) How could supporters of the Baltimore team argue that they had the most accurate technique? *(1 mark)*

(d) Why is it economically important to have an accurate measurement of the normal range for body temperatures? *(1 mark)*

4 Huntington's disease is an inherited condition which is caused by *a dominant allele*. The effects of the disease do not appear until the person with the allele is 30–40 years old.

See pages 210–13

(a) What is meant by:

(i) *allele*? *(1 mark)*

(ii) *dominant*? *(1 mark)*

(b) A man and his wife are both 45 years old. The man is suffering from Huntington's disease, but his wife is not a sufferer. They have one child who is now 14 years old.

(i) What system of the body is affected by Huntington's disease? *(1 mark)*

(ii) The man has both the H and h alleles. Draw a genetic diagram and use it to find the probability that the child will develop Huntington's disease.

Use the following symbols: H = allele for Huntington's disease

h = unaffected allele *(5 marks)*

[Higher]

GET IT RIGHT!

Read questions carefully or lose marks. For example, in question 4, part (b) (i), note the word 'system' and give the name of a system rather than that of an example of some organ or part of the system.

B3 | Further biology

What you already know

Here is a quick reminder of previous work that you will find useful in this unit:

- You can relate cells and cell functions to life processes in a variety of organisms.

- The products of digestion are absorbed into your bloodstream and transported throughout the body.

- The structure of your lungs plays a role in gas exchange.

- Smoking affects your lungs.

- Aerobic respiration involves a reaction in your cells between oxygen and glucose. The glucose is broken down into carbon dioxide and water.

- The reactants and products of respiration are transported throughout your body in the bloodstream.

- Plants need carbon dioxide, water and light for photosynthesis.

- The root hairs of plants play a role in absorbing water and minerals from the soil.

- The growth and reproduction of bacteria and viruses can affect your health.

The lungs play a vital part in gas exchange

RECAP QUESTIONS

1. a) What are the main functions of your lungs?

   b) If you smoke, your lungs often do not work as well. Explain how smoking can affect your lungs.

2. a) Why is a transport system so important in your body?

   b) What happens to the food you eat once it has been digested in your gut?

3. a) Explain why plants need carbon dioxide, water and light.

   b) How do they get the carbon dioxide that they need?

   c) Why are plant roots so important?

4. a) What is an infectious disease?

   b) What types of microorganisms cause infectious diseases?

   c) What are the differences between bacteria and viruses?

Making connections

All living organisms need water for the reactions to take place in their cells. Moving water into and out of the cells, using it for photosynthesis, to get rid of waste or to cool down. These are all important processes for animals and plants. In fact, almost all biology eventually depends on water!

Plants need water for photosynthesis. They have to take the water in from the soil through their roots. When water is in short supply, plants have some spectacular adaptations to help them survive!

When you breathe out, water evaporates from the surfaces of your lungs into the air and is lost. You can see this clearly in the winter when it condenses in the cold air as you breathe out. This loss happens all the time, day and night, summer and winter. So you have to take in enough water to replace it.

Fish take the oxygen they need from water. Their gills are specially adapted to make this possible. But whether fish live in fresh or salt water, they have to deal with the problem of water moving in to or out of the cells of their gills as it flows over them. Fish have several different adaptations to help them balance the water levels in their body.

The amount of water you take into your body varies enormously from one day to the next. Your body has to cope with whatever it is given. The concentration of your body fluids has to stay more or less the same whatever you drink and whatever you do. You use water to cool down, to remove waste and to carry materials around your body. All the chemical reactions of your cells take place dissolved in water. It's not surprising that you need your kidneys to maintain the water balance of your body!

Microorganisms are no different to other living things – water is vital for the chemical reactions in their cells to take place. That's why drying food preserves it. The microbes which cause decay simply can't grow.

ACTIVITY

Pet animals depend on their owners to keep them safe and well. Many types of pets, such as cats and dogs, cannot lose heat by sweating over most of their bodies. That's because of their thick layer of fur. They sweat through their feet and pant, cooling down as water evaporates from the surface of their mouths.

Design a poster or a leaflet for your local vet's surgery explaining why water is so important for animals. In particular, describe how to care for dogs and cats in hot weather.

Chapters in this unit

Exchange of materials   Transporting substances   Microbiology
                         around the body

B3 1.1

Active transport

LEARNING OBJECTIVES

1 What is active transport? [Higher]
2 Why is active transport so important? [Higher]

There are two main ways in which dissolved substances are moved into and out of cells. Substances move by diffusion, along a concentration gradient which must be in the right direction to be useful to the cells. Osmosis depends on a concentration gradient of water and a partially permeable membrane. Only water moves in osmosis. However, sometimes substances needed by your body have to be moved against a concentration gradient, or across a partially permeable membrane. The process is known as **active transport**.

Active transport

Active transport allows cells to move substances from an area of low concentration to an area of high concentration. So substances move against the concentration gradient. As a result, cells can absorb ions from very dilute solutions. It also makes it possible for them to move substances like sugars and ions from one place to another through the cell membranes.

a) How does active transport differ from diffusion and osmosis?

GET IT RIGHT!

Remember active transport takes place **against** a concentration gradient from low to high concentration and it requires energy from respiration.

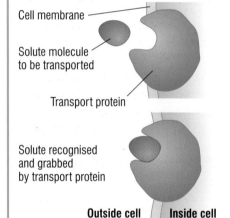

Cell membrane

Solute molecule to be transported

Transport protein

Solute recognised and grabbed by transport protein

Outside cell   Inside cell

Protein rotates in membrane and releases solute inside the cell (using energy)

Protein rotates back again (often using energy)

Figure 1 Sometimes it is worth using up energy when a resource is particularly valuable and its transport is really important!

It takes energy for the active transport system to carry a molecule across the membrane and then return it to its original position. (See Figure 1.) That energy comes from cellular respiration. Scientists have shown in a number of different cells that the rate of respiration and the rate of active transport are closely linked. (See Figure 2.)

In other words, if a cell is making lots of energy, it can carry out lots of active transport. These cells include root hair cells and your gut lining cells. Cells involved in a lot of active transport usually have lots of **mitochondria** to provide the energy they need.

b) Why do cells which carry out a lot of active transport have lots of mitochondria?

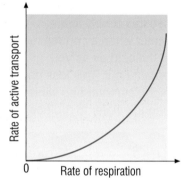

Figure 2 The rate of active transport depends on the rate of respiration

The importance of active transport

Active transport is widely used in cells. There are some situations where it is particularly important. For example, the mineral ions in the soil are usually found in very dilute solutions. These solutions are more dilute than the solution within the plant cells. By using active transport, plants can absorb these mineral ions, even though it is against a concentration gradient. (See Figure 3.)

Glucose is always moved out of your gut and kidney tubules into your blood, even when it is against a large concentration gradient.

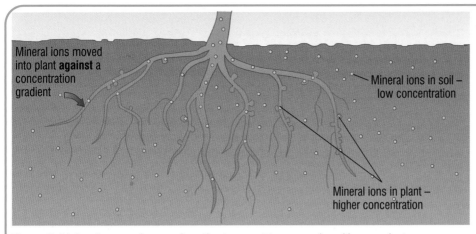

Mineral ions moved into plant **against** a concentration gradient

Mineral ions in soil – low concentration

Mineral ions in plant – higher concentration

Figure 3 It takes the use of energy in active transport to move mineral ions against a concentration gradient like this

Some marine birds and reptiles have a particular problem. They take in a great deal of salt in the sea water they drink. Their kidneys cannot get rid of it all.

The answer is special **salt glands**, which are usually found near the eyes and nostrils. Sodium ions are moved out of the body into the salt glands. The glands then produce a very strong salt solution – up to six times more salty than their urine!

The sodium ions have to be moved against a very big concentration gradient. So active transport is vital to the survival of these marine creatures.

Figure 4 Marine animals like this turtle live in very extreme conditions. The salt glands which some animals have as an adaptation for survival depend on active transport to move salt out of their bodies.

DID YOU KNOW?

People affected by the genetic disease cystic fibrosis (see page 212) produce thick, sticky mucus in their lungs, gut and reproductive systems. This fatal condition is the result of a mutation which affects a protein involved in the active transport system of the mucus-producing cells.

SUMMARY QUESTIONS

1   Copy and complete using the words below:

> **concentration   active transport   osmosis   against
> mitochondria   diffusion   energy**

…… and …… depend on a …… gradient in the right direction to work. Substances are moved …… a gradient by …… …… which uses …… produced by …… .

2   a) Explain how active transport works in a cell.
    b) Give some examples of a situation when a substance cannot be moved into a cell by osmosis or diffusion, and how active transport solves the problem.

3   The processes of diffusion and osmosis do not need energy to take place. Why does an organism have to provide energy for active transport and where does it come from?

4   a) Explain why cyanide is such an effective poison.
    b) Why is active transport so important for animals which live in the sea?

KEY POINTS

1   Substances are sometimes absorbed against a concentration gradient by active transport.

2   Active transport uses energy from respiration.

3   Cells can absorb ions from very dilute solutions and move molecules through cell membranes using active transport.

B3 1.2   Exchange of gases in the lungs

LEARNING OBJECTIVES

1 How are your lungs adapted to make gas exchange as efficient as possible?
2 What are your alveoli?

The movement of substances in and out of your body cells is extremely important. Many of your organ systems are specialised for exchanging materials. One of them is your breathing system, particularly your lungs.

The breathing system

Your body needs a constant supply of oxygen for cellular respiration. Breathing brings oxygen into your body and removes the waste carbon dioxide produced by your cells.

Your lungs are found in the upper part of your body – your chest or **thorax,** protected by your bony rib cage. Your lungs are separated from the digestive organs in the lower part of your body (your **abdomen**) by your **diaphragm** (a strong sheet of muscle). The job of your breathing system is to move air in and out of your lungs. It brings in oxygen-rich air and removes air containing waste carbon dioxide.

a) What is the thorax?

When you breathe in your ribs move up and out, and your diaphragm flattens from its normal domed shape. This pulls air into your lungs. When you breathe out your ribs move down and in. The diaphragm returns to its domed shape, forcing air out of your lungs again.

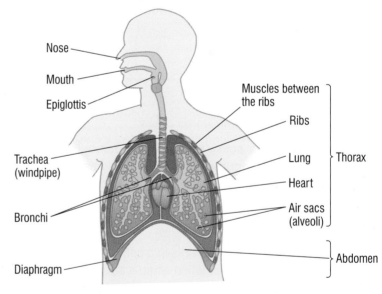

Figure 1 The breathing system supplies your body with vital oxygen and removes poisonous carbon dioxide

Exchange of gases in the lungs

Your lungs are specially adapted to make gas exchange more efficient. They are made up of clusters of **alveoli**. These tiny air sacs always have a very large surface area, which is kept moist. This is important for the most effective diffusion of the gases.

DID YOU KNOW...

Chronic obstructive pulmonary diseases (COPD) are often caused by smoking. They can result in the structure of the alveoli breaking down. The surface area of the lungs is reduced – so the person affected is always short of oxygen and feels breathless.

b) What is the function of the alveoli?

The alveoli also have a rich blood supply. This maintains a concentration gradient in both directions. Oxygen is constantly removed into the blood and more carbon dioxide is constantly delivered to the lungs. As a result, gas exchange takes place along the steepest concentration gradients possible, making it rapid and effective. The layer of cells between the air in the lungs and the blood in the capillaries is also very thin. This lets diffusion takes place over the shortest possible distance.

An analysis of the gases in inhaled and exhaled air shows clearly the differences in the quantities of some of the main gases.

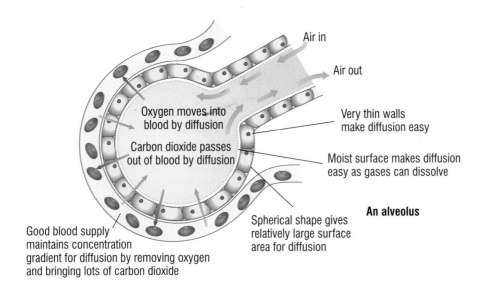

Figure 2 The alveoli are adapted so that gas exchange can take place as efficiently as possible in the lungs

Atmospheric gas	Air breathed in	Air breathed out
nitrogen	About 80%	About 80%
oxygen	20%	About 16%
carbon dioxide	0.04%	About 4%

KEY POINTS

1 Your breathing system takes air into and out of your body.
2 Oxygen from the air diffuses into your bloodstream and carbon dioxide diffuses out.
3 The alveoli of the lungs provide a very large, moist surface area with a rich blood supply and thin walls to make diffusion as effective as possible.

PRACTICAL

Comparing air breathed out and air breathed in

A detailed analysis of the air is not always possible, but you can carry out a relatively simple investigation. (See Figure 3.) It shows that the air breathed out (A) is different from the air breathed in (B). This experiment uses lime water as an indicator of the presence of carbon dioxide. The colourless solution turns cloudy when carbon dioxide is bubbled through it. The faster it turns cloudy, the greater the concentration of carbon dioxide present.

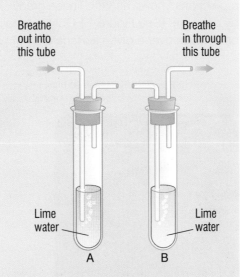

Figure 3 Comparing the level of carbon dioxide in inhaled and exhaled air

SUMMARY QUESTIONS

1 What is meant by the term gaseous exchange and why is it so important in your body?

2 How are the lungs adapted to allow gas exchange to take place as effectively as possible?

3 Draw a bar chart to show the difference in composition between inhaled and exhaled air. (Use the table above.)

B3 1.3 Exchange in the gut

Your gut is an area of your body where the exchange of materials is extremely important. The food you eat is broken down in your gut. It forms simple sugars such as glucose, amino acids, fatty acids and glycerol. But these products of digestion are of no use if they stay in your gut. They would simply be passed out of your body in your faeces.

Absorption in the small intestine

The molecules from food need to be made available to your body cells. In cells they provide fuel for respiration and the building blocks of all the tissues of your body. For this to happen they must move from the inside of your small intestine into your bloodstream. They do this by a combination of diffusion and active transport.

a) Why must the products of digestion get into your bloodstream?

This explains one reason why it is so important that your food is broken down into a soluble form during digestion. Only when the molecules are dissolved in water can diffusion take place.

The digested food molecules are then small enough to pass freely through the walls of the small intestine into the blood vessels. They move in this direction because there is a very high concentration of food molecules in the gut and a much lower concentration in the blood. In other words, they move into the blood along a steep concentration gradient.

b) Why is it so important that your food is broken down into smaller molecules?

The lining of the small intestine is folded into thousands of tiny finger-like projections known as **villi**. These greatly increase the uptake of digested food by diffusion. (See Figure 1.) Only a certain number of digested food molecules can diffuse over a given surface area of gut lining at any one time.

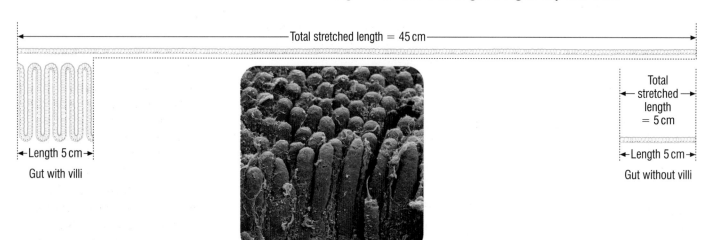

Figure 1 Without the villi of the small intestine we would be unable to absorb enough digested food to survive. They increase the surface area available for diffusion many times.

So increasing the surface area means there is more room for diffusion to take place.

Like the lungs, the lining of the small intestine has an excellent blood supply. This carries away the digested food molecules as soon as they have diffused from one side to the other. So a steep concentration gradient is maintained all the time, from the inside of the intestine to the blood. This in turn makes sure diffusion is as rapid and efficient as possible.

Diffusion isn't the only way in which dissolved food substances move from the gut into the blood. What happens as the food moves down the small intestine and the time since the last meal gets longer?

Glucose and other dissolved food molecules are moved from the small intestine into the blood by active transport, against the concentration gradient. (See page 222.) This makes sure that none of the digested food is wasted and lost in the faeces.

c) Why is it so important that the villi have a rich blood supply?

Exchange of materials in other organisms

Human beings are not the only organisms where an exchange of materials is important. Whether it is the gills of a fish or the kidneys of a desert rat, certain adaptations will always be seen:

- a large surface area to give plenty of opportunity for substances to diffuse,
- a rich blood supply to remove the substances, maintaining a steep concentration gradient and carrying them to where they are needed,
- moist surfaces for substances to dissolve,
- a short distance between the two areas so diffusion happens effectively. You can read more about this on the next two pages.

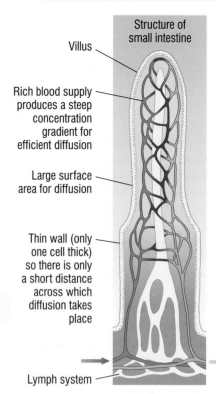

Structure of small intestine

Villus

Rich blood supply produces a steep concentration gradient for efficient diffusion

Large surface area for diffusion

Thin wall (only one cell thick) so there is only a short distance across which diffusion takes place

Lymph system

Figure 2 Thousands of finger-like projections in the wall of the small intestine – the villi – make it possible for all the digested food molecules to be transferred from your small intestine into your blood by diffusion

GET IT RIGHT!

Whenever you are looking at diffusion in a living organism, the surface area compared to the volume is very important. For efficient diffusion you need a big surface area!

SUMMARY QUESTIONS

1 Match A, B, C or D to the correct ending (1 to 4).

A Food needs to be broken down into small soluble molecules ……	1 …… by diffusion and active transport.
B The villi are ……	2 …… carry away the digested food to the cells and maintain a steep concentration gradient.
C Food molecules move from the small intestine into the bloodstream ……	3 …… so diffusion across the gut lining can take place.
D The small intestine has a rich blood supply to ……	4 …… finger-like projections in the lining of the small intestine which increase the surface area for diffusion.

2 Explain why a folded gut wall can absorb more nutrients than a flat one.

3 Places where materials are exchanged in the body always have a large, moist surface area, short distances across boundaries and a rich blood supply. Why is each of these features important for successful diffusion?

KEY POINTS

1 The villi in the small intestine provide a large surface area with an extensive network of capillaries. This makes them well adapted to absorb the products of digestion by diffusion and active transport.

2 In material exchanges, the surface area : volume ratio is always important – a big surface area is vital for successful diffusion.

B3 1.4

Exchange of materials in other organisms

Gas exchange in a fish

Fish cannot get oxygen directly from the water they live in because their bodies are covered in protective scales. Fortunately fish have evolved a very effective respiratory system which works really well in water.

Gills are made up of many thin layers of tissue with a rich blood supply. (See Figure 1 below.) The gills are thin so there is only a short distance for the gases to diffuse across. The surfaces are always moist as they work in water!

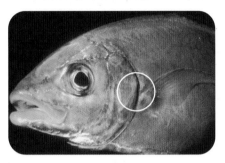
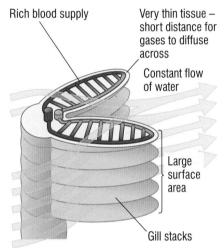

Rich blood supply

Very thin tissue – short distance for gases to diffuse across

Constant flow of water

Large surface area

Gill stacks

Figure 1 The gills of a fish – another example of an organ adapted for efficient gas exchange

In bony fish the gills are contained in a special gill cavity. Water is pumped over them constantly to maintain a concentration gradient. Fish such as sharks have to keep swimming all the time to keep water moving over their gills!

a) Why do fish need gills?

Unfortunately gills can't work in air – a fish out of water 'suffocates'. Without water surrounding them, the gill stacks all stick together. There simply isn't a big enough surface area available for the fish to get the oxygen it needs to survive from the air.

b) Why do fish die when they are taken out of water?

Tadpoles and frogs

Frogs are amphibians and have a very strange life history. The eggs hatch into tadpoles which spend all their time in water. Young tadpoles have frilly external gills with a large surface area and a rich blood supply. The tadpoles get all their oxygen by diffusion from the water through these gills. In the same way carbon dioxide diffuses out along a concentration gradient into the water.

When the tadpoles turn into frogs they spend a lot of time on the land – but they can still breathe in water!

c) Why would external gills be no use for an adult frog?

The external gills disappear and are reabsorbed into the body of the developing frog. We say that the tadpole undergoes *metamorphosis*. An adult frog has very moist skin with a rich blood supply, and under normal conditions most of its gas exchange takes place through the skin.

The mouth – which is very large and thin skinned – is also important for gas exchange. If it gets hot or the frog is being very active on land, it also has a pair of very simple lungs. These can be used to increase the surface area available for gas exchange to take place. When the frog is in the water, all the gas exchange takes place through the skin.

The respiratory system in insects

Many insects are very active so their muscles need a lot of oxygen. However, little or no gaseous exchange can take place through the tough outer covering of insects. To supply their needs, insects have an internal respiratory system which supplies oxygen directly to their cells and removes carbon dioxide. (See Figure 3.)

If you look along the side of an insect you can see the **spiracles**. These can open when the insect needs plenty of oxygen but close when they don't. This prevents water loss, rather like the guard cells of plants.

The spiracles lead into a system of tubes which run right into the cells of the tissues themselves. Most of the gas exchange takes place in the **tracheoles**. These tiny tubes are freely permeable to gases. They are very moist and air is pumped in and out of them by the insect to maintain a concentration gradient.

There is no blood supply in an insect. However, the tracheoles have a very large surface area and come into close contact with individual cells in the body of the insect. So they are very effective at gas exchange.

Figure 2 Tadpoles get all the oxygen they need from the water around them, while adult frogs can exchange gases with either water or air. They both have different adaptations for gas exchange. But the respiratory surfaces of both tadpoles and frogs have a big surface area, a rich blood supply, short diffusion routes and need to be moist.

Tracheoles are tiny tubes with a large surface area and moist lining. They are freely permeable to gases and pass right into the tissue of the insect, between the cells. Most gas exchange takes place here.

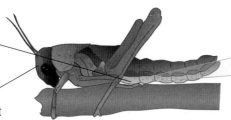

Spiracle – the opening through which air goes into and out of the insect. Often controlled so they can be opened and closed as needed.

Insect, e.g. locust

Trachea – largest tubes carrying air into the insect's body. Lined with rings of chitin, they are quite impermeable to gases, so little gaseous exchange takes place here.

Figure 3 The respiratory system of an insect has to do the same job as your breathing system. In spite of the very different design, there are many similar features which make gas exchange successful.

SUMMARY QUESTIONS

1 Copy and complete using the words below:

> **adaptations   surfaces   solute   short   surface area
> blood supply**

Wherever gas or …… exchange is important, certain …… will always be seen. These include a large …… ……, a rich …… ……, moist …… and …… diffusion distances.

2 a) Why do fish need a constant flow of water over their gills?
   b) Why are fish gills arranged in stacks?
   c) Why do you think the human breathing system does not work in water?

3 Draw and label a diagram of a tadpole and a frog, showing clearly the different ways in which gas exchange takes place.

4 What are the main features of the respiratory system of an insect and how are they important in successful gas exchange?

KEY POINT

1 Whatever the organism, gas and solute exchange depends on a large surface area, moist surfaces, short diffusion distances and a large concentration gradient.

229

B3 1.5

Exchange in plants

LEARNING OBJECTIVES

1 How are the leaves of plants adapted for gaseous exchange?
2 How are roots adapted for the efficient uptake of water and mineral ions?

Animals aren't the only living organisms that need to exchange materials. Plants rely heavily on diffusion to get the carbon dioxide they need for photosynthesis. Also important are osmosis to take water from the soil and active transport to obtain minerals from the soil. Plants have adaptations which make these exchanges as efficient as possible.

Gas exchange in plants

Plants need carbon dioxide and water for photosynthesis to take place. They get the carbon dioxide they need by diffusion through their leaves. The flattened shape of the leaves increases the surface area for diffusion. Most plants have thin leaves. This means the distance for the carbon dioxide to diffuse from the outside air to the photosynthesising cells is kept as short as possible.

What's more, the many air spaces inside the leaf allow carbon dioxide to come into contact with lots of cells. This provides lots of surface area for diffusion.

a) How are leaves adapted for efficient diffusion of carbon dioxide?

However, there is a problem. Leaf cells constantly lose water by evaporation. If carbon dioxide could diffuse freely in and out of the leaves, water vapour would also be lost very quickly. Then the leaves – and the plant – would die.

The leaf cells do not need carbon dioxide all the time. When it is dark, they don't need carbon dioxide because they are not photosynthesising. When light is a limiting factor on the rate of photosynthesis, the carbon dioxide produced by respiration can be used for photosynthesis. But on bright, warm, sunny days a lot of carbon dioxide needs to come into the leaves by diffusion.

So leaves are adapted to allow carbon dioxide in only when it is needed. They are covered with a waxy cuticle. This is a waterproof and gas-proof layer. Then all over the leaf surface there are small openings known as **stomata**. The stomata can be opened when the plant needs to allow air into the leaves so that carbon dioxide enters the cells. But they can be closed the rest of the time to control the loss of water. The opening and closing of the stomata is controlled by the guard cells.

b) Why don't leaves need carbon dioxide all the time?

Surface area
= 22 units²

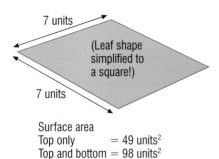

Surface area
Top only   = 49 units²
Top and bottom = 98 units²

Figure 1 The wide, flat shape of most leaves greatly increases the surface area for collecting light and exchanging gases, compared with more cylindrical leaves

PRACTICAL

Looking at stomata

You can look at stomata by coating the surface of a leaf with nail varnish. Allow the varnish to dry, peel off the layer of varnish and look at it under a microscope – the stomata will be revealed!

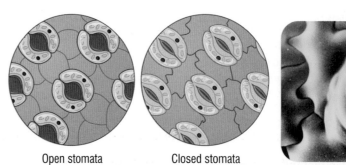

Open stomata   Closed stomata

Figure 2 Guards cells open and close the stomata to control the carbon dioxide going into the leaf

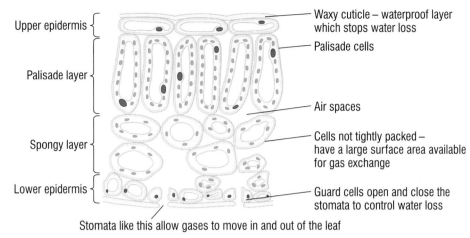

Waxy cuticle – waterproof layer which stops water loss

Upper epidermis

Palisade cells

Palisade layer

Air spaces

Spongy layer

Cells not tightly packed – have a large surface area available for gas exchange

Lower epidermis

Guard cells open and close the stomata to control water loss

Stomata like this allow gases to move in and out of the leaf

Figure 3 The arrangement of the cells inside a leaf, with plenty of air spaces and short diffusion distances, means that the carbon dioxide needed for photosynthesis reaches the cells as efficiently as possible

Uptake of water and mineral ions in plants

If you pull up a plant you will see a mesh of tiny white roots. These are adapted to enable plants to take water and minerals from the soil as efficiently as possible. Water is vital for plants. They need it to maintain the shape of their cells and for photosynthesis. They also need minerals to make proteins and other chemicals.

The roots themselves are thin, divided tubes with a large surface area. The cells on the outside of the roots near the growing tips also have their own adaptations. They increase the surface area for the uptake of substances from the soil. Known as root hair cells, they have tiny projections out from the cells which push out between the soil particles.

The membranes of the root hair cells also have microvilli. These increase the surface area for diffusion and osmosis even more. The water then has only a short distance to move across the root to the **xylem**, where it is moved up and around the plant.

Plant roots are also adapted to take in mineral ions using active transport. (See pages 222–3.) They have plenty of mitochondria to supply the energy they need. They also have all the advantages of a large surface area and the short pathways needed for the movement of water.

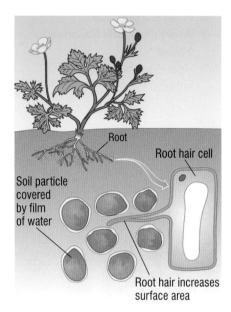

Root

Root hair cell

Soil particle covered by film of water

Root hair increases surface area

Figure 4 Many small roots, and the presence of microscopic root hairs on the individual root cells all increase diffusion of substances from the soil into the plant

SUMMARY QUESTIONS

1 How is a plant leaf adapted for the diffusion of carbon dioxide from the air?

2 a) What are stomata?
   b) What is their role in the plant?
   c) How are they controlled?

3 a) How are plant roots adapted for the absorption of water and minerals?

   b) How do the adaptations of plants for the exchange of materials compare with human adaptations in the lungs and the gut?

KEY POINTS

1 Plants have stomata which allow them to obtain carbon dioxide from the atmosphere.

2 Carbon dioxide enters the leaf by diffusion. Leaves have a flat thin shape and internal air spaces to increase the surface area available for diffusion.

3 Most of the water and mineral ions needed by a plant are absorbed by the root hair cells which increase the surface area of the roots.

B3 1.6

Transpiration

The top of a tree may be many metres from the ground, yet the leaves at the top need water just as much as the lower branches. So how do they get the water they need?

Water loss from the leaves

Plants have holes called stomata on the surfaces of their leaves. These stomata are opened to allow carbon dioxide into the plant for photosynthesis. But all the time that the stomata are open to allow carbon dioxide in, plants lose water vapour from the surface of their leaves. This loss of water vapour is what we call **transpiration**.

Stomata can be opened and closed by the guard cells which surround them. Losing water through the stomata is a side effect of opening them to let carbon dioxide in.

As water evaporates from the surface of the leaves, water is pulled up through the xylem to take its place. This constant movement of water molecules through the xylem from the roots to the leaves is known as the **transpiration stream**. It is driven purely by the evaporation of water from the leaves. So anything which affects the rate of evaporation will affect transpiration.

a) What is the transpiration stream?

The effect of the environment on transpiration

Firstly, conditions which increase the rate of photosynthesis will increase the rate of transpiration. Increased rates of photosynthesis mean more stomata are opened up to let carbon dioxide in. In turn, more water is lost by evaporation through the open stomata. So warm, sunny conditions increase the rate of transpiration.

Conditions which increase the rate of evaporation of water when the stomata are open will also make transpiration happen more rapidly. Hot, dry, windy conditions increase the rate of transpiration.

Figure 1 The transpiration stream can pull many litres of water up to 30 metres above the surface of the earth in a giant redwood tree like this one!

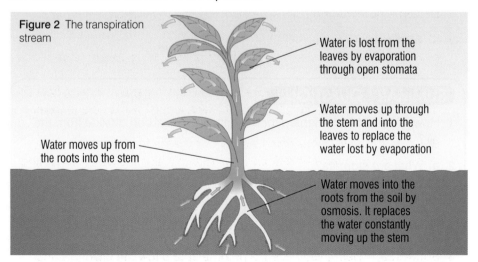

Figure 2 The transpiration stream

Water is lost from the leaves by evaporation through open stomata

Water moves up through the stem and into the leaves to replace the water lost by evaporation

Water moves up from the roots into the stem

Water moves into the roots from the soil by osmosis. It replaces the water constantly moving up the stem

b) Give three conditions that will increase the rate of transpiration from a leaf.

Controlling water loss

Plants have a constant problem – they need to open their stomata to photosynthesise but this means they are going to lose water. Most plants have a variety of adaptations which help them to photosynthesise as much as possible, while at the same time losing as little water as possible!

Most leaves have a waxy, waterproof layer (known as the **cuticle**) to prevent uncontrolled water loss. In very hot environments the cuticle may be very thick and shiny. Most of the stomata are found on the underside of the leaves. This means that they are not as exposed to the light and heat of the Sun, and reduces the time they are open.

If a plant begins to lose water faster than it is replaced by the roots, it can take some drastic measures. The whole plant may wilt. **Wilting** is a protection mechanism against further water loss. The leaves all collapse and hang down so the surface area available for water loss by evaporation is greatly reduced.

The stomata close, which stops photosynthesis and risks overheating. But this prevents most water loss and any further wilting. The plant will remain wilted until the temperature drops, the Sun goes in or it rains!

GET IT RIGHT!

Remember that plants have to make a compromise – when their stomata are open to get carbon dioxide for photosynthesis they have to pay the price through water loss by transpiration.

PRACTICAL

Evidence for transpiration

There are a number of experiments which can be done to investigate the movement of water in plants by transpiration. Many of them use a piece of apparatus known as a potometer.

A potometer can be used to show how the uptake of water by the plant changes with different conditions. This gives you a good idea of the amount of water lost by the plant in transpiration.

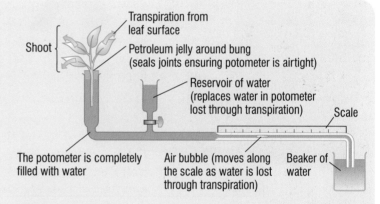

Figure 3 A potometer is used to show the water uptake of a plant under different conditions

SUMMARY QUESTIONS

1 Describe how water moves up a plant in the transpiration stream.

2 a) What is transpiration?
 b) What part of the leaves helps them to prevent losing water under normal conditions?
 c) If the top surfaces of the leaves were coated in petroleum jelly, how do you think it would affect the rate at which a plant takes up and loses water?
 d) If the bottom surfaces of the leaves were coated in petroleum jelly, how do you think it would affect the rate at which a plant takes up and loses water?
 e) What do you think would happen to the rate of transpiration if you turned a fan onto the leaves of the plant? Explain your answer.
 f) What does a potometer actually measure?

3 Suggest an investigation using a potometer to show the effect of either light or temperature on the rate of transpiration from a leafy stem.

KEY POINTS

1 The loss of water vapour from the surface of plant leaves is known as transpiration.
2 Water is lost through the stomata which are opened and closed by guard cells to let in carbon dioxide for photosynthesis.
3 Water is pulled up through the xylem from the roots to replace the water lost from the leaves in the transpiration stream.
4 Transpiration is more rapid in hot, dry, windy or light conditions.

Black lung!

OBITUARIES

and older brothers. The Jack Higgins was family have been well-known for his mining coal for 150 sense of humour and years and he was his skill in organising proud of his work. His the village fete. For enthusiasm for life will many years he was be sadly missed in the captain of the cricket village. He leaves a team until the lung wife and three disease, which finally children. killed him, made him hang up his boots four years ago. Jack started working down the mine when he was only 16 years old, following his father

The threat of 'black lung'

The disease 'black lung' is as threatening as its name. Black lung, or as it is properly known, coal worker's pneumoconiosis (CWP), is a form of chronic obstructive pulmonary disease (COPD) found only in mine workers.

The black dust produced as coal is mined gets breathed into the lungs. Normal amounts of dust are removed by the mucus produced in your breathing system. However, in a mine there is so much coal dust that your body cannot cope. Once the dust gets into the alveoli, an immune reaction is triggered. In the end the structure of the alveoli is lost. Instead of pink tissue containing millions of tiny air sacs, the lungs are blackened and have relatively few large air sacs.

The sufferer cannot get enough oxygen into their body, so they become tired, weak and less able to do things. They have to breathe pure oxygen to help them, until eventually even that isn't enough and they die.

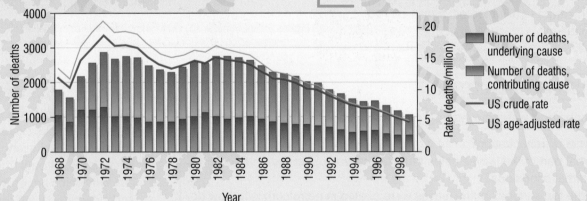

The number of people dying from CWP (coal worker's pneumoconiosis) in the USA is falling steadily. This is partly due to the fact that fewer people are working in the coal industry. Also, as people realised that coal dust in the mines caused the fatal disease, the level of dust in the mines has been controlled and miners are provided with masks to filter out the deadly particles.

ACTIVITY

Work in a group to put together a Powerpoint® presentation on coal worker's pneumoconiosis (black lung). Using this material and what you have learnt in this chapter as your starting point, make sure you cover the lungs, how they work, how CWP affects the lungs and prevents effective gaseous exchange and the social impact of diseases like this. You might include some history of the disease, and ways in which it can be prevented.

Cellular exchanges

The exchange of materials is vital to life as we know it. Diffusion, osmosis and active transport are necessary to move substances we need from one place to another – and to get rid of substances which would cause problems. Here are a few more of the different ways in which living organisms manage to transport substances into and out of their cells!

Keep it moving!

Bony fish, like cod and the goldfish you may keep as a pet, pump water over their gills. This helps to maintain a steep concentration gradient between the water and the blood flowing through the gill stacks. This in turn means that oxygen moves by diffusion into the blood and carbon dioxide moves out.

But most of the cartilaginous fish – like sharks – don't have a pumping mechanism. They also lack the swim bladders which make other fish buoyant. So if a shark stops swimming, it will start to sink to the bottom. Once it lands on the bottom, water is no longer flowing over the gills and it cannot get the oxygen it needs to live. So most sharks need to keep swimming to stay alive.

Paramecium and other single-celled freshwater organisms have a real problem because the concentration of solutes in their cells is greater than concentration in the water they live in. So water tends to flow into the cells by osmosis, which could make then swell up and burst – not good news! *Paramecium* deal with the problem using a contractile vacuole. Water is moved out of the cytoplasm into the vacuole against a concentration gradient by active transport. Once the vacuole is getting full, it is moved – again using energy from cellular respiration – to the surface of the cell where the water is released.

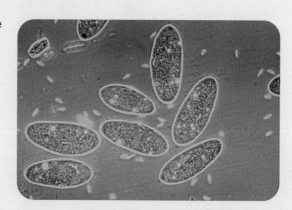

ACTIVITY

Design a wall poster – or three posters – on exchange mechanisms in animals and plants. Cover diffusion, osmosis and active transport. For each one find as many interesting examples as possible to show why the process is so important to living organisms.

SUMMARY QUESTIONS

1 Produce a table to compare diffusion, osmosis and active transport. Write a brief explanation of the advantages and disadvantages of all three processes in cells.

2 Look at the diagram of a fish gill below:

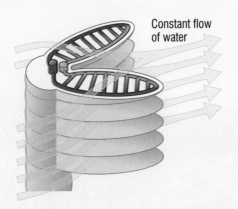

Constant flow of water

Explain carefully how it is adapted for the exchange of gases in water.

3 a) How are the lungs adapted to allow the exchange of oxygen and carbon dioxide between the air and the blood?

b) Explain what the experiment shown below tells us about inhaled and exhaled air.

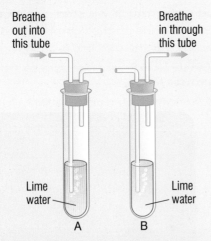

Breathe out into this tube

Breathe in through this tube

Lime water

Lime water

A   B

4 Compare the adaptations of a plant leaf for the exchange of carbon dioxide, oxygen and water vapour with the adaptations of the roots for the absorption of water and minerals.

5 Tell the story of water particles as they travel from the soil around the roots of a plant to the point when they reach the air surrounding the leaves on a sunny, windy day.

EXAM-STYLE QUESTIONS

1 The diagram shows two plant cells that have been immersed for the same time in different solutions.

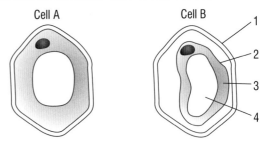

Cell A   Cell B
1
2
3
4

(a) Label the parts numbered **1, 2, 3** and **4** in cell **B**. (4)

(b) The contents of cell **B** have shrunk because water has moved out of the structure labelled **4**. What process is most likely to have caused this water to move out? (1)

(c) One cell has been immersed in distilled water and the other in a solution of sugar. Which cell has been immersed in distilled water? (1)

(d) The structure labelled **2** in cell **B** allows water to pass across it, but not other substances such as sugar. What name is given to structures that have this property? (1)

(e) Explain how cells in the same state as cell **A** can help to support the leaves of plants. (3)

2 (a) Copy out the table below and complete the boxes by writing 'Yes' or 'No' as appropriate. (5)

	Diffusion	Osmosis	Active transport
Occurs against a concentration gradient			
Needs energy			
Method by which oxygen is absorbed in the lungs			
Method by which digestive products are absorbed in the gut			
Method by which water is absorbed by plant roots			

(b) For the efficient transfer of materials by diffusion, the exchange surface needs to have certain features, such as a large surface area, that increase the rate of diffusion. Name two other features other than a large surface area that increase the rate of diffusion by an exchange surface. (2)

(c) In each of the following situations, name the particular structure that is adapted to produce a large surface area for the exchange of materials.

   (i) Absorption of oxygen into the blood in the lungs. (1)

   (ii) Absorption of glucose from the gut into the bloodstream. (1)

   (iii) Absorption of water by the roots of plants. (1)
   [Higher]

3 The diagram shows a small portion of the surface of a leaf as seen under a microscope.

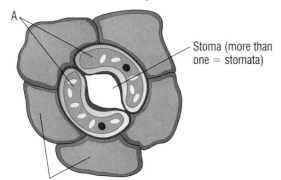

A

Stoma (more than one = stomata)

Leaf epidermis cells

(a) Name the structures labelled **A**. (1)

(b) Stomata are small pores in the leaf through which carbon dioxide and other gases can be exchanged between the inside of the leaf and the outside.

   (i) What process in the plant requires this carbon dioxide? (1)

   (ii) From what other process occurring within the plant cells might this carbon dioxide also be obtained? (1)

   (iii) By what process does this carbon dioxide enter the plant through stomata? (1)

(c) State two adaptations of leaves that produce a large surface area for the absorption of carbon dioxide. (2)

(d) Water vapour can be lost from the plant through stomata.

   (i) What name is given to this loss of water vapour? (1)

   (ii) Explain how this loss of water vapour can be controlled by the plant. (2)

   (iii) The loss of water vapour is affected by environmental conditions. State three changes in environmental conditions that would **decrease** the rate of water loss in plants. (3)

HOW SCIENCE WORKS

Alana was fascinated by the idea that she could get a measure of the concentration of the cytoplasm of some plant cells by a simple osmosis investigation. She prepared a range of sodium chloride solutions. She cut out some potato chips to more or less the same shape and size. She dried the outside of the chips and weighed them. Alana then put them into the different solutions and left them for half an hour. She then dried and weighed them again.

Alana's results are in the table below. L and R are the repeats

Concentration (mol/dm³)	Weight before (g)	Weight after (g)	Difference (g)	Mean difference (g)
L 0.0	8.584	8.873	+0.289	0.268
R 0.0	8.454	8.701	+0.247	
L 0.1	7.223	7.361	+0.138	0.144
R 0.1	8.048	8.198	+0.150	
L 0.2	8.157	8.059	−0.098	
R 0.2	8.236	8.198	−0.038	
L 0.3	7.720	7.433	−0.287	−0.266
R 0.3	8.590	8.344	−0.246	
L 0.4	7.032	6.616	−0.416	−0.467
R 0.4	7.798	7.361	−0.437	
L 0.5	8.286	7.789	−0.497	−0.487
R 0.5	7.399	6.922	−0.477	

a) Which important variable did Alana not control? (1)

b) Calculate the mean difference for the 0.2 (mol/dm³) concentration. (1)

c) Draw a graph for the mean differences shown in the different concentrations of sodium chloride solutions. Draw a line of best fit. (3)

d) Are there any anomalies to be found from the graph? If so, which results would you consider anomalous? (1)

e) What should be done with any anomalies? (1)

f) Which concentration could be considered to be similar to that of the cytoplasm in the potato cells? (1)

g) How accurate do you think this concentration is? (1)

h) How could you increase the accuracy of this concentration? (1)

B3 2.1 The circulatory system

LEARNING OBJECTIVES

1 How does your circulatory system work?
2 Where do substances enter and leave your blood?

You are made up of billions of cells and most of them are a long way from a direct source of food or oxygen. A transport system is absolutely vital to supply the needs of your body cells and remove the waste material they produce. This is the function of your blood circulation system. It has three elements – the pipes (**blood vessels**), the pump (the **heart**) and the medium (the **blood**).

A double circulation

You actually have not one transport system but two. Humans have a **double circulation**. One carries blood from your heart to your lungs and back again to exchange oxygen and carbon dioxide with the air. The other carries blood all around the rest of your body and back again.

A double circulation like this is very important in warm-blooded, active animals like ourselves. It makes our circulatory system very efficient. Fully oxygenated blood returns to the heart from the lungs and can then be sent off to the different parts of the body. This means more areas of your body can receive fully oxygenated blood quickly.

a) Why do we need a blood circulation system?

The blood vessels

You have three main types of blood vessels. They are adapted to carry out particular functions within your body, although they are all carrying the same blood.

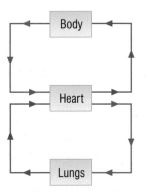

Figure 1 The two separate circulation systems supply the lungs and the rest of the body

Blood vessels	Distinctive features
Artery Thick walls — Small lumen — Thick layer of muscle and elastic fibres	Your **arteries** carry blood away from your heart to the organs of your body. This is usually oxygenated blood so it is bright red. They stretch as the blood is forced through them and go back into shape afterwards. You can feel this as a pulse where the arteries run close to the surface (like your wrist).
Vein Relatively thin walls — Large lumen — Often have valves	The **veins** carry blood towards your heart – it is usually low in oxygen and so is a deep purply-red colour. They do not have a pulse, but they often have valves to prevent the back-flow of blood as it moves from the various parts of the body back to the heart.
Capillary Walls a single cell thick — Tiny vessel with narrow lumen	In the organs of your body, between the arteries which bring blood from the heart and the veins which collect it up to take it back to the heart, your blood flows through a huge network of **capillaries**. No cell in your body is more than 0.05 mm away from a capillary! Capillaries are narrow with very thin walls so the substances needed by your body cells, such as oxygen and glucose, can easily pass out of your blood and into your cells by diffusion. In the same way substances produced by your cells, such as carbon dioxide, pass easily into the blood through the walls of the capillaries.

b) Substances can only enter and leave the blood in the capillaries. Why is this?

The heart as a pump

Your heart is made up of two pumps (for the double circulation) which beat together about 70 times each minute. The walls of your heart are almost entirely muscle, supplied with oxygen by the coronary blood vessels.

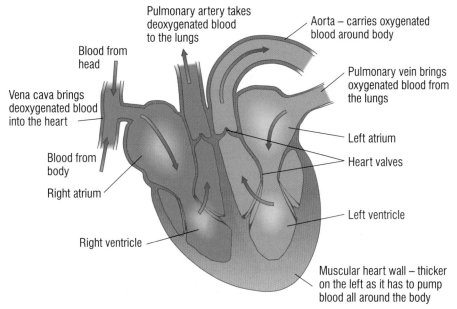

Pulmonary artery takes deoxygenated blood to the lungs

Aorta – carries oxygenated blood around body

Blood from head

Pulmonary vein brings oxygenated blood from the lungs

Vena cava brings deoxygenated blood into the heart

Left atrium

Heart valves

Blood from body

Right atrium

Left ventricle

Right ventricle

Muscular heart wall – thicker on the left as it has to pump blood all around the body

Figure 2 The structure of the human heart is perfectly adapted for the job it has to do, pumping blood to your lungs and your body. The two sides of the heart fill and empty at the same time to give a strong, co-ordinated heart beat.

DID YOU KNOW...

The noise of the heart beat we can hear through a stethoscope is actually the sound of the valves of the heart working in the surging blood, preventing it from flowing backwards.

FOUL FACTS

Because the blood in the arteries is under pressure, it is very dangerous if an artery is cut. The blood spurts out rapidly every time the heart beats. If a large artery is cut the blood may spurt a couple of metres or more.

KEY POINTS

1 The body transport system consists of the blood vessels, the heart and the blood.
2 Human beings have a double circulation.
3 The heart works as a pump, moving blood around the body.
4 The three main types of blood vessels are the arteries, veins and capillaries.

SUMMARY QUESTIONS

1 a) Draw a diagram which explains the way the arteries, veins and capillaries are linked to each other and to the heart.
  b) Label the diagram and explain what is happening in the capillaries.

2 Blood in the arteries is usually bright red because it is full of oxygen. This is not true of the arteries leaving the heart for the lungs. Why not?

B3 2.2 Transport in the blood

LEARNING OBJECTIVES

1 What is blood made up of?
2 How are red blood cells adapted to carry oxygen around your body?

Figure 1 Blood plasma is a yellow liquid which transports everything you need – and need to get rid of – around your body to the right places

Your blood is a complex mixture of cells and liquid which carries a huge range of substances around your body. The liquid part of your blood is called the plasma. It carries red blood cells, white blood cells and **platelets**.

The white blood cells are part of your immune system which is your defence against disease. The platelets are involved in the clotting of your blood. But it is the blood plasma and the red blood cells which are involved in the transport of materials around your body.

a) What are the roles of the white blood cells and the platelets in your blood?

The blood plasma as a transport medium

Your **blood plasma** is a yellow liquid – the red colour of whole blood comes from your red blood cells. The plasma transports all of your blood cells and a number of other things around your body. Carbon dioxide produced in the organs of the body is carried in the plasma back to the lungs.

Similarly **urea**, a waste product formed in your liver from the breakdown of proteins, is carried in the plasma to your kidneys. In the kidneys the urea is removed from your blood to form urine. (See page 246).

All the small, soluble products of digestion pass into the blood from your gut. They are carried in the plasma around your body to the organs and individual cells which need them.

b) What is transported in your blood plasma?

Red blood cells

There are more **red blood cells** than any other type of blood cell in your body. You have about 5 million red blood cells in each 1 mm³ of your blood – and the average person has between 4.7–5.0 litres of blood in their body! The red blood cells pick up oxygen from your lungs and carry it to the tissues and cells where it is needed. Your red blood cells have a number of adaptations which make them very efficient at their job:

- they have a very unusual shape – they are *biconcave discs*. This means they are concave – pushed in – on both sides. This gives them an increased surface area : volume ratio over which the diffusion of oxygen can take place,
- they are packed full of a special red pigment called **haemoglobin**, which can carry oxygen,
- they do not have a nucleus – this makes more space to pack in molecules of haemoglobin.

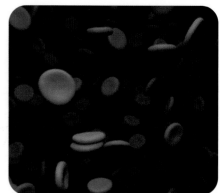

Figure 2 Red blood cells have a unique shape which helps them pick up and carry as much oxygen as possible

The formation and breakdown of oxyhaemoglobin

Haemoglobin is a large protein molecule folded around four iron atoms. In a high concentration of oxygen, such as in the lungs, the haemoglobin reacts with oxygen to form **oxyhaemoglobin**. This is bright scarlet in colour, which is why the blood in most of your arteries is bright red.

In areas where the concentration of oxygen is lower, such as the cells and organs of the body, the reaction reverses. The oxyhaemoglobin splits to give haemoglobin and oxygen. The oxygen then diffuses into the cells where it is needed. Haemoglobin is purply-red – the colour of the blood in your veins.

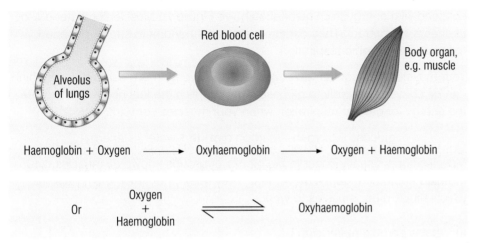

Figure 3 This reversible reaction makes active life as we know it possible by carrying oxygen to all the places where it is really needed

Because haemoglobin is based on iron, if your diet lacks iron, your body cannot make enough red blood cells and you suffer from anaemia. People who are anaemic are pale and lack energy. That's because they cannot carry enough oxygen around the body for their needs

DID YOU KNOW…

One red blood cell contains about 250 million molecules of haemoglobin, which allow it to carry 1000 million molecules of oxygen.

SCIENCE @ WORK

Scientists get involved all sorts of different jobs. For example, some religious groups don't allow blood transfusions even to save lives. So scientists have developed an artificial blood which can be used during surgery and after blood is lost in an accident.

SUMMARY QUESTIONS

1 Copy and complete using the words below:

   transported   glucose   red blood cells   urea   blood
   lungs   plasma   oxygen

   Substances are …… around your body in your …… . Dissolved food molecules such as …… and waste substances such as …… are carried in the ……, while …… is carried from the …… to the cells by your …… …… …… .

2 a) Why is it not accurate to describe the blood as a red liquid?
   b) What actually makes the blood red?
   c) Give three important functions of the blood plasma.

KEY POINTS

1 Your blood is the main transport medium of your body.
2 Your blood plasma transports dissolved food molecules, carbon dioxide and urea.
3 Your red blood cells are adapted to transport oxygen from your lungs to the organs of your body.
4 Red blood cells are biconcave discs which have no nucleus and are packed with the red pigment haemoglobin.
5 Oxygen is carried by haemoglobin which becomes oxyhaemoglobin in a reversible reaction.

B3 2.3

The effect of exercise on the body

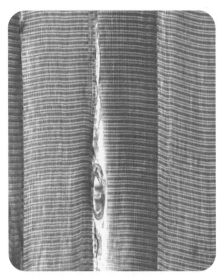

Figure 1 All the work done by the muscles is based on these very special protein fibres which contract when they work and relax afterwards

Your muscles use a lot of energy. They move you about and help support your body against gravity. Your heart is made of muscle, and the movement of food along your gut depends on muscles too.

Muscle tissue is made up of protein fibres which contract when they are supplied with energy from respiration. (See Figure 1.) Muscle fibres need a lot of energy to contract. They contain many mitochondria to supply the energy they need by aerobic respiration.

Your muscles also contain **glycogen** stores. Glycogen is a carbohydrate which can be converted rapidly to glucose. This supplies the fuel needed to provide the energy for cellular respiration when your muscles contract.

$$\text{glucose} + \text{oxygen} \rightarrow \text{carbon dioxide} + \text{water} \; (+ \text{energy})$$

Muscles fibres usually occur in big blocks or groups known as muscles. Your muscles contract to cause movement. They relax when their role is finished which allows other muscles to work.

a) What is aerobic respiration?

The response to exercise

Even when you are resting your muscles use up a certain amount of oxygen and glucose. This is because some of your muscles fibres are constantly contracting to keep you in position against the pull of gravity. Muscles are also involved in your life processes such as breathing and circulation of the blood.

But when you begin to exercise, your muscles start contracting harder and faster. As a result they need more glucose and oxygen to supply their energy needs. During exercise the muscles also produce increased amounts of carbon dioxide. This needs to be removed for them to keep working effectively.

b) Why do you need more energy when you exercise?

So during exercise, when muscular activity increases, several changes take place in your body:

- Your heart rate increases and the arteries supplying blood to your muscles dilate. These changes increase the blood flow to your exercising muscles. This in turn increases the supply of oxygen and glucose and increases the removal of carbon dioxide.
- Your breathing rate increases and you breathe more deeply. These changes mean that not only do you breathe more often, but you also bring more air into your lungs each time you breathe. This increases the amount of oxygen brought into your body and picked up by your red blood cells. The oxygen is carried to the exercising muscles. It also means that more carbon dioxide can be removed from the blood in the lungs and breathed out.

c) Why do you produce more carbon dioxide when you are exercising hard?

DID YOU KNOW?

The maximum rate to which you should push your heart is usually calculated as approximately 220 beats/minute minus your age. When you exercise, you should ideally get your heart rate into the range between 60% to 90% of your maximum!

The benefits of exercise

Your heart and lungs benefit from regular exercise. Both the heart and the lungs become larger. They both develop a bigger and very efficient blood supply. This means they function as effectively as possible at all times, whether you are exercising or not.

	Before getting fit	After getting fit
Amount of blood pumped out of the heart during each beat (cm^3)	64	80
Heart volume (cm^3)	120	140
Breathing rate (no. of breaths per minute)	14	12
Pulse rate	72	63

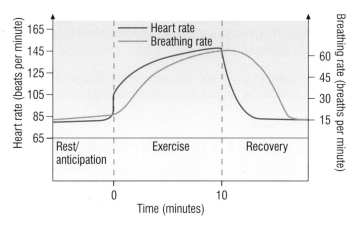

Figure 2 During exercise the heart rate and breathing rate increase to supply the muscles with what they need and remove the extra waste produced

Figure 3 Not everyone can be as fit as a top-class athlete – but doing exercise helps keep your heart and lungs healthy at whatever level you take part

GET IT RIGHT!

Exercise results in an increased rate of delivery of oxygen to the muscles – not just 'more oxygen'.

Be clear about the difference between the rate and the depth of breathing.

PRACTICAL

Testing fitness

A good way of telling how fit you are is to measure your resting heart rate and breathing rate. The fitter you are, the lower they will be! Then see what happens when you exercise. The increase in your heart rate and breathing rate and how fast they return to normal is another way of finding out how fit you are – or aren't!

SUMMARY QUESTIONS

1 Using Figure 2, describe the effect of exercise on the heart rate and the breathing rate of a fit person and explain why these changes happen.

2 Plan an investigation into the fitness levels of your class mates. Describe what you would do and how you would record and analyse your results. What pattern would you expect to see?

KEY POINTS

1 The energy that is released during respiration is used to enable muscles to contract.

2 When you use your muscles you need glucose and oxygen to be supplied at a faster rate. The rate at which carbon dioxide is removed from muscle tissues needs to increase too.

3 Body responses to exercise include an increase in heart rate, and increase in breathing rate and depth of breathing. The arteries supplying blood to the muscles dilate and the glycogen stores in the muscle are converted to glucose to use as fuel for respiration.

4 Regular exercise benefits the muscles, heart and lungs.

B3 2.4

Anaerobic respiration

LEARNING OBJECTIVES

1 Why do muscles use anaerobic respiration to obtain energy?
2 Why is less energy released by anaerobic respiration than aerobic respiration? [Higher]
3 What is an oxygen debt? [Higher]

Your everyday muscle movements are made possible with energy produced by aerobic respiration. However, during vigorous exercise your muscle cells may become short of oxygen. In spite of your increased heart and breathing rates, the blood simply cannot supply oxygen fast enough. When this happens the muscle cells can still obtain energy from glucose. But they have to do it by a type of respiration which does not use oxygen – **anaerobic respiration**.

Muscle fatigue

When you have been using your muscle fibres for vigorous exercise for a long time they can become fatigued. This means they stop contracting efficiently. At this stage they are usually very short of oxygen, so they switch to anaerobic respiration.

However, anaerobic respiration is not as efficient as aerobic respiration. In anaerobic respiration the glucose molecules are not broken down completely. So far less energy is released than during aerobic respiration.

The end products of anaerobic respiration are **lactic acid** and water, instead of the carbon dioxide and water produced by aerobic respiration.

Anaerobic respiration:

$$\text{glucose} \rightarrow \text{lactic acid} \, (+\, \text{energy})$$

a) How does anaerobic respiration differ from aerobic respiration?

DID YOU KNOW?

Training at altitude so your blood carries more oxygen is legal. There are other ways of increasing your red blood cell count to avoid oxygen debt (see next page) which are not. Sometimes athletes remove some of their own blood, store it, and then, just before a competition, transfuse it back again (blood doping). Others use hormones to stimulate the growth of more red blood cells. Both of these methods give an athlete extra red blood cells to carry more oxygen to the working muscles so they can run faster or compete better. Both of them are illegal.

PRACTICAL

Making lactic acid

Carry out a single repetitive action such as stepping up and down or lifting a weight or a book from the bench to your shoulder time after time. You will soon feel the effect of a build up of lactic acid in your muscles.

● How can you tell when your muscles have started to respire anaerobically?

Figure 1 Repeated movements can soon lead to anaerobic respiration in your muscles – particularly if you're not used to it

Oxygen debt

If you have been exercising hard, you often carry on puffing and panting for some time after you stop. The length of time you remain out of breath depends on how fit you are. But why do you keeping breathing faster and more deeply when you have stopped using your muscles?

The waste lactic acid you produce during anaerobic respiration presents your body with a problem. You cannot simply get rid of lactic acid by breathing it out as you can with carbon dioxide. As a result, when the exercise is over lactic acid has to be broken down to produce carbon dioxide and water. This needs oxygen, and the amount of oxygen needed to break down the lactic acid is known as the **oxygen debt**.

Even though your leg muscles have stopped, your heart rate and breathing rate stay high to supply extra oxygen until you have paid off the oxygen debt. The bigger the debt, the longer you will puff and pant!

Oxygen debt repayment:

lactic acid + oxygen → carbon dioxide and water

b) What is an oxygen debt?

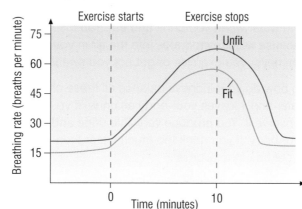

Figure 2 Hard exercise means everyone has to pay off their oxygen debt – but if you are fit you can pay it off faster!

Figure 3 Training hard is the simplest way to avoid anaerobic respiration as it enables your body to get more oxygen to your muscles and to remove carbon dioxide more efficiently

SUMMARY QUESTIONS

1 Define the following words:

    aerobic respiration; anaerobic respiration; oxygen debt; blood doping [Higher]

2 a) If you exercise very hard or for a long time your muscles begin to ache and do not work so effectively. Explain why.
    b) If you exercise very hard you often puff and pant for some time after you stop. Explain what is happening. [Higher]

3 Look at Figure 2.

    a) Explain what is happening to both people.
    b) Why is the graph for the unfit person different from the graph for the fit person?
    c) What could the unfit person do to change their body reactions to be more like those of the fit person? [Higher]

KEY POINTS

1 If muscles work hard for a long time they become fatigued and don't contract properly. If they don't get enough oxygen they will respire anaerobically.
2 Anaerobic respiration is respiration without oxygen. Glucose is broken down to form lactic acid, water and a small amount of energy. [Higher]
3 After exercise, oxygen is still needed to break down the lactic acid which has built up. The amount of oxygen needed is known as the oxygen debt. [Higher]

B3 2.5   The human kidney

Your kidneys are one of the main organs which help to maintain homeostasis, keeping the conditions inside your body as constant as possible.

What are the functions of your kidney?

Your kidneys are very important in your body for homeostasis. For example, you produce urea in your liver when you break down excess amino acids. These excess amino acids come from protein in the food you eat and from the breakdown of your worn out body tissues. Urea is poisonous, but your kidneys filter it out of your blood and get rid of it in your urine.

a) What is urea?

Your kidneys are also vital in the water balance of your body. If the concentration of your body fluids change, water will move in to or out of your cells by osmosis. (See page 144.) This could damage or destroy the cells. You gain water when you drink and eat. You lose water constantly from your lungs because water evaporates into the air in your lungs and is breathed out. Whenever you exercise or get hot you sweat and lose more water.

So how do your kidneys balance all these changes? They remove any excess water and it leaves your body as urine. If you are short of water your kidneys conserve it. You produce very little urine and most water is saved for use in your body. If you drink too much water then your kidneys produce lots of urine to get rid of the excess.

The ion concentration of your body is very important. You take in mineral ions with your food. The amount you take in varies. Sometimes you eat very little.

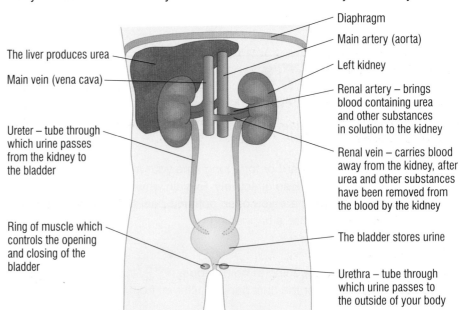

The liver produces urea

Main vein (vena cava)

Ureter – tube through which urine passes from the kidney to the bladder

Ring of muscle which controls the opening and closing of the bladder

Diaphragm

Main artery (aorta)

Left kidney

Renal artery – brings blood containing urea and other substances in solution to the kidney

Renal vein – carries blood away from the kidney, after urea and other substances have been removed from the blood by the kidney

The bladder stores urine

Urethra – tube through which urine passes to the outside of your body

Figure 1 The kidney, a very important organ of homeostasis, is involved in controlling the loss of water and mineral ions from the body as well as getting rid of urea. The kidneys have a very rich blood supply.

But if you eat processed food which is high in salt, you take in a big load of minerals. Some are lost through your skin when you sweat. Again your kidneys are most important in keeping a mineral ion balance. They remove excess mineral ions (particularly salt) which are lost in the urine.

b) Why do your kidneys work hard after you eat a lot of processed food?

How do your kidneys work?

Your kidneys filter your blood and then take back (re-absorb) everything your body needs. So sugar (glucose), amino acids, mineral salts and urea all move out of your blood into the kidney tubules by diffusion along a concentration gradient. The blood cells are left behind – they are too big to pass through the membrane of the tubule.

Then **all** of the sugar is reabsorbed back into the blood by active transport. But the amount of water and the dissolved mineral ions which are reabsorbed varies. It depends on what is needed by your body. This is known as **selective reabsorption**.

> Both sugar and dissolved mineral ions move back into the blood both by diffusion along a concentration gradient and by active transport. Active transport is used to move them against a concentration gradient. This makes sure no sugar is left in the urine and the right quantity of dissolved mineral ions is reabsorbed.

The amount of water reabsorbed depends on what your body needs. It is controlled by a very sensitive feedback mechanism. Urea is lost in your urine. However, some of it leaves the kidney tubules and moves back into your blood by diffusion along a concentration gradient.

Your kidneys have a very rich blood supply and they produce urine all the time. It trickles into your bladder where it is stored until the bladder is full and you choose to empty it!

What does urine contain?

Your urine contains the waste urea along with excess mineral ions and water not needed by your body. The exact quantities vary depending on what you have taken in and given out. For example, on a hot day if you drink little and exercise a lot you will produce very little urine. This will be concentrated and relatively dark yellow. On a cool day if you drink a lot of liquid and do very little you will produce a lot of dilute, almost colourless urine.

SUMMARY QUESTIONS

1 Urea is one of the main waste products of your body. Describe carefully
   a) how it is formed,
   b) why it has to be removed,
   c) how it is removed from your body.

2 Explain how your kidneys would maintain the water and mineral balance of your blood on
   a) a cool day when you stayed inside and drank lots of cups of tea,
   b) a hot sports day when you ran three races and had forgotten your drink bottle.

HIGHER

DID YOU KNOW?

Water, glucose, urea and salt are all colourless. So why is urine yellow? **Urobilins** are yellow pigments which come from the breakdown of haemoglobin in your liver. They are another waste product excreted by your kidneys in the urine along with everything else – and make it yellow!

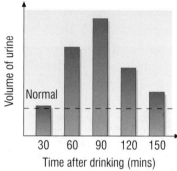

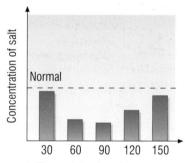

Figure 3 These data show how your kidneys respond when you drink a lot. They show volume of urine produced and the concentration of salt in the urine after a student drank a large volume of water.

KEY POINTS

1 A healthy kidney produces urine by filtering the blood. It then reabsorbs **all** of the sugar, and the mineral ions and water needed by your body.

2 Excess mineral ions and water along with urea are removed in the urine.

3 Sugar and dissolved ions can be actively reabsorbed against a concentration gradient. [Higher]

B3 2.6

Dialysis – an artificial kidney

Your kidneys can be damaged and destroyed by infections. Some people have a genetic problem which means their kidneys fail, and sometimes the kidneys are damaged during an accident. Whatever the cause, untreated failure of both your kidneys can lead to death. Toxins, such as urea, build up in the body and the salt and water balance of your body is lost.

Dialysis

For centuries kidney failure meant certain death, but now we have two effective methods of treating kidney failure. We can carry out the function of the kidney artificially using **dialysis** or we can replace the failed kidneys with a healthy one in a **kidney transplant**. (See pages 250–1.)

a) Why is kidney failure such a threat to life?

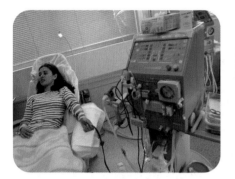

Figure 1 These 'artificial kidney machines' have not only saved countless lives, but allowed sufferers from kidney failure to lead full, active lives

The machine which carries out the functions of the kidney is known as a **dialysis machine**. It relies on the process of dialysis to clean the blood. In a dialysis machine a person's blood leaves their body and flows between partially permeable membranes. On the other side of these membranes is the dialysis fluid. The concentration of the solutes in the dialysis fluid makes sure that unwanted substances pass out of the blood by diffusion. These include urea and excess mineral ions. However, glucose and other useful substances remain in the blood. (See Figure 2.)

Without functioning kidneys, the concentration of urea and mineral ions builds up in the blood. Treatment by dialysis restores the concentrations of these dissolved substances to normal levels. Then as the patient carries on with normal life, urea and other substances build up again. So the dialysis has to be repeated at regular intervals.

It takes around eight hours for dialysis to be complete. So people with kidney failure have to remain attached to a dialysis machine for many hours several times a week. They also have to manage their diets carefully. This helps keep their blood chemistry as stable as possible.

b) What process does dialysis depend on?

How dialysis works

During dialysis, it is vital that patients lose the excess urea and mineral ions which have built up in their systems. It is equally important that they do not lose useful substances such as glucose and useful mineral ions from their blood.

The loss of these substances is prevented by the careful control of the dialysis fluid. The dialysis fluid contains the same concentration of glucose and mineral ions as normal blood plasma so there is *no* net movement of glucose and useful mineral ions out of the blood. It also contains normal plasma levels of mineral ions so excess ions are lost from the blood along a concentration gradient, but no more.

SCIENCE @ WORK

Scientists have developed a way of carrying out dialysis within the body of the patient. The dialysis fluid is put into the body cavity and removed hours later. People can do this for themselves at home without a machine. They rely on a huge support team. This includes the scientists who developed the technique, the doctors who treat the patients, the nurses who train and manage the patients and the technical support staff who deliver the bags of dialysis fluid and fetch the patients into hospital fast if things go wrong.

In contrast the dialysis fluid contains no urea, so there is a strong concentration gradient from the blood to the fluid for this substance and all the urea leaves the blood. The whole process of dialysis depends on diffusion along concentration gradients which have to be maintained by the flow of fluid. There is no active transport.

Many people go to hospital to receive dialysis, but in 1964 home dialysis machines were made available for the first time. This meant that at least some people with kidney failure could set up their dialysis in their own home. Even with all our modern technology, dialysis machines are still quite large – certainly much bigger than the kidneys they replace!

Dialysis has some disadvantages. You have to follow a very carefully controlled diet and there are regular, long sessions connected to a dialysis machine. Over many years, the balance of substances in the blood can become more difficult to control however careful the dialysis. But for many people with kidney failure, dialysis means life rather than death, because we have successfully copied the action of the living kidney in the body.

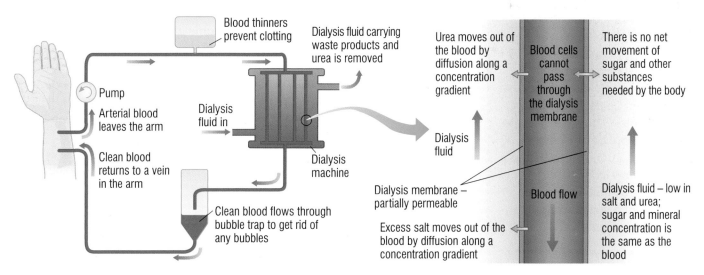

Figure 2 A dialysis machine relies on simple diffusion to clean the blood, removing the waste products which would damage the body as they built up. Changing the concentration of substances in the dialysis fluid allows unwanted substances to be removed from the blood while the concentration of useful substances remains unchanged.

SUMMARY QUESTIONS

1 Why is kidney failure potentially life threatening?

2 Produce a flow chart to explain how a dialysis machine works.

3 a) Why do people with kidney failure have to control their intake of protein and salt?
   b) Why can patients with kidney failure eat and drink what they like during the first few hours of dialysis?

4 a) Explain the importance of dialysis fluid containing no urea and normal plasma levels of salt, glucose and minerals.
   b) Both blood and dialysis fluid are constantly circulated through the dialysis machine. Explain why the constant circulation of dialysis fluid is so important.

KEY POINTS

1 People suffering from kidney failure may be treated by regular sessions on a kidney dialysis machine or by having a kidney transplant.

2 In a dialysis machine, the concentration of dissolved substances in the blood is restored to normal levels.

3 The levels of useful substances in the blood are maintained, while urea and excess salt pass out from the blood into the dialysis fluid.

B3 2.7   Kidney transplants

LEARNING OBJECTIVES

1 Why are kidney transplants sometimes rejected?
2 Which is better – dialysis or a kidney transplant?

SCIENCE @ WORK

Scientists are working hard to solve the problem of the lack of organ donors. Some are working on **xenotransplantation**, producing genetically engineered pigs and other animals with hearts and kidneys which can be used for human transplants. Other scientists are hoping **stem cell** research (see page 204) will enable them to grow new kidneys on demand, so that no-one dies waiting for a suitable organ to become available.

If your kidneys fail, your life can be saved using a dialysis machine. However once your body has been returned to health and your blood is kept balanced by regular dialysis, another treatment becomes possible – a **kidney transplant**.

What is a kidney transplant?

If your kidneys have failed they may be replaced in a transplant operation by a single healthy kidney from a **donor**. The donor kidney is joined to the normal blood vessels in the groin of the person getting the new kidney (the **recipient**). If all goes well, it will function normally to clean and balance the blood. One kidney is quite capable of keeping your blood chemistry in balance and removing your waste urea for a lifetime.

However, there are some difficulties to overcome in transplanting a kidney from one person to another. The main problem is that your new kidney comes from a different person, so the antigens on the cell surfaces will be different to yours. This means there is a risk that the donor kidney will be rejected by your immune system. When this happens your immune system destroys the new organ. Everything is done to make sure the new kidney is not rejected but it is always a risk.

a) There is just one situation where there is no risk of a new kidney being rejected. What do you think that might be?

There are a number of ways of reducing the risk of rejection. The match between the donor and the recipient is made as close as it can be. Whenever possible donor kidneys with a 'tissue type' similar to the recipient will be used so their antigens are very similar. For example, they will come from people with the same blood group. Also the recipient of the new kidney will be given drugs that suppress their immune response and stop it working (**immunosuppressant drugs**) – for the rest of their lives. As immunosuppressant drugs get better, the need for a really close tissue match is getting less important.

The down-side of these drugs is that the patients cannot deal with infectious diseases very well. They have to take great care if they become ill in any way. Most people feel this is a small price to pay for a new, working kidney and the immunosuppressant drugs are improving all the time.

Transplanted organs don't last forever – the average transplanted kidney works for around nine years. Once the organ starts to fail the patient has to return to dialysis until another suitable kidney is found.

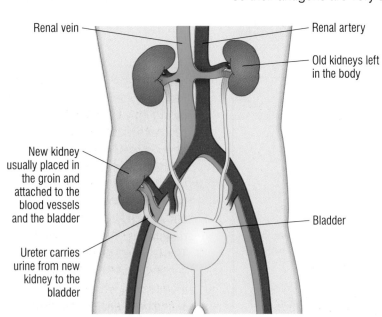

Renal vein

Renal artery

Old kidneys left in the body

New kidney usually placed in the groin and attached to the blood vessels and the bladder

Bladder

Ureter carries urine from new kidney to the bladder

Figure 1 A donor kidney is placed in the body where it takes over the functions of the organs which have failed

Dialysis v transplants

The great advantage of receiving a kidney transplant is that you no longer have to live like someone with kidney failure. You can eat what you like and are free from the restrictions which come with regular dialysis sessions. An almost completely normal life is the dream of everyone waiting for a kidney transplant.

The disadvantages are mainly to do with the risk of rejection. You have to take medicine every day of your life in case the kidney is rejected. You also need regular check-ups to see if your body has started to reject the new organ. And of course, the biggest disadvantage is that you may never get the chance of a transplant at all.

Dialysis is much more readily available than donor organs, so it is there when your kidneys fail. It enables you to lead a relatively normal life, although you are tied to a special diet and regular sessions on the machine.

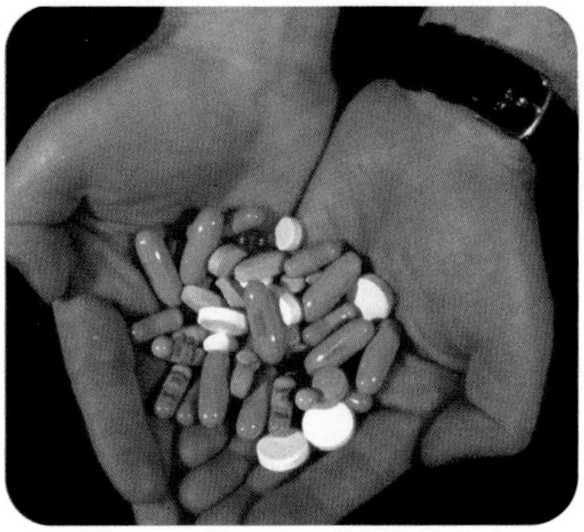

Finding the donors

One of the biggest problems in the kidney transplant programme is the lack of donor kidneys. The main source of kidneys is from people who die suddenly and unexpectedly either from road accidents or from strokes and heart attacks. In the UK, organs can only be taken from people if they carry an organ donor card or are on the on-line register giving permission for their organs to be used in this way – or if the bereaved relatives give consent.

Because many of us do not carry donor cards, there are never enough donor kidneys to go around. What's more, as cars become safer, fewer people die in traffic accidents and become potential donors. At any one time there are thousands of people having kidney dialysis who would love to have a kidney transplant but who never get the opportunity.

Figure 3 This young man has been given a new lease of life by a kidney transplant. Not everyone who suffers from kidney failure is so lucky, due to a lack of donors.

Figure 2 These are some of the immunosuppressant drugs taken by the recipients of kidney transplants. They can have unpleasant long-term side effects. However, most people feel that it is a relatively small price to pay for the hope of a normal life.

DID YOU KNOW?

Dialysis is more expensive than a transplant. It costs an average of £30 000 a year to treat a patient with dialysis. In comparison a kidney transplant costs £20 000 in the first year (including the surgery and hospital stay) and £6500 a year afterwards for the anti-rejection drugs.

SUMMARY QUESTIONS

1 How does a kidney transplant overcome the problems of kidney failure?

2 Sometimes – in very rare cases – a healthy kidney is taken from a live donor to be given to someone with kidney failure. Usually the live donor is a close family member – the parent, brother or sister of the recipient. Live donor kidney transplants have a higher rate of success than normal transplants from dead, unrelated donors.

a) Suggest two reasons why live transplants have a higher success rate than normal transplants.

b) Why do you think that live donor transplants are relatively rare?

3 Produce a table to compare the advantages and disadvantages of treating kidney failure with dialysis or with a kidney transplant. Which treatment do you think is preferable and why?

KEY POINTS

1 In a kidney transplant diseased or damaged kidneys are replaced with a healthy kidney from a donor.

2 The donor kidney may be rejected by the recipient's immune system. To try and prevent rejection the tissue types of the donor and the recipient are matched as closely as possible and immunosuppressant drugs are used.

B3 2.8   Transporting substances around the body – the past and the future

Understanding the circulatory system
A complicated story!

ACTIVITY

The History of the Heart

Choose one of the scientists mentioned here and plan a display for a Museum of Medicine on him. Decide what objects you might like to display, and write out information cards or produce a PowerPoint presentation which could accompany your display. Remember that to make it interesting you need little details about the person as well as the story of their scientific discovery. Use web resources such as www.timelinescience.org to help you.

1 The Ancient Chinese model

The circulation of the blood was understood in China by the second century BC at the latest – about two thousand years before it was accepted in Europe!

The ancient Chinese thought there were two separate circulations of fluids in the body with the heart acting as a pump. They even made a model heart with bellows and bamboo tubes!

2 Galen of Pergamum ca 130–ca 200 AD

Galen was a Greek doctor who lived in the second century AD who taught his students that there were two distinct types of blood. 'Nutritive blood' was made by the liver and consumed by the organs and 'vital blood' was made by the heart and moved through the arteries to carry the 'vital spirits'. Galen believed that the heart acted not to pump blood, but to suck it in from the veins.

Galen carried out lots of dissections and vivisections on animals. He was the first person to show that experiments were important in medicine. In spite of this he got the circulation completely wrong – but his ideas were accepted for over a thousand years!

3 Ibn-El-Nafis 1208–88

Ala'El-Deen Ibn-El-Nafis was a brilliant Arab doctor. He worked out the correct anatomy of the heart and lungs and the way the blood flowed through them. He was also the first person known to record the blood supply to the heart itself.

Around 300 years after his death, some of his writings were translated into Latin and became available in Europe. Soon afterwards some European scientists and doctors began to make the same discoveries!

4 Michael Servetus 1511–53

Michael Servetus was a Spanish doctor who was trained in Paris. He accurately described the circulation of the blood from the heart to the lungs and back again. Unfortunately he wrote about it as part of some revolutionary religious arguments which upset Protestants and Catholics alike, so his findings were ignored.

Servetus met a grisly fate. First he was burned to death as a heretic by the Protestants. A few months later he was executed again by the Catholic Inquisition. As he was already dead, they executed an effigy of him!

5 William Harvey 1578–1657

William Harvey was court physician to both King James I and King Charles I. At the same time he carried out lots of research on the blood flow in the human body. Harvey questioned the teachings of Galen and investigated them scientifically.

He carried out lots of experiments. For example, when Harvey removed the beating heart from a living animal, it continued to beat, acting as a pump, not a sucking organ.

Harvey's notes show that he developed his ideas about the circulation as early as 1615 but he didn't publish his findings until 1628. Why did he wait so long? To rebel against the ideas of Galen could quickly end the career of any doctor, and this is what Harvey risked. After his work was published, many doctors and scientists rejected him and his findings. Some of his patients left him. Controversy raged for twenty years – yet today Harvey's work is considered to be one of the most important contributions in the history of medicine.

I sold one of my kidneys last year. The money has paid for my children to go to school and my wife to have some medicine.

Our faiths all support organ donation – there is no greater gift to give someone than life itself.

Our only comfort in those awful days after the accident was the thought that although we had lost our daughter, other people would get the chance of life. We've been told that two different people had successful kidney transplants and they used her liver and heart as well…

We are desperate for our son to get a new kidney – he really struggles with dialysis. The awful thing is that someone else has to die for him to live a full life. If someone offered to sell us a kidney, I really think we would buy it

It's not that I have anything against kidney transplants, but I don't think stem cell research or transplanting parts of other animals into people should be allowed.

I don't want to think about dying. I think all this business about donor cards is just morbid!

ACTIVITY

Most people agree that kidney transplants are a good thing. However there are still some ethical issues involved. In some parts of the world organs are bought and sold. Some of the new technologies which could increase the number of organs available raise their own ethical questions. Some people simply don't want to think about it.

a) Write two short speeches, one in favour of organ donation and the other raising issues about it.

b) At the moment you have to sign up to be an organ donor. In some countries everyone is automatically an organ donor unless they opt out. This provides many more organs for transplantation. It is suggested that this method is used in the UK. What do you think – should things stay as they are or should they change?

Write a brief paragraph explaining your personal point of view.
Then design a poster or leaflet supporting kidney transplants to be handed out at blood donation sessions, in hospitals and at doctor's surgeries.

SUMMARY QUESTIONS

1 Here are descriptions of two heart problems and how they may be overcome. In each case use what you know about the heart and the circulatory system to explain the problems caused by the condition and how the treatment helps.

a) Sometimes babies are born with a 'hole in the heart' – there is a gap in the central dividing wall of the heart. They may look blue in colour and have very little energy. Surgeons can close up the hole.

b) The blood vessels supplying blood to the heart muscle itself may become clogged with fatty material. The person affected may get chest pain when they exercise or even have a heart attack. Doctors may be able to replace the clogged up blood vessels with bits of healthy blood vessels taken from other parts of the patient's body.

2 a) Explain how red blood cells are adapted for carrying oxygen around your body.

b) In each of the following examples, explain the effect on the blood and what this will mean to the person involved:
   i) an athlete running a race after acting as a blood donor and giving blood,
   ii) someone who eats a diet low in iron.

3 It is often said that taking regular exercise and getting fit is good for your heart and your lungs.

a) Make bar charts to show the data given on page 243.

b) Use the information on your bar charts to explain exactly what effect increased fitness has on
   i) your heart, and
   ii) your lungs.

4 a) Explain the importance of dialysis fluid containing no urea and normal plasma levels of salt, glucose and minerals for the successful treatment of the blood.

b) Both blood and dialysis fluid are constantly circulated through the dialysis machine. Explain why the constant circulation of dialysis fluid is so important.

c) What are the main similarities and differences between a working kidney and a dialysis machine?

EXAM-STYLE QUESTIONS

1 The table below shows which parts of the blood transport different substances from one part of the body to another. Some words have been replaced by the letters **A–G**. State the word or words represented by these letters. (7)

Part of the blood	Substance transported	Transported from	Transported to
A	Carbon dioxide	Various organs	B
Blood plasma	Soluble products of digestion	C	Various organs
Blood plasma	D	E	Kidneys
F	G	Lungs	Various organs

2 The diagram shows the rate of blood flow to various parts of the body when a person is at rest and when they are exercising.

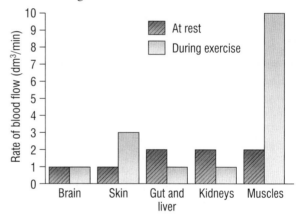

(a) What is the volume of blood that is pumped each minute to the muscles when:
   (i) at rest? (1)
   (ii) during exercise? (1)

(b) (i) What is the total volume of blood that is pumped each minute to **all** of the organs when the person is exercising? (1)
   (ii) What is the percentage increase in blood pumped to all the organs when the person is exercising compared to the volume pumped when at rest? (1)

(c) Suggest reasons for the changes in the rate of blood flow to the following organs during exercise compared to the flow when at rest.
   (i) muscles (2)
   (ii) skin (2)
   (iii) gut and liver (2)

(d) Variation in the distribution of blood to organs is one change that occurs during exercise. State two other changes that take place within the body when exercise is taking place. (2)

3 The passage below about anaerobic respiration has some words replaced by the numbers **1**, **2**, **3** and **4**. Match the words **A**, **B**, **C** and **D** with the appropriate numbers in each sentence of the passage.

A energy   **B** glucose
C oxygen   **D** water

Anaerobic respiration is the partial breakdown of ....**1**.... to produce lactic acid.

Compared to aerobic respiration much less ....**2**.... is released.

Anaerobic respiration results in an ....**3**.... debt.

This debt must be repaid in order to convert the lactic acid to carbon dioxide and ....**4**.... [Higher]

4 The kidneys remove wastes from the liquid portion of the blood.

(a) Name the solution of waste that is stored in the bladder. (1)

(b) Describe how a healthy kidney produces this solution of waste. (5)

(c) If a person's kidney fails to function properly they may die because toxic substances cannot be removed from the blood. One way in which they may be kept alive is by using a dialysis machine. During dialysis, blood passes through branched or coiled tubes on the other side of which is dialysis fluid.

(i) Suggest a reason why the tubes containing blood are branched or coiled. (1)
(ii) What special property does the membrane of the tubes possess? (1)
(iii) Name one substance that diffuses from the blood into the dialysis fluid. (1)
(iv) Why do useful substances such as glucose and some mineral ions not also pass out of the blood? (1)

(d) Another form of treatment is to replace a diseased kidney with one transplanted from a donor. The new kidney may however be rejected by the immune system of the person receiving it. State three ways in which rejection of the transplanted kidney may be prevented. (3)

HOW SCIENCE WORKS

Kidney dialysis machines are designed to remove harmful substances such as urea and excess salts and water from the body. They also maintain a healthy level of potassium and sodium ions. People who need to use a dialysis machine will visit a treatment centre about three times a week for about 3 to 5 hours.

Their blood supply is connected to the machine. Every month the patient's blood is checked to ensure that the machine is working properly. The blood is tested for its urea content (URR test). The results have to be above 64%. Another test checks the amount of blood being filtered compared with the amount of fluid in the body (Kt/V test). This needs to be more than 1.1.

Look at this chart for a patient and answer the questions that follow.

Test	Target	Jan	Feb	Mar	April	May	June	July	Aug
Kt/V	≥ 1.2	1.1	1.15	1.2	1.23	1.24	1.2	1.2	1.2
URR	≥ 65	60	62	64	65	66	65	65	65
Weight (kg)	80	81	80	82	81	79	80	81	81

a) What was the range for the Kt/V test? (1)
b) What was the pattern for the Kt/V test? (3)
c) How does the Kt/V test results compare to those for the URR test? (2)
d) Can you say that there is a causal link between the two sets of test results? (1)
e) For how many months were the patient's tests both satisfactory? (1)
f) The doctors were concerned at the beginning of the year that the test results were a little low for this patient. What should they have done to get a better idea of what was happening? (1)
g) Urologists say that the two tests really measure the same thing. Why then is it a good idea to do both tests? (1)
h) What are the economic issues related to kidney dialysis? (2)
i) What are the social issues related to kidney dialysis? (2)

B3 3.1 | Growing microbes

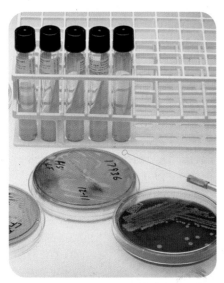

Figure 1 Culturing microorganisms like bacteria makes it possible for us to see what they are like and what they need to grow

Microbiology is the study of microorganisms. These are tiny living organisms such as bacteria, viruses, fungi and protoctista which are usually too small to be seen with the naked eye. To see and understand them properly you need to use a microscope. There are many different reasons why we study microorganisms. They play a vital role in decay and the recycling of nutrients in the environment. (See pages 168–9.) They can cause disease and they are enormously useful to people.

Many microorganisms can be grown in the laboratory. This allows us to learn a lot more about them. We can find out which nutrients they need to survive and which chemicals will kill them. We can also discover which microbes can be useful to us and which cause deadly disease.

a) What are the four main types of microorganisms?

What do microbes need to grow?

If you want to find out more about microorganisms you need to **culture** them. In other words, grow very large numbers of them so that you can see the colony as a whole.

To culture microorganisms you must provide them with everything they need. This usually involves providing a culture medium containing carbohydrate to act as an energy source. Along with this, various mineral ions and in some cases extra, supplementary protein and vitamins are included.

The nutrients are often contained in an **agar** medium. Agar is a substance which dissolves in hot water and sets to form a jelly. You pour hot agar containing all the necessary nutrients into a **Petri dish**. Then leave it to cool and set before you add any microorganisms. The other way to provide nutrients to grow microorganisms is as a broth in a culture flask. Whichever way you do it, you usually need to provide warmth and oxygen as well.

b) What do bacteria need to grow?

Safety precautions in the lab

You have to take great care when you culture microorganisms. This applies even when the microbes you want to grow are completely harmless. There is always the risk that a mutation may take place resulting in a new and dangerous pathogen. Also your culture may be contaminated by disease-causing pathogens which are present in the air, soil or water around you.

Not only do you want to avoid contamination by any dangerous miroorganisms, you also need to keep any pure strains of bacteria that you are growing free from other microorganisms. There are always millions of microbes – some harmful and some not – in the air, on the lab surfaces and on your own skin. So whenever you are culturing microorganisms, you must carry out very strict health and safety procedures.

PRACTICAL

Growing microbes safely

Sterilise the Petri dishes and the nutrient agar before they are used. This kills unwanted microorganisms. You sterilise them using heat, often in a special oven known as an autoclave. The autoclave uses steam at high pressure.

Then inoculate the sterile nutrient agar with the microorganism you want to grow following the steps shown below:

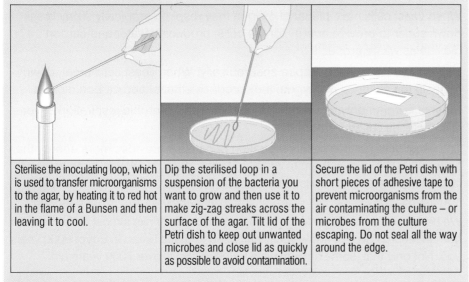

| Sterilise the inoculating loop, which is used to transfer microorganisms to the agar, by heating it to red hot in the flame of a Bunsen and then leaving it to cool. | Dip the sterilised loop in a suspension of the bacteria you want to grow and then use it to make zig-zag streaks across the surface of the agar. Tilt lid of the Petri dish to keep out unwanted microbes and close lid as quickly as possible to avoid contamination. | Secure the lid of the Petri dish with short pieces of adhesive tape to prevent microorganisms from the air contaminating the culture – or microbes from the culture escaping. Do not seal all the way around the edge. |

Figure 4 Culturing microorganisms in the lab

Once you have taken these steps, you need to incubate the sealed Petri dishes to allow the microorganisms to grow. In schools and college labs the maximum temperature at which cultures should be incubated is 25°C. This relatively low temperature greatly reduces the likelihood of pathogens growing which might be harmful to people. In industrial conditions, bacterial cultures are often grown at higher temperatures to promote rapid growth of the microbes. Once you have observed the microbes you have grown, you need to re-sterilise your Petri dishes complete with their cultures by heating them to 100°C and then throwing them away.

SUMMARY QUESTIONS

1 Why do we culture microorganisms in the lab?

2 Why do you think that all bacteria need a carbohydrate energy source as part of their nutrient medium?

3 Explain why cultures are not incubated at 37°C in a school lab.

4 Look at the graph in Figure 3.

   a) What does this tell you about the growth of bacteria in ideal conditions?

   b) A Petri dish provides ideal conditions for bacteria as they start to grow, but the ideal conditions do not last forever. What might limit the growth of the bacteria in a culture on a Petri dish?

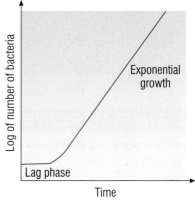

Figure 3 If you give bacteria the right conditions they can grow and divide very rapidly – which is why it is relatively easy to culture them in the lab

FOUL FACTS

You are surrounded by disease-causing bacteria all the time. If you cultured bacteria at 37°C – human body temperature – there would be a very high risk of growing some very nasty pathogens indeed.

Make sure you can explain why we culture bacteria at 25°C rather than 37°C in the lab at school.

KEY POINTS

1 Microorganisms can be grown in an agar culture medium with a carbohydrate energy source and various minerals, vitamins and proteins.

2 You need to take careful safety measures and use sterilised equipment to grow uncontaminated cultures of microorganisms and to avoid the growth of harmful pathogens.

B3 3.2

Food production using yeast

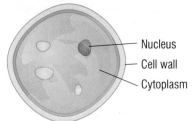

- Nucleus
- Cell wall
- Cytoplasm

Figure 1 Yeast cells – these microscopic organisms have been useful to us for centuries

One of the microorganisms which is most useful to people is yeast. The yeasts are single celled organisms. Each yeast cell has a nucleus, cytoplasm, and a membrane surrounded by a cell wall. The main way in which yeasts reproduce is by asexual budding – splitting in two to form new yeast cells.

When yeast cells have plenty of oxygen they respire aerobically. They break down sugar to provide energy for the cells, producing water and carbon dioxide as waste products.

However yeast can also respire anaerobically. When yeast cells break down sugar in the absence of oxygen, they produce ethanol and carbon dioxide.

Ethanol is commonly referred to as alcohol. The anaerobic respiration of yeast is sometimes called *fermentation*.

The yeast cells need aerobic respiration because it provides more energy than anaerobic. This allows them to grow and reproduce. However, once there are large numbers of yeast cells, they can survive for a long time in low oxygen conditions. They will break down all the available sugar to produce ethanol.

We have used yeast for making bread and alcoholic drinks almost as far back as human records go. We know yeast was used to make bread in Egypt 6000 years ago. Not only that, some ancient wine found in Iran is over 7000 years old.

a) What are the advantages to yeast of being able to use both aerobic and anaerobic respiration?

PRACTICAL

Using yeast to make bread

When you make basic bread, you mix yeast with sugar to provide it with an energy source for respiration. Just 1 gram of yeast contains about 25 billion cells! Then you mix the yeast and sugar with water and flour. Kneading the mixture makes sure the yeast is evenly spread throughout the dough and improves the texture. Then you leave the mixture somewhere warm.

As the yeast grows and respires it produces carbon dioxide, making the bread rise. When you bake the bread the bubbles of gas expand in the high temperature, giving the cooked bread a light texture.
The yeast cells are killed during the cooking process.

b) What by-product of yeast respiration makes bread rise?

Figure 2 When you make bread the dough is usually left to rise at least twice, to give the yeast a chance to make as much carbon dioxide as possible

Making alcoholic drinks

When fruit fall to the ground and begin to decay, wild yeasts on the skin break down the fruit sugar. They form ethanol and carbon dioxide. These fermented fruits can cause animals to become drunk when they eat them. This is probably how our ancestors discovered alcohol!

We now use this same reaction in a controlled way to make beers and wine. In both cases the yeast must be supplied with carbohydrates to act as an energy source for respiration.

Beer making depends on a process called **malting**. You soak barley grains in water and keep them warm. As germination begins, enzymes break down the starch in the barley grains into a sugary solution. You then extract the sugary solution produced by malting and use it as an energy source for the yeast. The yeast and sugar solution mixture is then fermented to produce alcohol. You can add hops at this stage to give flavour. Finally you allow the beer to clear and develop its flavour fully before putting it in barrels, bottles or tins to be sold.

In contrast, *wine making* uses the natural sugar found in fruit such as grapes as the energy source for the yeast. You press the grapes and mix the juice with yeast and water. You then let the yeast respire anaerobically until most of the sugar has been used up. At this stage you filter the wine to remove the yeast and put it in bottles. It will be stored for some time to mature before it is sold. Most wine sold commercially is made from grapes. However, wine can actually be made from almost any fruit or vegetable. The yeast isn't at all fussy about where the sugar it uses comes from!

Interestingly, alcohol in large amounts is poisonous to yeast as well as to people. This is why the alcohol content of wine is rarely more than 14%. Once it gets much higher, it kills all the yeast and stops the fermentation!

SUMMARY QUESTIONS

1 Produce word equations to show the difference between aerobic and anaerobic respiration in yeast.

2 a) In breadmaking, why is the yeast mixed with sugar before it is added to the flour and water?
  b) Why is the bread dough kept warm while it is rising?
  c) Why are cooked rolls or loaves bigger than the risen dough which is placed in the oven?

3 a) Make a flow chart to explain the process of making beer.
  b) Explain how wine making is similar to and different from brewing beer.

Figure 3 Brewing at Ringwood Brewery in Hampshire. In tanks like these billions of yeast cells respire anaerobically, turning sugar into alcohol every day of the year. Who would guess, looking at the finished product, where it has all come from?!

GET IT RIGHT!

Remember: yeast can respire aerobically in bread making but *must* respire anaerobically to make alcoholic drinks.

KEY POINTS

1 Yeast is a single-celled organism which can respire aerobically producing carbon dioxide and water. This reaction is used in breadmaking to make the dough rise.
2 Yeast can also respire anaerobically producing ethanol and carbon dioxide in a process known as fermentation.
3 The fermentation reaction of yeast is used to produce ethanol in the production of beer and wine.

B3 3.3   Food production using bacteria

Figure 1 In the UK we get most of our milk from cows, but around the world a number of different animals including camels, horses, sheep and goats are used for milking

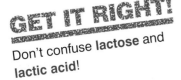

Don't confuse **lactose** and **lactic acid**!

People began to domesticate animals quite early in history. They soon realised that the milk that female animals made for their babies could be used as food for us too! However, there is one big drawback in using milk as part of the diet. It very rapidly goes off, starts to smell and tastes disgusting! However, it didn't take people long to find ways of changing the milk into foods which lasted much longer than milk itself. These changes depended on the action of microorganisms.

Yoghurt has long been a staple part of the diet in the Middle East but has only become popular in the UK quite recently. Cheese, on the other hand, has been around for a very long time almost all over the world.

a) How many different animals can you think of that are used to provide milk for people?

Making yoghurt

Traditionally, yoghurt is fermented whole milk but now we can make it from semi-skimmed milk, skimmed milk and even soya milk. Yoghurt is formed by the action of bacteria on the **lactose** (milk sugar) in the milk.

b) Why is yoghurt so much thicker than milk?

Figure 2 In just a few years yoghurts have become one of the most widely eaten dairy products in the UK!

Cheese-making

Like yoghurt-making, cheese-making depends on the reactions of bacteria with milk. These change the texture and taste and also preserve the milk. Cheese-making is very successful at preserving milk, and some cheeses can survive for years without decay.

Just as in yoghurt-making, you add a starter culture of bacteria to warm milk. The difference is in the type of bacteria added. The bacteria in cheese making also convert lactose to lactic acid, but they make much more of the lactic acid. As a result the solid part (the curds) are much more solid than the yoghurt ones.

Enzymes are also added to increase the separation of the milk. When it has completely curdled you can separate the curds from the liquid whey. Then you can use the curds for cheese-making. The whey is often used as animal feed.

Next you cut and mix the curds with salt along with other bacteria or moulds before you press them and leave them to dry out. The bacteria and moulds which you add at this stage of the process are very important. They affect the development of the final flavour and texture of the cheese as it ripens. The ripening may take months or years depending on the cheese being made.

Figure 3 Curds are formed by the action of one set of bacteria on the milk. Then the final flavour and texture of the cheese may well depend on other bacteria added at this stage of the process.

KEY POINTS

1 Bacteria are used in making both yoghurt and cheese.
2 In the production of yoghurt a starter culture of bacteria acts on warm milk. Lactose is converted to lactic acid in a lactic fermentation reaction. This changes the texture and taste of the milk to make yoghurt.
3 In cheese-making a different starter culture is added to warm milk giving a lactic fermentation which results in solid curds and liquid whey. The curds are often mixed with other bacteria or moulds before they are left to ripen into cheese.

SUMMARY QUESTIONS

1 Produce a flow chart to summarise the production of yoghurt from milk.

2 Produce a flow chart to summarise the production of cheese from milk.

3 Compare the processes of yoghurt-making and cheese-making and summarise the differences between them.

B3 3.4   Large-scale microbe production

It is one thing to grow microorganisms in the laboratory on a small scale in a Petri dish. In those conditions it is relatively easy for you to provide the food, oxygen and warmth the microorganisms need to grow. The cultures are not usually kept for long enough for the build-up of waste products to be a problem.

But it becomes a very different story when we need to grow microorganisms on an industrial scale. Increasingly we want to make use of materials made by the microbes. These may be drugs, like antibiotics, or food. This means we need to keep a very large fermentation going for a long time so we can harvest the products.

Scaling up brings its own problems. Imagine moving a culture of microorganism from an agar plate to vessels which might hold $1\,000\,000\,dm^3$ of medium. This is a very big step to manage successfully.

a) Which drugs are commonly made by microbes?

Fermenters

In ideal conditions the numbers of bacteria can double every 20 minutes. However, ideal conditions are rare and what usually happens in a bacterial culture is shown in Figure 1.

As the numbers of microorganisms begin to rise, conditions change. The food is used up. The metabolism of all the millions of microorganisms causes the temperature to rise. Oxygen levels fall as it is used up in respiration.

The carbon dioxide waste from respiration can also alter the pH of the culture. If the pH changes, the activity of the enzymes in the culture can be affected so it stops growing or dies. Other waste products may begin to build up and poison the culture. In industrial fermentations, very large microbe cultures are involved. So problems like these can develop very rapidly.

When we grow microbes on an industrial scale we use large vessels known as **fermenters**. These industrial fermenters are designed to overcome the problems which stop a culture from growing well. They react to changes, keeping the conditions as stable as possible. This in turn means we can get the maximum yield. Industrial fermenters usually have:

● an oxygen supply to provide oxygen for respiration of the microorganisms,
● a stirrer to keep the microorganisms in suspension. This maintains an even temperature and makes sure that oxygen and food are evenly spread throughout the culture,
● a water-cooled jacket which removes the excess heat produced by the respiring microorganisms. Any rise in temperature is used to heat the water which is constantly removed and replaced with more cold water,
● measuring instruments which constantly monitor factors such as the pH and temperature so that changes can be made if necessary.

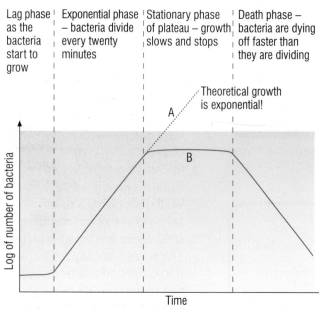

| Lag phase as the bacteria start to grow | Exponential phase – bacteria divide every twenty minutes | Stationary phase of plateau – growth slows and stops | Death phase – bacteria are dying off faster than they are dividing |

Theoretical growth is exponential!

A

B

Log of number of bacteria

Time

Figure 1 In real life rather than ideal conditions, all sorts of factors make bacterial growth slow down

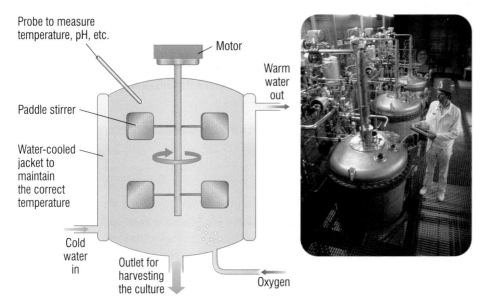

Probe to measure temperature, pH, etc.

Motor

Paddle stirrer

Warm water out

Water-cooled jacket to maintain the correct temperature

Cold water in

Outlet for harvesting the culture

Oxygen

Figure 2 The design of fermenters is improving all the time as new ways are developed to keep conditions inside the fermenter as stable as possible

b) Why does a change in pH in a fermenter have such a serious effect on the microorganisms?

Mycoprotein production

Recently a completely new food based on fungi has been developed. It is known as **mycoprotein**, which means 'protein from fungus'. It is produced using the fungus *Fusarium*. This grows and reproduces rapidly on a relatively cheap sugar syrup (made from waste carbohydrate) in large specialised fermenters. It needs aerobic conditions to grow successfully. Then it can double its mass every five hours!

The fungal biomass is harvested and purified. Then it is dried and processed to make mycoprotein. This is a pale yellow solid with a faint taste of mushrooms. On its own it has very little flavour.

However, mycoprotein can be given a range of tastes and flavours to make it similar to many familiar foods. It is a high-protein, low-fat meat substitute. So it is used by vegetarians and people who want to reduce the fat in their diet plus people who just like the taste!

When mycoprotein was first developed people thought a world food shortage was on its way. They were looking for new ways to make protein cheaply and efficiently. The food shortage never happened, but the fungus based food continued. It is versatile, high in protein and fibre, low in fat and calories and so is widely used in the developed world.

Figure 3 Mycoprotein can be made to look like meat, chicken and fish. You can have mycoprotein pies and burgers. It is also sold relatively unflavoured for people to use in their own way. Mycoprotein can even be ground into flour that can be used to make snacks and a variety of sweet things.

DID YOU KNOW?

The protein content of mycoprotein is similar to that of prime beef!

KEY POINTS

1 Microorganisms can be grown on a large scale in vessels known as fermenters to make useful products such as antibiotics and mycoprotein food.

2 Industrial fermenters usually have a range of features to make sure the fermentation takes place in the best possible conditions for a maximum yield of the product.

3 The fungus *Fusarium* is grown on sugar syrup in aerobic conditions to produce mycoprotein foods.

SUMMARY QUESTIONS

1 a) Describe how curve B differs from curve A in Figure 1.
  b) What causes the difference between the growth pattern in ideal conditions and the growth pattern in real conditions?

2 How might an industrial fermenter be adapted to ensure the best possible conditions for the growth of a particular microorganism?

3 Why do the following factors tend to change during a fermentation process?
  a) Temperature.   b) Oxygen levels.   c) pH levels.

B3 3.5   Antibiotic production

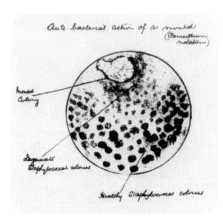

Figure 1 The keen eyes of Alexander Fleming noticed the clear areas on his plates. He realised he had made a discovery of enormous potential.

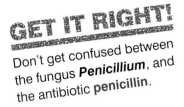

GET IT RIGHT!

Don't get confused between the fungus **Penicillium**, and the antibiotic **penicillin**.

Figure 2 Alexander Fleming, Howard Florey and Ernst Chain each received a well-deserved Nobel Prize for their work on the miracle drug penicillin. This reflected the fact that it wasn't just the original discovery that was important. It wasn't until penicillin could be made on a large scale at plants like this early production unit in the UK that the impact of the drug on our health could really be seen.

Penicillin is one of the best known medicines in the world. It has revolutionised medicine since it was first manufactured.

The discovery of penicillin

In 1928, Alexander Fleming was a young researcher at St Mary's Medical School, London. He left some of the plates on which he was culturing bacteria uncovered near an open window. When he remembered to look at them he found bacteria were growing on the surface of his dishes as he expected. But Fleming also noticed spots of mould growing, and around them were clear areas of agar. The bacteria were no longer growing. Whatever had blown in and started growing on his plates was producing a chemical which killed the bacteria.

a) Which parts of Fleming's experimental technique would not be allowed in your school labs?

Fleming found that the microorganism which had invaded his Petri dishes was a common mould called *Penicillium notatum*. He set about trying to extract the substance which killed bacteria but found it almost impossible with the technology available at that time.

He managed to get a tiny amount and used it to treat an infected wound. He called his extract **penicillin**. But it was very difficult to extract, and very unstable once he had got it. So Fleming decided he wouldn't be able to get useful amounts of penicillin from his mould. He saved his cultures and returned to other areas of research.

b) Why did Fleming give up on penicillin?

During the Second World War the need for a drug to kill bacterial infections became ever more urgent. Howard Florey and Ernst Chain were working at Oxford University in a desperate search to find an antibacterial drug. They turned to Fleming's mould and finally managed to extract enough penicillin to show what it could do.

After successful animal trials, they tried to save the life of a London policeman dying of a blood infection. The dying man made an amazing recovery. However, the supply of penicillin ran out, the infection regained its hold and he died. More months of work produced enough penicillin to save the life of a boy dying of a bacterial infection. But what was needed was enough of the drug to treat thousands of wounded and sick soldiers. British factories were dedicated to the war effort, so Chain and Florey turned to the American pharmaceutical industry for help in developing a manufacturing process.

Fleming's original mould was incredibly difficult to grow in large cultures. Then a mould growing on a melon in a market was found to yield 200 times more penicillin than the original. What's more it grew relatively easily in deep tanks, making large-scale production possible. By 1945, we were producing enough penicillin each year to treat 7 million people.

Modern penicillin production

The production of penicillin did not stop with the end of the war. We now use modern strains of *Penicillium* mould which give even higher yields. We grow the mould in a sterilised medium. It contains sugar, amino acids, mineral salts and other nutrients. It is made from soaking corn in water (corn steep liquor mould is grown in huge 10 000 dm³ fermenters).

We use huge 10 000 dm³ fermenters which have strong paddles to keep stirring the broth. That's because the penicillin mould needs lots of oxygen to thrive. Sterile air is blown in to provide the oxygen. We control the temperature by a cooling jacket which surrounds the fermenter.

During the first 40 hours of the fermentation the mould grows rapidly, using up most of the nutrients in the broth. It is only after most of the nutrients have gone that the mould begin to make penicillin. This is why there is a 40-hour lag period from the start of the fermentation to the start of the production of penicillin. We have to provide enough food to allow lots of mould to grow. Then limit the supplies so that it produces the penicillin we need!

Over a period of about 140 hours broth is removed regularly and small amounts of nutrients are added. This allows us to get the maximum yield of the drug, which is then extracted from the broth. It is purified and turned into medicines – almost 30 000 tonnes a year! Although we now have many other antibiotics, penicillin is still used all over the world.

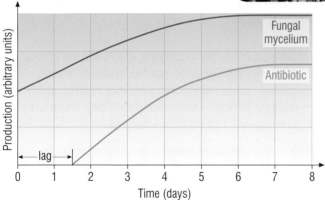

Figure 3 The production of penicillin is geared to the growth of the *Penicillium* mould, which doesn't start making the chemical we need until it starts to run out of food!

KEY POINTS

1 The antibiotic penicillin was discovered by Alexander Fleming. The method of mass production was the work of Howard Florey and Ernst Chain.
2 Penicillin is made by growing the mould *Penicillium* in a fermenter.
3 The medium contains sugars and other nutrients and has a good supply of oxygen.
4 The mould only starts making penicillin after most of the nutrients are used up.

SUMMARY QUESTIONS

1 Produce a timeline to show the stages in the discovery and development of penicillin.

2 Why has the development of penicillin had such an impact on the quality of human life?

3 What have been the difficulties of producing penicillin on a commercial scale?

B3 3.6   Biogas

Figure 1 Biogas generators like this have made an enormous difference to many families by producing cheap and readily available fuel

FOUL FACTS

Biogas is produced naturally in sewers and rubbish dumps. In the days before electricity, biogas was taken from the London sewers and used as fuel for the gas lamps which lit the steets.

Everyone needs fuel of some sort but there is only a finite amount of fossil fuels like coal, oil and gas to use. Even wood and peat are getting scarcer. Around the world, we all need other, renewable forms of fuel. The generation of **biogas** from human and animal waste is becoming increasingly important in both the developing and the developed world.

What is biogas?

Biogas is a flammable mixture of gases formed when bacteria break down plant material or the waste products of animals in anaerobic conditions. It is mainly **methane** but the composition of the mixture varies. It depends on what is put into the generator and which bacteria are present:

Type of gas	Percentage in the mixture by volume
methane	50–80
carbon dioxide	15–45
water	5
other gases including hydrogen	0–1
hydrogen sulfide	0–3

a) What is the main component of biogas?

Biogas generators

Around the world millions of tonnes of faeces and urine are made by animals like cows, pigs, sheep and chickens. We produce our fair share of waste materials too! Also, in many places, plant material grows very rapidly. Both the plant material and the animal waste contain carbohydrates. They make up a potentially enormous energy resource – but how can we use it?

The bacteria involved in biogas production work best at a temperature of around 30°C. So biogas generators tend to work best in hot countries. However, the process generates heat (the reactions are *exothermic*). This means that if you put some heat energy in at the beginning to start things off, and have your generator well insulated to prevent heat loss, biogas generators will work anywhere.

Under ideal conditions, 10 kg of dry dung can produce 3 m³ of biogas. That will give you 3 hours of cooking, 3 hours of lighting or 24 hours of running a refrigerator. Not only that, but you can use the waste from your generator as a fertiliser!

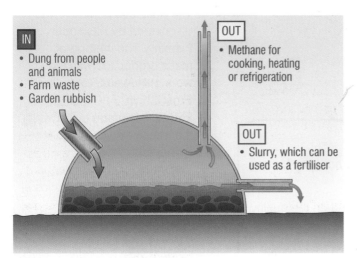

IN
• Dung from people and animals
• Farm waste
• Garden rubbish

OUT
• Methane for cooking, heating or refrigeration

OUT
• Slurry, which can be used as a fertiliser

Figure 2 Biogas generators take in waste material or plants and biogas and useful fertilisers come out the other end

b) What is an exothermic reaction?

Scaling up the process

At the moment most biogas generators around the world operate on a relatively small scale. They supply the energy needs of one family, a farm or at most a whole village.

What you put into your small generator has a big effect on what comes out. Biogas units are widely used in China. There are well over 7 million biogas units, which produce as much energy as 22 million tonnes of coal. Waste vegetables, animal dung and human waste are the main raw materials. These Chinese digesters produce excellent fertiliser but relatively low-quality biogas.

In India, there are religious and social taboos against using human waste in biodigesters. As a result only cattle and buffalo dung is put into the biogas generators. This produces very high quality gas, but much less fertiliser.

There are also different sizes and designs of biogas generators. The type chosen will depend on local conditions. For example, many fermenters are sunk into the ground, which provides very good insulation. Others are built above ground, which may be easier and cheaper but offers less insulation. If the night-time temperatures fall a long way it could cause problems.

Many countries are now looking at biogas generators and experimenting with using them on a larger scale. The waste material we produce from sugar factories, sewage farms and rubbish tips could be used to produce biogas. We have some problems to overcome with scaling the process up, but early progress looks promising.

Biogas could well be an important fuel for the future for all of us. It would help us to get rid of much of the waste we produce as well as providing a clean and renewable energy supply.

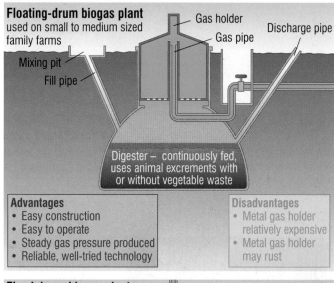

Floating-drum biogas plant used on small to medium sized family farms

Gas holder
Gas pipe
Discharge pipe
Mixing pit
Fill pipe
Digester – continuously fed, uses animal excrements with or without vegetable waste

Advantages
- Easy construction
- Easy to operate
- Steady gas pressure produced
- Reliable, well-tried technology

Disadvantages
- Metal gas holder relatively expensive
- Metal gas holder may rust

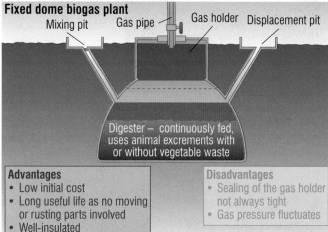

Fixed dome biogas plant

Mixing pit
Gas pipe
Gas holder
Displacement pit
Digester – continuously fed, uses animal excrements with or without vegetable waste

Advantages
- Low initial cost
- Long useful life as no moving or rusting parts involved
- Well-insulated

Disadvantages
- Sealing of the gas holder not always tight
- Gas pressure fluctuates

Figure 3 Different types of biogas generators have different advantages and disadvantages – it is a case of choosing the right design for the conditions available

SUMMARY QUESTIONS

1 Why is it important that there are many different bacteria present in a biogas generator rather than a culture containing just one or two microbes?

2 Some types of biogas generators are set up with a large amount of plant material like straw and a starter mixture of bacteria, and left to produce gas. These **batch digesters** are very effective. Once gas generation begins to drop, the generator is emptied and cleaned out and the process starts again.

   a) Using a generator like this, how could you be sure of a continuous supply of gas for cooking?
   b) What are the advantages of a batch type digester over the types shown in Figure 3?

3 What are the advantages of a) a floating drum, and b), fixed dome biogas plant, over a batch type digester?

KEY POINTS

1 Biogas – mainly methane – can be produced by anaerobic fermentation of a wide range of plant products and waste materials that contain carbohydrates.

2 Many different organisms are involved in the breakdown of material in biogas production.

B3 3.7   More biofuels

LEARNING OBJECTIVES

1 How can yeast produce fuel for your car?
2 What is the environmental impact of biofuels?

Figure 1 The Sun and the rain in areas like the Caribbean allow plants like this sugar cane to photosynthesise and grow very rapidly. The next step is to convert them into usable fuel.

GET IT RIGHT!

At the moment the advantages and disadvantages of using biofuels depends on the space, climate and economy of a country – but from an environmental point of view everyone benefits.

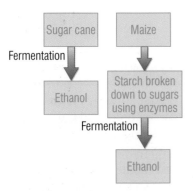

Figure 2 The starch in maize needs to be broken down by enzymes before yeast can use it as fuel for anaerobic respiration. Although it takes more steps to produce ethanol from maize than from sugar cane, maize can be grown in many more countries around the world.

In tropical countries plants grow fast. Sugar cane grows about 4 to 5 metres in a year. It has a juice which is very high in carbohydrates, particularly sucrose. Maize (which we often refer to as sweet corn) is another fast grower. We can break the starch in the maize kernels down into glucose using the enzyme **carbohydrase**. (See page 186.) But can we convert the carbohydrates we grow into clean and efficient fuels?

Ethanol-based fuels

If sugar-rich products from cane and maize are fermented anaerobically with yeast, the sugars are broken down incompletely to give **ethanol** and water. You can extract the ethanol from the products of fermentation by **distillation**, and you can then use it in cars as a fuel.

Car engines need special modification to be able to use pure ethanol as a fuel, but it is not a major job. Many cars can run on a mixture of petrol and ethanol without any problems at all.

a) Why is it cheaper to produce ethanol-based fuels from sugar cane than from maize?

The advantages and disadvantages of ethanol as a fuel

In many ways ethanol is an ideal fuel. It is efficient and it does not produce toxic gases when you burn it. It is much less polluting than conventional fuels which produce carbon monoxide, sulfur dioxide and nitrogen oxides. In addition, you can mix ethanol with conventional petrol to make a fuel known as gasohol. This is being done increasingly in the USA and it reduces pollution levels considerably. However, there is still some pollution from the petrol part of the mix.

Using ethanol as a fuel is known as **carbon neutral**. This means there is no overall increase in carbon dioxide in the atmosphere when you burn ethanol. The original plants removed carbon dioxide from the air during photosynthesis. When you burn the ethanol, you simply return it.

The biggest difficulty with using plant-based fuels for our cars is that it takes a lot of plant material to produce the ethanol. As a result, the use of ethanol as a fuel has largely been limited to countries with enough space and a suitable climate to grow lots of plant material as fast as possible.

Brazil was the trailblazer for ethanol as a fuel when oil prices shot up in the 1970s. The Brazilians grew their own 'green petrol' and slashed the money paid out on oil imports which were crippling the economy of the country. They were very successful, and in the 1980s 90% of the cars produced in Brazil had ethanol-powered engines.

However, when oil prices dropped again the Brazilian government couldn't afford to subsidise ethanol as a fuel. As a result they began to move back to petrol driven cars.

Now people all over the world are worried about the environmental problems caused by the burning of fossil fuels. Interest in clean alternatives such as ethanol is soaring. Brazil is again taking a lead, supporting countries such as India with advice on technology for producing ethanol from plants.

b) What is meant by the term 'carbon neutral'?

In America, the use of gasohol – a mixture of 90% petrol and 10% ethanol – is increasing all the time. A lot of the ethanol is fermented and distilled from maize grown within the USA itself. However as the use of ethanol grows the Americans are also importing ethanol from places like Brazil and the Caribbean. In 2004, the USA imported 160 million gallons of ethanol fuel on top of its own production!

The main problem is finding enough ethanol. If we Europeans added 5% ethanol to our fuel it would reduce carbon dioxide emissions – but we would need 7.5 billion litres of ethanol a year. That's more than half of the total production level in Brazil!

The methods of ethanol production we use at the moment leave large quantities of unused cellulose from the plant material. To make ethanol production work financially in the long term, we need to find a way to use this cellulose. We might develop biogas generators which can break down the excess cellulose into methane, another useful fuel.

Genetically engineered bacteria or enzymes may be able to break down the cellulose in straw and hay and make it available for yeast to make more ethanol. We don't know exactly what the future will hold, but it seems likely that ethanol-based fuel mixe s will be part of it!

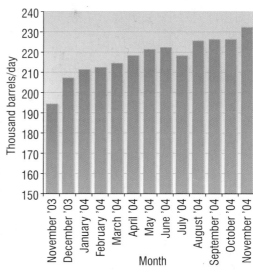

Figure 3 Increasing demand for gasohol in the USA has lead to increasing production of ethanol from maize, as the data clearly shows

SCIENCE @ WORK

Scientists are attempting to find ways of producing economically viable quantities of ethanol from plants which grow fast and well in Europe. They have tried pine trees and beet but have not yet been very successful. Now they are looking at fast growing grasses!

SUMMARY QUESTIONS

1 Make a table to summarise the advantages and disadvantages of ethanol as a fuel for cars.

2 Use the data in Figure 3 to help you answer the following questions:

   a) What was the increase in ethanol production in the USA between
      i)   November and December 2003,
      ii) October and November 2004?
   b) What was the percentage increase in USA ethanol production between November 2003 and November 2004?
   c) Much of the ethanol made is mixed with normal petrol to make a more ecologically sound motor fuel. If each barrel of gasohol sold contains 10% ethanol, how many barrels of gasohol could have been produced in January 2004?

KEY POINTS

1 Ethanol based fuels can be produced by the anaerobic fermentation of sugar cane juices and from glucose derived from maize starch by the action of the enzyme carbohydrase.

2 Ethanol is distilled from the fermentation products and can be used as a fuel in motor vehicles on its own or mixed with petrol to produce gasohol.

An historical perspective on microbes

Lazzaro Spallanzani

Theodor Schwann

Louis Pasteur

For centuries people believed in *spontaneous generation*. They thought that living things appeared from nothing in water or on food as it decayed.

In 1765, Lazzaro Spallanzani set up two sets of containers full of broth. One set was left open to the air. The other was boiled to kill anything living and then sealed. Only the second set remained sterile. Spallanzani claimed this showed spontaneous generation didn't happen. His opponents said boiling killed a 'vital principle' in the air so nothing could live there!

In 1836/7, Theodor Schwann and Franz Schulze showed that when air goes through red-hot glass tubes, or through strong sulfuric acid, the amount of oxygen stays the same but living things are killed. They passed this sterile air through boiled broth and nothing grew. Then when they exposed the broth to normal air, moulds and other microbes soon grew. Opponents said they had destroyed something important in the air – it might not have been living things.

In the 1850s Louis Pasteur first proved that microorganisms (yeast) are responsible for fermentation. Later he carried out a series of experiments with swan-necked flasks. These showed clearly that the moulds and bacteria which appeared in liquids were tiny organisms that had come from the air. Spontaneous generation did not exist.

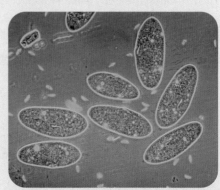

ACTIVITY

1 This account gives you a taste of some of the controversy which surrounded the scientists who started working with microorganisms in the past. Louis Pasteur is the most famous of them. Find out as much as you can about him and produce plans for a brief biography of Pasteur to be part of a science programme aimed at young people. Make sure you explain what he did and why it was so important. Try and find something about the sort of person he was as well, to make it even more interesting!

Grow your own!

ACTIVITY

2 When students work with microorganisms in the school labs for the first time they have some funny ideas …! Make a poster or a safety leaflet which could be given to students before they start working with microorganisms for the first time. You need to make sure they understand the need for care and good hygiene, both to get good results and to keep everyone safe.

She said we'd be growing bacteria – bet we can't see much!

I hope we don't all get ill from it. My Mum'll see the head if I get ill

I'm going to grow something really gruesome like plague

A microbe miracle?

Imagine you work for a food company. A research team has come to see you. They have been working on a new form of food which can be manufactured safely, quickly and cheaply. The new foodstuff can be used in different ways, and it is high in vitamins, fibre and protein but low in fat. The production of this new food helps to solve some environmental problems by getting rid of waste from human activities. Your task is to look at the details of the food and its production, and decide how you might market it.

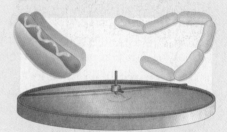

This food is based on a microorganism which grows on human sewage, supplied very cheaply by local authorities. The microorganism turns sewage into protein with 30% efficiency. It also acts as an effective treatment for the sewage. Once the microbes have been harvested the liquid which remains in the production tanks can be pumped safely into rivers or the sea. The microorganism product is a creamy colour with very little taste or smell. It can be processed to give it a wide range of textures and tastes, both savoury and sweet.

ACTIVITY

3 a) Make a list of the points in favour of the new food and the points against it.

b) Discuss these points and use them to decide how you will overcome any problems.

c) Think of a name for your new food. Then make an outline plan of the ways you will try and advertise it. Produce a draft advert to go in a magazine.

Fuel for school?

You go to school all through the year. The school is heated and lit; there is hot water to wash your hands and the whole system uses an awful lot of energy. Yet your school also produces an enormous amount of waste of various sorts.

ACTIVITY

4 Put together a presentation to explain the benefits of installing a biogas generator at your school. Your presentation will be seen by the parents, the governors and the local education authority. You need to explain how your idea will work – where the fuel supply for the generator might come from and how the biogas produced might be used – and all the benefits it could bring. Be prepared to answer questions and push your ideas as hard as possible.

SUMMARY QUESTIONS

1 a) Produce a set of instructions for setting up and growing a culture of soil bacteria in the school lab.

b) Explain why all the safety precautions are necessary.

2 a) You can leave bread dough in a fridge for hours without it rising. Put it somewhere warm and it starts to rise again. Bread usually rises fully in an hour or two. In a cool room it may take as long as twenty-four hours. Explain how these differences come about.

b) Temperature is vital for successful beer and wine making. Why is it so important?

c) To make sparkling wine or champagne, a small amount of yeast is left in the bottle. What effect does this have and what is the gas that makes the bubbles in the drink?

3 Write a brief report on 'Bacteria in the dairy industry'. Use what you know about yoghurt and cheese-making, and find out about the growing market in 'probiotic' yoghurts and drinks. Try to answer the question: What are 'friendly bacteria' and does eating them really do you any good?

4 Why are microorganisms so useful in industrial processes? Make a table to summarise the advantages and disadvantages of using these living organisms in the food and drugs industries.

5 Write a letter from a mother to her friend in the late 1940s, describing the recovery of her child from a serious throat infection thanks to the use of penicillin.

6 a) Suggest ways in which people might improve the quality of the biogas produced in their fermenters.

b) Suggest reasons for and against the use of fermenters and biogas in the UK.

7 Write a letter to your local authority explaining why you think they should look into the idea of running all their vehicles – buses, emergency vehicles, etc. – on ethanol or gasohol. Explain the potential value of ethanol in helping to prevent the greenhouse effect and global warming.

EXAM-STYLE QUESTIONS

1 The passage below describes how a student grows a culture of microorganisms on an agar plate. Petri dishes and a culture medium containing agar, carbohydrate, protein and mineral ions are heated to 120°C for 15 minutes. The culture medium is poured into the Petri dishes and left to set. A wire inoculating loop is passed through a flame until red hot, allowed to cool and then dipped into a container of microorganisms. The loop is then streaked across the medium in the Petri dish. The Petri dish is sealed with adhesive tape and incubated at a temperature not exceeding 25°C.

In each case, give one reason why the following procedures were carried out:

(a) Carbohydrate was included in the culture medium.

(1)

(b) The culture medium and Petri dishes were heated to 120°C for 15 minutes. (1)

(c) The wire inoculating loop was cooled before being used to transfer microorganisms. (1)

(d) The Petri dish was sealed with adhesive tape. (1)

(e) The temperature at which the microorganisms are grown was not allowed to exceed 25°C. (1)

2 In countries such as China, India and Nepal, biogas generators, similar to the one shown below, are common. Although they may cost over £300 to set up they are an energy-efficient way of producing biogas in the long term.

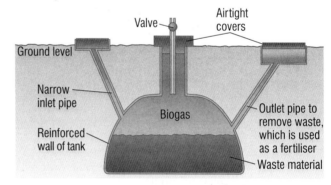

(a) Biogas is a mixture of different gases. Which gas makes up the largest proportion of this mixture? (1)

(b) Give one use of biogas. (1)

(c) Suggest two types of waste material that could be added to the container in order to generate biogas.

(2)

(d) Explain how the waste material is converted into biogas. (3)

(e) Suggest a reason why
    (i) airtight covers are used. (1)
    (ii) the wall of the tank is reinforced. (1)

(f) Biogas generators have both advantages and disadvantages over other forms of energy generation.
    (i) Suggest two advantages of a biogas generator. (2)
    (ii) Suggest one disadvantage of a biogas generator. (1)

3 The diagram shows a fermenter that can be used to produce the antibiotic penicillin using the mould *Penicillium*.

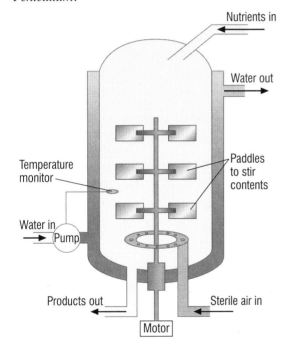

(a) Name two nutrients that a mould like *Penicillium* requires. (2)

(b) (i) Suggest a reason why air is bubbled through the fermenter. (1)
    (ii) Why must this air be sterile? (1)

(c) (i) Explain why it is necessary to pump water around the fermenter. (2)
    (ii) Suggest a reason why the temperature monitor is attached to the pump circulating the water around the fermenter. (1)

(d) Suggest **two** reasons why the contents of the fermenter are continually stirred. (2)

HOW SCIENCE WORKS

A class of students were set the task of finding the best temperature at which to manufacture yoghurt. They checked out the recipe for yoghurt. All it said was that after preparing the milk and the culture they should be left in a warm cupboard until the yoghurt is set.

Each group followed the same recipe. They were each assigned a temperature at which to keep their yoghurt.

a) Suggest a suitable range and number of different temperatures to use. (2)

b) If you were not at all sure about the answer to question a), what could you do to find out? (1)

c) What would be a suitable end point? That is, how will everybody know when the yoghurt is ready? (1)

d) What would be a suitable number of repeats? Give reasons for your answer. (1)

e) Construct a suitable table for the results. (3)

f) Suggest how the results might be displayed. (1)

g) This information could be very useful to anyone wanting to be involved in the yoghurt industry.

Suggest some of the ways in which this information could be exploited by different companies. (2)

h) Explain why you think that you might have some difficulty convincing an industrialist of the reliability of your data. (1)

EXAMINATION-STYLE QUESTIONS

1 Complete the table by choosing the word from the list in the box that matches the description in the table. Write the chosen word next to the relevant letter in the 'process' column in the table. *(5 marks)*

See pages 222–33

| active transport | breathing | dialysis | diffusion | digestion |
| osmosis | respiration | transpiration | | |

Process	Description
A	Loss of water from plant leaves
B	Movement of air in and out of the lungs
C	Movement of particles of a substance from high to low concentration
D	Movement of substances from low to high concentration using energy from respiration
E	Movement of water from dilute to concentrated solution through a partially permeable membrane

2 In the following questions, write down one answer from the list that follows the question.

See pages 238–49

(a) Through which vessels does blood flow from the heart to the organs? *(1 mark)*

      arteries   capillaries   veins

(b) There are two separate circulations in the body, one goes to the organs. Where does the other go to? *(1 mark)*

      intestines   kidneys   lungs

(c) In which part of the blood is carbon dioxide transported? *(1 mark)*

      plasma   red blood cells   white cells

(d) What is the name of the pigment in red blood cells that transports oxygen? *(1 mark)*

      glycogen   haemoglobin   lactic acid

(e) Blood plasma transports urea to the kidneys from which organ? *(1 mark)*

      liver   lungs   muscles

GET IT RIGHT!

Never look for a pattern in responses where you choose from a list. It is just as likely that it is always the first word or item in the list as any other combination. Often, as in this case, the choices are simply presented in alphabetical order.

3 The table compares the volume of blood pumped and the number of heartbeats of individuals at rest and during strenuous exercise. The individuals were of two types – trained athletes and untrained non-athletes.

See pages 242–5

State	Training level	Volume of blood pumped out of the heart at each beat /cm³	Number of heartbeats per minute
At rest	Untrained	65	70
	Trained	90	52
During strenuous exercise	Untrained	100	210
	Trained	155	190

(a) Using the data for individuals at rest, describe the effects of training on the

    (i) volume of blood pumped at each heartbeat. *(1 mark)*

    (ii) the number of heartbeats per minute. *(1 mark)*

    (iii) Why is it important that we know the results for each person tested before we can be really confident in any conclusions drawn? *(1 mark)*

(b) (i) **Cardiac output** is the volume of blood pumped by the heart in one minute. Calculate the cardiac output for an untrained individual during strenuous exercise. Show your working. *(2 marks)*

    (ii) Calculate the percentage increase in the number of heartbeats per minute of an untrained individual during strenuous exercise compared to at rest. Show your working. *(2 marks)*

(c) Explain why it is necessary for blood flow to increase during strenuous exercise. *(2 marks)*

(d) If strenuous exercise continues for a long period, the blood may be unable to supply sufficient oxygen or glucose to the muscles. The muscles however continue working for some time longer. In these circumstances:

    (i) what alternative source of glucose (carbohydrate) do the muscles use? *(1 mark)*

    (ii) how is energy still produced by the muscles? *(2 marks)*

4 When people suffer from kidney failure, they may be treated with a dialysis machine. The patient's blood is passed through the machine where the composition of the blood is adjusted.

See pages 248–9

(a) Name a waste substance, carried in the blood, which is removed by the dialysis machine. *(1 mark)*

(b) Doctors sometimes give these patients dialysis treatment, rather than a kidney transplant. Suggest **four** reasons for this. *(4 marks)*

5 In the 1880s, Louis Pasteur investigated the reasons for food going rotten. To check one of Pasteur's experiments, a student set up the two flasks shown in the diagram.

See page 270

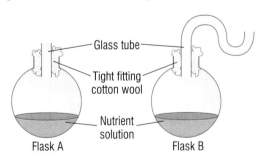

After three days, the nutrient solution in both flasks had gone rotten.

(a) What makes the nutrient solution go rotten? *(1 mark)*

(b) The student then set up two more similar flasks. This time, she boiled the nutrient solution in both flasks for ten minutes.

    (i) Why did she boil the nutrient solution? *(1 mark)*

    (ii) Why did the nutrient solution in flask **A** go rotten? *(1 mark)*

    (iii) Why did the nutrient solution in flask **B not** go rotten? *(2 marks)*

GET IT RIGHT!

Always be as precise as possible in the use of terms, especially when writing about micoorganisms. The term 'germ' is far too vague and unscientific. It should not be used in examinations.

Glossary

A

Abdomen The lower region of the body. In humans it contains the digestive organs, kidneys, etc.

Accuracy An accurate measurement is close to the true value.

Acid A substance that produces hydrogen ions when it dissolves in water.

Acid rain Rain that is acidic due to dissolved gases, such as sulphur dioxide, produced by the burning of fossil fuels.

Activation energy The minimum amount of energy needed for a given chemical reaction to take place.

Active site The site on an enzyme where the reactants bind.

Active transport The movement of substances against a concentration gradient and/or across a cell membrane, using energy.

Adaptations Special features which make an organism particularly well suited to the environment where it lives.

Aerobic Using oxygen.

Aerobic respiration The process by which food molecules are broken down using oxygen to release energy for the cells.

Agar The nutrient jelly on which many microorganisms are cultured.

Alcohol The commonly used name for drinks containing ethanol.

Alcoholics People who are addicted to alcohol.

Alleles A version of a particular gene.

Alveoli The tiny air sacs in the lungs which increase the surface area for gaseous exchange.

Amino acids The building blocks of protein.

Anaerobic respiration Cellular respiration in the absence of oxygen.

Anomalous A measurement that is well away from the pattern shown by other results.

Anorexia nervosa A mental disorder linked to an unrealistic body image and a need for control.

Antibiotics Drugs which destroy bacteria inside the body without damaging human cells.

Antigens The unique proteins on the surface of a cell. They are recognised by the immune system as 'self' or 'non-self'.

Aqueous solution A solution with water as the solvent.

Arthritis A painful and debilitating disease affecting the joints.

Association When two variables change together, but they are both linked by a third variable. E.g. lack of carbon dioxide in soil and poor growth of plants: both could be linked to too much water in the soil.

B

Bar charts Used when the independent variable is categoric and the dependent variable is continuous.

Bias The influence placed on scientific evidence because of: wanting to prove your own ideas; supporting the person who is paying you; political influence; the status of the experimenter.

Binge drinking Short bouts of very heavy drinking.

Biogas Gas made by the action of microorganisms on the remains of living organisms.

Biological detergents Washing detergents that contain enzymes.

Biodiesel Diesel fuel made from plant materials.

Biomass The amount of biological material in an organism.

Biomass fuel Fuel from animal waste or cut-down plants.

Bladder The organ where urine is stored until it is released from the body.

Blood The liquid which is pumped around the body by the heart. It contains blood cells, dissolved food, oxygen, waste products, mineral ions, hormones and other substances needed in the body or needing to be removed from the body.

Blood plasma The clear yellow liquid part of the blood which carries dissolved substances and blood cells around the body.

Blood vessels The tubes which carry blood around the body, i.e. arteries, veins and capillaries.

C

Carbohydrases Enzymes which speed up the breakdown of carbohydrates.

Carbon cycle The cycling of carbon through the living and non-living world.

Carbon neutral A process which uses as much carbon dioxide as it produces.

Carcinogen A chemical which is known to cause cancer.

Carriers People who have a single recessive allele for a genetic disease.

Catalyst A substance that speeds up the rate of another reaction but is not used up or changed itself.

Categoric variable These tell us the name of the variable, e.g. oak tree, beech tree, ash tree.

Causal link One change in a variable has caused a change in another variable. You can only be reasonably certain of this when you have valid and reliable evidence. E.g. increasing light intensity causes an increase in the rate of photosynthesis.

Cell membrane The membrane around the contents of a cell which controls what moves in and out of the cell.

Central nervous system (CNS) The central nervous system made up of the brain and spinal cord where information is processed.

Chance When there is no scientific link between the two variables. E.g. increased sea temperatures and increased diabetes.

Chlorophyll The green pigment contained in the chloroplasts.

Chloroplasts The organelles in which photosynthesis takes place.

Cholesterol A substance made in the liver and carried around the body in the blood. High blood cholesterol levels seem to be linked to a high risk of heart disease.

Chromosomes Thread-like structures carrying the genetic information found in the nucleus of a cell.

Cirrhosis of the liver A disease which is often the result of heavy drinking over a long period of time.

Clones Offspring produced by asexual reproduction which are identical to their parent organism.

Combustion The process of burning.

Compost heap A site where garden rubbish and kitchen waste are decomposed by microorganisms.

Concentration gradient The gradient between an area where a substance is at a high concentration and an area where it is at a low concentration.

Conclusion A conclusion considers the results and states how those results match the hypothesis. The conclusion must not go beyond the data available.

Constrict To narrow.

Continuous variable A continuous variable can be any numerical value, e.g. your own weight.

Control groups Often used when there are a large number of control variables that cannot be kept constant. E.g. when testing a drug on thousands of different people, half will be given the drug and half will be given a similar treatment that does not contain the drug (placebo).

Control variable These are the variables that might affect your result and therefore must be kept the same for a valid investigation. E.g. concentration of the enzyme in an investigation of the effect of temperature.

Controlled An experiment is controlled when all variables that might affect your result (apart from the independent variable) have been kept constant.

Culture Growing microorganisms in the laboratory.

Cuticle The waxy covering of a leaf (or an insect) which reduces water loss from the surface.

Cystic fibrosis A genetic disease that affects the lungs, digestive and reproductive systems. It is inherited through a recessive allele.

Cytoplasm The water-based gel in which the organelles of all living cells are suspended.

D

Data Measurements or observations of a variable. Plural of datum.

Daughter cells The cells produced by cell division.

Decompose To split up or break down organisms or waste material.

Decomposers Microorganisms that break down waste products and dead bodies.

Denatured Enzymes that are denatured have their protein structure broken down and can no longer catalyse a reaction.

Dependent variable The variable that you are measuring as a result of changing the independent variable, e.g. the volume of CO_2 produced.

Detritus feeders See **decomposers**.

Diabetes A condition in which it becomes difficult or impossible for your body to control the levels of sugar in your blood.

Dialysis The process of cleansing the blood through a dialysis machine when the kidneys have failed.

Dialysis machine The machine used to remove urea and excess mineral ions from the blood when the kidneys fail.

Diaphragm The sheet of muscle which divides the thorax from the abdomen.

Differentiated Specialised for a particular function.

Diffusion The net movement of particles of a gas or a solute from an area of high concentration to an area of low concentration (along a concentration gradient).

Dilate To widen.

Directly proportional A graph will show this if the line of best fit is a straight line through the origin.

Discrete variable These are numerical, but can only be whole numbers, e.g. numbers of layers of insulation.

Distillation A process which separates the components of a mixture on the basis of their different boiling points.

DNA Deoxyribose nucleic acid, the material of inheritance.

Dominant The characteristic that will show up in the offspring even if only one of the alleles is inherited.

Donor The person who gives material from their body to another person who needs healthy tissues or organs, e.g. blood, kidneys. Donors may be alive or dead.

Double circulation The separate circulation of the blood from the heart to the lungs and then back to the heart and on to the body.

E

E number A number given to a food additive in order to identify it.

Economic How science affects the cost of goods and services. E.g. developing a new drug might increase the cost of treatment.

Ecosystem All the animals and plants living in an area, along with things which affect them such as the soil and the weather: also the interaction between many different types of living organisms and the non-living features of their home.

Effector organs Muscles and glands which respond to impulses from the nervous system.

Electron microscope An instrument used to magnify specimens using a beam of electrons.

Emulsifier A substance which stops the two liquids in an emulsion separating.

Emulsify To physically break down large droplets into smaller droplets.

Emulsion A mixture of tiny droplets of one liquid in another liquid.

Environmental How science affects our natural surroundings. E.g. killing badgers to stop disease in cows.

Enzyme Protein molecules which act as biological catalysts. They change the rate of chemical reactions without being affected themselves at the end of the reaction.

Enzyme substrate complex The combination of the enzyme and the substrate at the active site.

Ethanol A chemical found in alcoholic drinks and biofuels such as gasohol, chemical formula C_2H_5OH.

Ethical Whether it is 'right' or 'wrong' to do something. E.g.

experimentation on animals to develop new drugs.

Evidence Scientific evidence should be reliable and valid. It can take many forms. It could be an observation, a measurement or data that somebody else has obtained.

Evolution The process of slow change in living organisms over long periods of time as those best fitted to survive breed successfully.

Extinction The process by which animals become extinct – the permanent loss of all the members of a species from the face of the Earth.

F

Fair test Only the independent variable is affecting your dependent variable, all other variables are kept the same.

Fatty acids Building blocks of lipids.

Fermenters The large vessels used in commercial fermentation processes.

Food additive A substance added to food to improve its flavour, texture or shelf-life.

Fossil fuel Coal, oil or gas or any other fuel formed long ago from the fossilised remains of dead plants or creatures.

Fructose syrup A sugar syrup.

G

Gametes Sex cells.

Gasohol A mixture of petrol (gasoline) and ethanol.

Genes A short section of DNA carrying genetic information.

Genetic engineering/genetic modification A technique for changing the genetic information of a cell.

Genetic diseases Diseases which are inherited.

Genetic disorders See **genetic diseases**.

Global warming Warming of the

Earth due to greenhouse gases in the atmosphere trapping infra-red radiation from the surface.

Glucagon Hormone involved in the control of blood sugar levels.

Glucose A simple sugar.

Glycerol Building block of lipids.

Glycogen Carbohydrate store in animals, including the muscles, liver and brain of the human body.

Greenhouse gases Gases such as carbon dioxide in the atmosphere that absorb infra-red radiation from the Earth's surface.

H

Haemoglobin The red pigment which carries oxygen around the body.

Heart The muscular organ which pumps blood around the body.

High-density lipoproteins (HDLs) Chemicals which carries cholesterol in the blood and lower the risk of heart disease.

Homeostasis The maintenance of constant internal body conditions.

Hormones Chemical messages secreted by special glands and carried around the body in the blood.

Hydroponics Growing plants in water enriched by mineral ions rather than soil.

Hypothesis Using theory to suggest explanations for observations, e.g. 'I think that the plants are smaller because they do not have enough water.'

I

Immune system The body system which recognises and destroys foreign tissue such as invading pathogens.

Immunisation Giving a vaccine which allows immunity to develop without exposure to the disease itself.

Immunosuppressant drugs Drugs which suppress the immune system of the recipient of a transplanted organ to prevent rejection.

Independent variable The variable that you have decided to change in an investigation, e.g. the temperature of the enzyme in an investigation to find out the effect of temperature on enzyme activity.

Insoluble Unable to dissolve in a given solvent.

Insulin Hormone involved in the control of blood sugar levels.

Interval measurements The values of your independent variable that you choose within the range e.g. $10\,cm^3$; $20\,cm^3$; $30\,cm^3$; $40\,cm^3$; $50\,cm^3$.

Isomerase An enzyme which converts one form of a molecule into another.

K

Kidneys Organs which filter the blood and remove urea, excess salts and water.

Kidney transplant Replacing failed kidneys with a healthy kidney from a donor.

L

Lactic acid One product of anaerobic respiration. It builds up in muscles with exercise. Important in yoghurt and cheese making processes.

Lactic fermentation Fermentation using bacteria to break down the lactose in milk and produce lactic acid.

Lactose The main sugar found in milk.

Light microscope An instrument used to magnify specimens using lenses and light.

Lime water Solution of calcium hydroxide, used to test for carbon dioxide.

Limiting factors Factors which limit the rate of a reaction, e.g. photosynthesis.

Line graphs Used when the

independent and the dependent variables are both continuous.

Line of best fit Used to show the underlying relationship between the independent and the dependent variables. It should fit the pattern in the results and have roughly the same number of plots on each side of the line. It could be a straight line or a curve. Remember to ignore any anomalies!

Linear These are straight line graphs that can be positive (as the concentration increases so too does the oxygen produced) or negative (as the concentration increases the oxygen produced decreases).

Link due to association When two variables change together, but they are both linked by a third variable. E.g. lack of carbon dioxide in soil and poor growth of plants, both could be linked to too much water in the soil.

Link due to chance When there is no scientific link between the two variables. E.g. increased sea temperatures and increased diabetes.

Lipids Fats and oils.

Liver A large organ in the abdomen which carries out a wide range of functions in the body.

Low-density lipoproteins (LDLs) Chemicals which carry cholesterol in the blood and raises the risk of heart disease.

M

Magnesium A metallic element. Magnesium ions are needed by plants to make chlorophyll.

Malting The process, important in making beer, of soaking barley grains in water and keeping them warm so germination begins and enzymes break down the starch in the barley grains into a sugary solution.

Mean Add up all of the measurements and divide by how

many measurements there are. Don't forget to ignore any anomalous results.

Meiosis The two-stage process of cell division which reduces the chromosome number of the daughter cells. It is involved in making the gametes for sexual reproduction

Menstrual cycle The reproductive cycle in women controlled by hormones.

Metabolic rate The rate at which the reactions of your body take place, particularly cellular respiration.

Methane The main, flammable component of biogas, chemical formula CH_4.

Microbiology The study of microorganisms.

Microorganism Bacteria, viruses and other organisms which can only be seen using a microscope.

Mitochondria The site of aerobic cellular respiration in a cell.

Mitosis Asexual cell division where two identical daughter cells are formed.

Model Description of a theory or theories that suggests further ideas that could test those theories. E.g. 'plum pudding' model of the atom that was tested and found not to be correct. A better model was then suggested.

Mutation A change in the genetic material of an organism.

Mycoprotein A form of food derived from fungi grown in giant fermenters on sugar solution from waste carbohydrates.

N

Natural selection The process by which evolution takes place. Organisms produce more offspring than the environment can support so only those which are most suited to their environment – the 'fittest' – will survive to breed and pass on their useful characteristics.

Net Overall.

Neurones Basic cells of the nervous system which carry minute electrical impulses around the body.

Nicotine Colourless, poisonous substance which is the addictive drug in tobacco smoke.

Nitrates Mineral ions needed by plants to form proteins.

Nitrogen Inert gas making up around 80% of the Earth's atmosphere.

Nucleus (of a cell) An organelle found in many living cells containing the genetic information.

O

Opinion Opinions are personal judgements. Opinions can be formed from scientific evidence or non-scientific ideas.

Ordered variable Variables that can be put into an order, e.g. small, large, huge lumps of rock.

Organ A group of different tissues working together to carry out a particular function.

Organ systems A group of organs working together to carry out a particular function.

Organelles Membrane-bound structures in the cytoplasm of a cell which carry out particular functions.

Osmosis The net movement of water from an area of high concentration (of water) to an area of low concentration (of water) along a concentration gradient.

Ova The female sex cells, eggs.

Ovaries Female sex organs which contain the eggs and produce sex hormones during the menstrual cycle.

Ovulation The release of a mature egg from the ovary in the middle of the menstrual cycle.

Oxyhaemoglobin The molecule formed when haemoglobin binds to oxygen molecules.

P

Pancreas An organ which produces the hormone insulin and many digestive enzymes.

Partially permeable Allowing only certain substances to pass through.

Pathogens Microorganisms which cause disease.

Pay-back period (or time) Length of time for the savings from an improvement to match the actual cost of the improvment.

Penicillin The first broad spectrum antibiotic to be produced commercially.

Percentage yield The percentage of product formed in a chemical reaction compared with the maximum possible amount of product that could be formed.

Petri dish The flat glass dishes often used to culture microorganisms in the laboratory.

Phloem The living transport tissue in plants which carries sugars around the plant.

Photosynthesis The process by which plants make food using carbon dioxide, water and light energy.

pH scale A scale running from 0 to 14 that describes the degree of acidity of a solution.

Pituitary gland Small gland in the brain which produces a range of hormones controlling body functions.

Platelets Fragments of cells in the blood which are vital for the clotting mechanism to work.

Pollution The contamination of air, water or soil by substances which are harmful to living organisms.

Population A group of individuals of the same species living in the same habitat.

Precision Where your repeat results are very close to each other. This is related to the smallest scale division on the measuring instrument used.

Predator An animal which preys on other animals for food.

Prediction A hypothesis that can be used to design an investigation e.g. I predict that if I increase the amount of water given to plants there will be an increase in the mass of the plants.

Protease An enzyme which breaks down proteins.

Protein synthesis The process of building up protein molecules from amino acids on the surface of a ribosome.

Puberty The stage of development when the sexual organs and the body become adult.

Pyramid of numbers A model of feeding relationships based on the numbers of organism at each level of a food chain.

R

Random changes Changes that cannot be predicted.

Random error Measurements when repeated are rarely exactly the same. If they differ randomly then it is probably due to human error when carrying out the investigation.

Range The maximum and minimum values.

Recessive The characteristic that will show up in the offspring only if both of the alleles are inherited.

Recipient The person who receives a donor organ.

Red blood cells Blood cells which contain the red pigment haemoglobin. They are biconcave discs in shape and they give the blood its red colour.

Reflexes Rapid automatic responses of the nervous system which do not involve conscious thought.

Reliable Describes data we can trust. E.g. others get the same results.

Reliability The trustworthiness of data collected.

Renewable energy Energy from sources that never run out, including wind energy, wave energy, tidal energy, hydroelectricity, solar energy and geothermal energy.

Respiration The process by which food molecules are broken down to release energy for the cells.

Ribosomes The site of protein synthesis in a cell.

S

Salt glands Special glands which enable some animals to remove excess salt from their bodies.

Sankey diagram Diagram to show the energy transfer through a system.

Saturated A hydrocarbon which contains as many hydrogen atoms as possible in each molecule.

Selective breeding Choosing parents with a desired characteristic for breeding.

Sense organ Collection of special cells known as receptors which respond to changes in the surroundings (e.g. eye, ear).

Sensitivity The smallest change that an instrument can measure, e.g. 0.1 mm.

Sewage treatment plant A site where human waste is broken down using microorganisms.

Sex chromosomes The chromosomes which carry the information about the sex of an individual.

Social issues How science influences and is influenced by its effects on our friends and neighbours. E.g. building a wind farm next to a village.

Spiracles The openings in the outer coat of an insect which allow air in and out of the breathing system.

Statins Drugs which lower the blood cholesterol levels and improve the balance of HDLs to LDL.

Stem cell research Research into the stem cells found in embryonic tissue and in some adult tissues.

Stimulus A change which causes a response in the body.

Soluble Able to dissolve in a given solvent.

Solute The solid which dissolves in a solvent to form a solution.

Solvent A liquid in which some solids will dissolve.

Specialised Adapted for a particular function.

Sperm The male sex cells.

Stem cells Undifferentiated cells with the potential to form a wide variety of different cell types.

Stomata Openings in the leaves of plants (particularly the underside) which allow gases to enter and leave the leaf. They are opened and closed by the guard cells.

Sugars Simple carbohydrates.

Sustainable development Using natural resources in a way which also conserves them for future use.

Synapses The gaps between neurones where the transmission of information is chemical rather than electrical.

Systematic error If the data is inaccurate in a constant way, e.g. all results are 10 mm more than they should be. This is often due to the method being routinely wrong.

T

Tar Thick, black chemical found in tobacco smoke which can cause cancer.

Technology Scientific knowledge can be used to develop equipment and processes that can in turn be used for scientific work.

Territory An area where an animal lives and feeds which it may mark out or defend against other animals.

Testes Male sex organs which produce sperm and sex hormones.

Theory A theory is not a guess or a fact. It is the best way to explain why something is happening. E.g. Sea levels are rising, and the global warming theory is the best way to describe why they are. Theories can be changed when better evidence is available.

Thermoregulatory centre The area of the brain which is sensitive to the temperature of the blood.

Thorax The upper (chest) region of the body. In humans it includes the rib cage, heart and lungs.

Tissue A group of specialised cells all carrying out the same function.

Tracheoles Minute breathing tubes in insects with a large surface area which penetrate right through the tissues.

Transpiration The loss of water vapour from the leaves of plants through the stomata when they are opened to allow gas exchange for photosynthesis.

Transpiration stream The movement of water through a plant from the roots to the leaves as a result of the loss of water by evaporation from the surface of the leaves.

Trial run Carried out before you start your full investigation to find out the range and the interval measurements for your independent variable.

U

Unsaturated A hydrocarbon which contains a carbon–carbon double bond.

Urea The waste product formed by the breakdown of excess amino acids in the liver.

Urine The liquid formed by the kidneys.

Urobilins Yellow pigments that come from the breakdown of haemoglobin in the liver, and which colour the urine yellow.

V

Valid Describes an investigation that successfully gathers the data needed to answer the original question. Data may not be valid if you have not carried out a fair test.

Valid data Evidence that can be reproduced by others and answers the original question.

Villi The finger-like projections from the lining of the small intestines which increase the surface area for the absorption of digested food into the blood.

W

Wilting The process by which plants droop when they are short of water or too hot. This reduces further water loss and prevents cell damage.

X

Xenotransplantation Transplanting tissues or organs from one species to another, e.g. pig organs into people.

Xylem The non-living transport tissue in plants, which transports water around the plant.

Y

Yield The amount of product formed in a chemical reaction.

Z

Zero error A systematic error, often due to the measuring instrument having an incorrect zero. E.g. forgetting that the end of the ruler is not at zero.

Index

blood plasma 240–1, 248
blood pressure 40, 41, 54
blood sugar 33, 137, 196–7
blood transfusions 241
blood vessels 194, 195, 238–9
    disease of 42, 54
body temperature 33, 83, 165, 179, 194–5, 198, 228
bone marrow, stem cells in 204, 214
bony fish 228, 235
brain 25, 26, 28
    and drugs/alcohol 50, 52, 53
brain, thermoregulatory centre in 194
Brazil, production of biofuels 268–9
bread making 258
breathing 32, 192, 193
breathing system 224–5
    effects of exercise 242–3
    problems of 234
building investigations 8–9
burettes 10
burning
    and acid rain 122
    and global warming 124

C

cambium cells 154
cancer 52, 54, 56
cannabis 48, 49, 50, 51, 58–9
capillaries 194, 195
capillaries 227, 238–9
carbohydrases 182, 186
carbohydrates 154, 156, 170, 182, 196
    see also cellulose; glycogen; starch; sugars
carbon cycle 170–1
carbon dioxide 131, 170–1, 180
    in biogas 266
    in blood plasma 240
    and burning fuel 122
    emissions 268–9
    in photosynthesis 136, 137, 150, 152, 159, 230–1
    production of 242–3
    in respiration 136, 180, 192–3, 225
    in transpiration 232–3
    as waste 170, 192–3
carbon monoxide 55
carbon neutral fuels 268–9
carcinogens 54
cardio-vascular disease *see* heart disease
carnivores 86, 90–1
cars
    pollution from 123
cartilaginous fish 235
catalysts 176
categoric variables 4, 13
cattle/cows 98–9, 125

causal links 5, 15, 17
cell division 202–3, 206–7
cells 62, 94, 98–9, 100, 101, 107
    see also eggs; sperm; white blood cells
cells (biological) 136, 138–47, 149
    in leaves 150, 151
    respiration in 138, 139, 180–1
    sex cells (gametes) 206–7
        see also ova; sperm
    specialised cells 140–1, 202–3
    stem cells 199, 202, 204–5, 214
cells, transport of substances 222–3
cellulose 138, 139, 154, 269
Chain, Ernst 264
chance links 15
cheese making 260, 261
China, production of biogas 267
chlorophyll 138, 150, 156, 157
chloroplasts 138, 139, 150, 151
cholesterol 42–3, 44
chromosomes 94, 95, 202, 206–7, 208, 210
chronic obstructive pulmonary disease (COPD) 224,
    234
climate change *see* greenhouse effect
cloning 203, 205
cloning/clones 96, 98–101, 104, 117
coal 122
coal worker's pneumoconiosis (CWP) 234
cocaine 48, 49, 50, 51
cold-blooded animals 165
Colton, Frank 18
combustion/burning 170, 171
competition 80, 86–9, 115
concentration 142–3, 144–5, 147
conclusions 2, 14–15
cone cells 140
consumers 162, 163
continuous variables 4, 13
contraception 18–19, 30, 31, 34
contraceptives 18–19
control groups 8
control variables 5, 7
controls in investigations 2, 5, 7, 8
cooling down 194, 195
culture medium 256
curds 261
cuticles, plants 233
cyanide 223
cystic fibrosis 137, 212, 213, 223
cytoplasm 138, 139, 140

D

data 5, 7, 9, 12–15
    accuracy/precision of 9–10, 11
    drawing conclusions from 14–15

Acknowledgements

Action +/Glyn Kirk 51.3; Alamy fotolincs 129.3, /Perihelion 170.1, /Photofusion Picture Library 212.1, /Ritterbach/f1online 159bl; Ann Fullick 116tl, 164.2, 173bm, 173br; Axon Images 156.1; Bananastock T (NT) 22tl; Clare Marsh/John Birdsall 33r; Corbis V84 (NT) 57; Corbis V98 (NT) 48.1; Corbis/John Sparks 20l, /Pete Saloutos/zefa 24.1, /Tom Brakefield 117mr; Corel 11 (NT) 39.3, 194.1b; Corel 12 (NT) 221tl; Corel 21 (NT) 5.3; Corel 27 (NT) 181.2; Corel 46 (NT) 84.2; Corel 50 (NT) 81tl; Corel 60 (NT) 110.2; Corel 73 (NT) 147bl, 235tr; Corel 92 (NT) 165.3; Corel 103 (NT) 147tr; Corel 148 (NT) 98.1; Corel 178 (NT) 122.2; Corel 219 (NT) 52.1; Corel 263 (NT) 228.1; Corel 337 (NT) 32.1, 109.3; Corel 344 (NT) 173tr; Corel 357 (NT) 194.1a; Corel 459 (NT) 166.1; Corel 465 (NT) 50.1; Corel 467 (NT) 126.1, 155.3; Corel 511 (NT) 39.2; Corel 588 (NT) 41.2; Corel 599 (NT) 94.1; Corel 603 (NT) 159ml; Corel 638 (NT) 100bl; Corel 657 (NT) 128.2; Corel 671 (NT) 158bl; Corel 706 (NT) 108.1; Corel 713 (NT) 154ml; Corel 753 (NT) 91tr; Corel 765 (NT) 261.3b; Corel 780 (NT) 198br; David Buffington/Photodisc 67 (NT) 32.2; Diabetes Research Organisation/by permission of Dr Allan Kirk 251.2; Digital Stock 10 (NT) 234l; Digital Vision 4 (NT) 80tl; Digital Vision 5 (NT) 87.2; Digital Vision 7 (NT) 120.1, 130; Digital Vision 15 (NT) 122.1; Empics/Malcolm Croft/PA News 30.2; Frink/Digital Vision LU (NT) 142.1, 235l; Gerry Ellis/Digital Vision JA (NT) 7.2, 82.1, 86.1b; Holt Studios International Ltd/Nigel Cattlin 157.2a, 157.2b; Illustrated London News V2 (NT) 270bl; Image 100 22 (NT) 137br; Ingram ILP V1 CD2 (NT) 42.1, 221bck; Ingram ILP V1 CD5 (NT) 91br; James Cook 81br, 81tr; Karl Ammann/Digital Vision AA (NT) 83.4; Keith Brofsky/Photodisc 59 (NT) 149r; Kilmer McCully 45m; Martyn F. Chillmaid 1m, 1ml, 1mr, 7.3, 20r, 20tl, 121.4, 177.1, 184.2a, 184.2b, 184.2c, 221m; Mary Evans Picture Library 118l; Photodisc 6 (NT) 137tl; Photodisc 10 (NT) 192.1c; Photodisc 19 (NT) 8.1, 162.1, 268.1; Photodisc 21 (NT) 39.4, 115.3; Photodisc 29 (NT) 20m, 112.1, 128.1, 221tr; Photodisc 31 (NT) 125.3, 127.2; Photodisc 38A (NT) 28.1; Photodisc 40 (NT) 35; Photodisc 44 (NT) 3.1, 112.2, 164.1; Photodisc 45 (NT) 186.2, 192.1b; Photodisc 50 (NT) 24.3; Photodisc 51 (NT) 232.1, 244.1,; Photodisc 54 (NT) 19.3, 34l, 70.1; Photodisc 59 (NT) 34r, 240.1; Photodisc 67 (NT) 38.1, 137m, 243.3, 261.2b; Photodisc 71 (NT) 192.1a, 192.1d; Photodisc 72 (NT) 221br; Photodisc 75 (NT) 234m; Photodisc 76 (NT) 245.3; Photodisc 83 (NT) 240.2; Randy Allbritton/Photodisc 72 (NT) 209.3; Rex Features/SIPA 169.2, /Sipa Press 41.3; Ringwood Brewery 259.3a, 259.3b; Roslin Institute 100.1; Ryan McVay/Photodisc 79 (NT) 136tl; Science Photo Library 63.3, 146tl, 252bl, 270ml, 270l, /A. Barrington Brown 95.4, /Adam Hart-Davis 72.1, 85.3, /Alfred Pasieka 62.1, /Andrew Syred 64.2, 235br,

270tr, /Anthony Mercieca 83.3, /Astrid and Hanns-Frieder Michler 137mr, /Bryan Peterson/AGSTOCKUSA 172tr, /BSIP VEM 42.2, /Beranger 248.1, /Carlos Dominguez 158mr, /CNRI 12.1, 53.2a, 53.2b, 210.1; /Cordelia Molloy 45t, 67.3, 90m, 104t, 150.1, 155.2, 172tl, 186.1, 188m, /Cristina Pedrazzini 96.1, /Curt Maas/Agstock 167.2, /Darwin Dale 179.4, /David Hall 266.1, /David Munns 189bl, /David Nunuk 89.2, 157.3, /Div. of Computer Research and Technology, National Institute of Health 95.3, /Dr Gopal Murti 146ml, /Dr Jeremy Burgess 21bl, 121.3, 230.2; /Dr Kari Lounatmaa 62.2, /Dr P. Marazzi 185.3, /Dr R. Dourmashkin 143.4, /Eamonn McNulty 27, /Erika Craddock 168.1, /Eye Of Science 169.4, 207.3, 226.1, /Francoise Sauze 113.3, /Gary Parker 97.3, /Gregory Dimijian 81bl; /Gusto 263.3, /J. C. Revy 5.2, 139.3a, 177.2, /James King-Holmes 208.1, 220l, /Jim Zipp 229.2b, /John Cole 198tr, /John Howard 167.3b, /John Kaprielian 6.1, /John M. Burnley 229.2a, /Josh Sher 30.1, /Kenneth W. Fink 114.2, /Kent Wood 64.1, /M. I. Walker 203.3, /Manfred Kage 242.1, /Maria E. Bruno Petriglia 217tr, /Mark Clarke 197.2, /Martin Bond 269.3, /Martyn F. Chillmaid 4.1, /Matt Meadows, Peter Arnold Inc. 256.1, /Mauro Fermariello 104b, 260.1, /Maximilian Stock Ltd 263.2, 265.3, /NASA 50.2a, 50.2b, /Omikron 146mr, /Pascal Goetgheluck 86.1a, /Paul Whitehill 66.1, /Phillipe Plailly 103.2, 109.2, /Robert Brook 103.3, /Rod Planck 90br, /Rosenfeld Image Ltd 261.2a, /SCIMAT 258.1, /Scott Sinklier/Agstock 167.3a, 173tl, /Sheila Terry 252ml, /Simon Fraser 129.4, /Simon Fraser/University of Newcastle Upon Tyne 204.1, /St Mary's Hospital School 67.2, 264.1, /Steve Gschmeissner 94.2, 139.3b, 146b, /TH Foto-Werbung 18.1, /Tony Craddock 54.1, /USA Library Of Medicine 69.3, /Will and Deni McIntyre 215br; Stephen Frink/Digital Vision LU (NT) 147br, 147tl, 221bl, 223.4; Stockbyte 9 (NT) 43.3; Stockbyte 28 (NT) 127.3; Stuart Sweatmore 33.3, 273tr; The Syndics of Cambridge University Library 111.3; Topfoto.co.uk, David R Frazier/The Image Works 261.3a, /PAL 264.2, /National Pictures 49.3, /The ArenaPAL Library 71.3, /The Image Works 40.1, /UPPA Ltd 187.3; Trustees, Royal Botanic Garden, Kew 117bl, 117ml; USA Transplant Games 2004 251.3; Wearset 258.2a, 258.2b, 258.2c

Picture research by Stuart Sweatmore, Science Photo Library and johnbailey@ntlworld.com.